Technology on Trial

Uppsala Studies in History of Science 1

Editor: Tore Frängsmyr

To the memory of
Marja from Slacktarbo and *Pär Jönsdotter Karin,*
girls who worked in the windlasses at
the Dannemora Mines in 1726 and never experienced
the "public utility of science"

SVANTE LINDQVIST

Technology on Trial

The Introduction of
Steam Power Technology
into Sweden, 1715–1736

UPPSALA 1984

ALMQVIST & WIKSELL INTERNATIONAL, STOCKHOLM, SWEDEN

Doctoral Thesis at Uppsala University 1984

ABSTRACT

Lindqvist, S., 1984. Technology on Trial: The Introduction of Steam Power Technology into Sweden, 1715–1736. *Uppsala Studies in History of Science 1*. 392 pp. ISBN 91-86836-00-5 (paper), 91-22-00716-4 (bound).

The steam engine came to Sweden as something of a paradox: both as a solution and at the same time as a threat. It promised to meet industry's need for mechanical energy, but it also threatened the limited resources of thermal energy. Part I is a study of this issue in Sweden as a whole during the eighteenth century.

Part II is a study of how knowledge of the Newcomen engine was transferred to Sweden in 1715–1725 by Swedish travellers and foreign adventurers, and how information about the new technology was evaluated by the Board of Mines. In particular, the technology assessment by Emanuel Swedenborg in 1725 is discussed.

Part III is a study of the first attempt to build and operate a Newcomen engine in Sweden: Mårten Triewald's engine at the Dannemora Mines in 1726–1730. A dispute between the mineowners and Triewald led to a lawsuit in which the new technology was brought to trial in 1731, and finally rejected in 1736. The concluding chapter is a discussion of critical factors in technology transfer, of which some were specific to this particular failure and some were general difficulties always attending the movement of new technologies across national boundaries.

The present study illustrates the roles of science and technology in early eighteenth-century Sweden. It is also an enquiry into the nature of technological knowledge, its structure and sociology.

Svante Lindqvist, Institutionen för idé- och lärdomshistoria, Uppsala universitet, Box 256, S-751 05 Uppsala, Sweden.

ISSN 0282-1036
ISBN 91-86836-00-5 (paper)
ISBN 91-22-00716-4 (bound)

Graphic design Jerk-Olof Werkmäster

Printed in Sweden by Almqvist & Wiksell Tryckeri, Uppsala 1984

Contents

Part III: Transfer of technology

Acknowledgements

Catharina Åhlund-Lindqvist must be mentioned first and not last, contrary to all academic traditions. My greatest debt is to my wife for her daily companionship, loyalty and humour.

Work on this project was begun in 1974 under the guidance of Torsten Althin at the Royal Institute of Technology, Stockholm. He encouraged me to continue my studies under Sten Lindroth, Professor of the History of Ideas and Science at Uppsala University. My thanks for their inspiring teaching can no longer reach either of them.

Gunnar Eriksson, Professor Lindroth's successor, has been my tutor for the last four years, and he has gently unlocked the writer's cramp induced by these two awe-inspiring figures. Tore Frängsmyr, Professor of the History of Science at Uppsala, has been of continuous support since I began my studies there, and he has done me the honour of accepting my dissertation in his series. Most parts of the manuscript have been presented at seminars in the History of Ideas and Science at Uppsala, and Anders Lundgren and Sven Widmalm have been particularly helpful with their careful scrutiny and critical comments.

The Royal Institute of Technology made this work possible by creating a position as successor to Torsten Althin in 1978 which has enabled me to divide my time during the last seven years between teaching the history of technology in Stockholm and my own studies in Uppsala. This was an administrative innovation designed to bridge the gap between the faculties of engineering and humanities and I am deeply grateful for it. The Royal Institute of Technology has also generously financed translation of the manuscript, photography and travel, and met other costs in connection with this work. At the Royal Institute of Technology, I am indebted to Sune Berndt, Gunnar Brodin, Erling Dahlberg, Albert Danielsson, Sven Lundgren, Ulla Malm, Jan Nygren, Bengt Rundqvist, Björn Sandberg, Lennart Sandin and Gunnar Sohlenius. I would also like to acknowledge my gratitude to the late Rector, Anders Rasmuson.

A number of friends in Sweden deserve special mention. Nicolai Herlofson has been my adviser and benefactor since engineering school. Stephan Schwarz has given constant support and daily inspiration over the years. Jan Hult's enthusiasm and initiative have sustained my faith in the possibility of establishing the history of technology as an academic discipline in Sweden. Kjell Peterson has been my companion on journeys between Stockholm and Uppsala, which has made them into enjoyable pre- and post-seminars. Sven Rydberg has been of great support, both as a friend and as the author of an excellent study on Swedish travellers in England during the eighteenth century. Wilhelm Odelberg has gracefully and generously shared his encyclopaedic knowledge and his advice has been of great value. Per Sörbom has been an animated travelling companion on several journeys in pursuit of history of technology to places as different as Roanoke, Va., and Höganäs. Olle Edqvist has been the helpful combination of loyal friend and severe critic, and I have learned much from him about technology and developing countries. For many years the meetings of the Symposium Committee of the National Museum of Science and Technology, Stockholm, under the presidency of Nils-Eric Svensson, have provided me with a stimulating milieu: a perpetual symposium in themselves.

I am also indebted for their comments and help to the following persons: Göran Andolf, Per-Olof Bjällhag, Göran Blomberg, Gunnar Broberg, Ralph Edenheim, Ulla Ehrensvärd, Allan Ellenius, Rune Ferling, Pierre Guillet de Monthoux, Gunnar Hoppe, Karin Johannisson, Inge Jonsson, Åke Kromnow, Jan Larsson, Torsten Lindström, Arne Losman, Marie Nisser, Åke Norrgård, Göran Philipson, Bo Sahlholm, Inga-Britta Sandqvist, Georg Schackne, Einar Stridsberg, Carl-Gustaf Wachtmeister, Klaus Wohlert, and Gustaf Östberg.

Like most historians of technology, I have not had the advantage of daily institutional contact with colleagues in my own discipline. International contacts have therefore been of great importance. In the USA, I am particularly obliged to Eugene S. Ferguson, Melvin Kranzberg, and Edwin T. Layton, Jr, for their encouragement and support. Eugene S. Ferguson has done me the great favour of reading and commenting on the manuscript despite the fact that I imposed my own tight schedule on him. Our discussions and correspondence have taught me much about the difficulties involved in writing on the history of technology. That I have been aware of these difficulties may be apparent, but I am not sure that I can pay him the compliment of showing that I have learned how to overcome them. I am also indebted to Brooke Hindle, David A. Hounshell, Thomas P. Hughes, Jonathan Liebenau, Patrick M. Malone, Bronwyn M. Mellquist, Carroll W. Pursell, Sheldon Rothblatt, Merritt Roe Smith, and John F. Staudenmaier, S. J. In addition I would like to acknowledge my debt to the late Derek J. de Solla Price for many stimulating conversations in New Haven and Stockholm.

In England, I wish particularly to thank John S. Allen and Richard L. Hills for their help over several years. My thanks are also due to James H. Andrew, Angus Buchanan, Stephen Buckland, John Butler, Donald S. L. Cardwell, Graham J. Hollister-Short, Alex Keller, Rodney J. Law, Jennifer Tann, Alan Smith and George Watkins. R. M. Gard, County Archivist at the Northumberland County Record Office in Newcastle-upon-Tyne, has been exceedingly helpful over the years. I would also like to acknowledge my gratitude to the late Torsten Berg and his family.

Abroad, I wish also to express my appreciation of the help given by Otto Mayr and Wolfhard Weber in the Federal Republic of Germany; Edo G. Loeber in Holland; Helmer Dahl and Gunnar Nerheim in Norway; Georges Hansotte in Belgium; Alexandre Herlea, Jocelyn de Noblet and Jacques Payen in France.

Sven och Dagmar Saléns Stiftelse is a foundation whose grant in 1978 made it possible for me to give my work an international perspective, both by travelling abroad and by receiving foreign visitors in Sweden. A grant made in 1982 to the Royal Institute of Technology by another foundation, *Knut och Alice Wallenbergs Stiftelse*, enabled premises to be built and fitted out for our Department for the History of Technology directly adjoining the magnificient collection of older technical literature in the Library of the Institute. This created a most pleasing setting in which to work.

The publication of this book has been possible thanks to generous financial support from the following foundations: *Konung Gustaf VI Adolfs fond för svensk kultur, Axel och Margaret Ax:son Johnsons Stiftelse, Anders Diös fond för svensk hembygdsforskning vid Kungl. Gustav Adolfs Akademien, Kungl. Patriotiska Sällskapet, Kungl. Tekniska Högskolans i Stockholm fond för vetenskap och högre utbildning, Nordenskjöldska Swedenborgsfonden vid Kungl. Vetenskapsakademien, Stiftelsen Lars Hiertas Minne*, and *Torsten Althins Minnesfond vid Föreningen Tekniska Museets Vänner;* as well as from *Svenskt Stål AB*, the present owners of the Dannemora Mines.

Bernard Vowles, of Västra Frölunda, has translated the major part of the manuscript and wrestled with my English in those parts where I have ventured forth on my own. I am grateful to him for his interest and our many discussions. Anne Ræder has helped me to compile the bibliography from the footnotes, and Anna Hult has assisted me with the index. Bengt Vängstam has taken most of the photographs. My friends Mats Fröier and Sven Widmalm have done me an enormous service by their painstaking reading of the proofs.

With such a galaxy of benefactors, colleagues and friends to help me, I am inclined to blame any remaining faults on them.

Stockholm, June 1984

Svante Lindqvist

1. Introduction

BACKGROUND

In 1979, the Royal Swedish Academy of Engineering Sciences (IVA) published a study of the role of technology in Sweden's industrial development over the last hundred years.[1] One of the questions raised in this study was: to what extent is the technological competence of Swedish industry a result of domestic contributions, and to what extent is it due to technologies acquired from abroad? The conclusion was that the advanced technological position of Sweden has been achieved without any significant domestic contribution during the post-war era, and is due to the successful development and commercial application of imported technology.[2] In 1982, Erland Waldenström, a former President of the Academy, said in a paper on the importance to Swedish industry of an awareness of the history of technology:

We know that Sweden accounts for a very small part of the engineering research and innovation that takes place in the world, perhaps one per cent or so. By far the greater part of our economy is based on technologies imported from abroad. How this technology transfer takes place and the chief factors that influence it are therefore of fundamental importance to our economic progress. [...] To throw a scientific light on these matters it ought to be useful to approach them historically—to discover what really did take place in the past and to draw conclusions therefrom.[3]

A general definition of *technology transfer* is "the movement of a technology from one country to another or from one industry or application to another".[4] This definition covers three different phenomena, i.e. three different kinds of movement:

- Established or new technologies from industrialized nations to developing nations
- New technologies from one industrialized nation to another
- New technologies from one industry or application to another within a nation

The last of these phenomena is more often designated "diffusion of innovation", and the term "technology transfer" will be used here to describe only the first two, viz. the movement of technologies across national boundaries. Technology transfer has been a matter of interest to economists for a number of years, and many policy studies have been devoted to an appraisal of the effectiveness of efforts to stimulate economic growth by technology transfer. This reflects a growing awareness that the process of transfer is as important to technological development as—or even more important than—the actual innovation.

Most empirical studies are case studies of technology transfer in the vaguely defined period from, say, the end of the Second World War to the present day that we regard as our own time. Historical studies of technology transfer, on the other hand, have mainly attracted the interest of other historians, and much remains to be done to relate the findings of these historical studies to the work of the economists and policy makers of today.[5] Some may object that historical studies are irrelevant to the urgent problems of economic growth and social change facing modern societies. They may argue that historical studies are of no practical value, since conditions in the past were different. But that is, in fact, just what makes them so useful.

Technology transfer across national boundaries often involves moving a well-established technology to a totally different environment. There are innumerable variables—technical, economic, social, cultural, political, geographical, etc.—that may influence this process. Modern studies of technology transfer take into account only a limited number of these factors. Many factors are taken for granted without discussion, even though they may be highly relevant to the case being studied. Historical studies of technology transfer have their empirical value in that they increase our knowledge of this wide range of factors and their relative significance. Such studies deepen our understanding of the complexities that are involved when a technology is moved across cultural boundaries. To take into account the findings of historical studies when considering the nature of technology transfer is thus merely to increase the number of parameters that are examined. The conclusions drawn from such a wide variety of empirical data will be of more universal value than those drawn from studies limited to our own time.

THE PROBLEM

David S. Landes, writing about the introduction of the Newcomen engine in the eighteenth century, has said:

the early engines were generally employed only where coal was extremely cheap—as in collieries; or in mines too deep for other techniques, as in Cornwall; or in those occasional circumstances—the naval drydock at Saint Petersburg for example—where cost was no object.[6]

This was also true in Sweden, and events in Sweden during the eighteenth century reflected those in Europe as a whole. Newcomen engines were built at the mines of Dannemora and Persberg, which were too deep for other techniques; in Höganäs, which was the only colliery in Sweden; and in the dry docks of Karlskrona, where cost was no object.[7] None of the Newcomen engines built in Sweden during the eighteenth century were successful, however, and it was not until near the end of the century that the steam engine acquired any practical and economic utility in Swedish industry. The first attempt to build a Newcomen engine in Sweden was made in 1726–1728, only fifteen years after it had been successfully introduced in the collieries of England. This was the engine built by Mårten Triewald (1691–1747) at the Dannemora Mines, and well-known today from his book *Kort beskrifning, om eld- och luft-machin wid Dannemora grufwor* (A short description of the fire and air engine at the Dannemora Mines), published in 1734.[8] This book is one of the few sources of information concerning the invention and early history of the Newcomen engine, and it has often been quoted in general histories of technology. The engraving of the Dannemora engine in Triewald's book has been reproduced almost as often as the woodcuts in Agricola's *De re metallica*.

The Dannemora engine figures prominently in the popular history of Swedish technology,[9] and is frequently mentioned in the literature.[10] But knowledge of the details of this unsuccessful attempt to introduce steam power technology into Sweden has generally been superficial. Weaknesses in the design of the engine are always asserted to have been the cause of the failure, i.e. it is assumed to have been a *technical* failure. Its parts, so the explanation goes, were not robust enough for the jerky action of the engine. All of these pronouncements may be traced back to a verdict by the engineer and mining historian Carl Sahlin, in an article in *Teknisk Tidskrift* in 1918.[11] The source of Sahlin's portrayal and assessment of the Dannemora engine was a history of the Dannemora company published by Johan Wahlund in 1879.[12] But this book is not a history in the modern sense of the word, and deals only with the story of the individual iron mines in excerpts from minutes of the company's board meetings and other documents. The section relating to the engine consists of a handful of disconnected quotations from the one-sided reports of the Mine Bailiff to the Mine Inspector.[13] Not only do these quotations give an incomplete and unobjective picture of the course of events, they also, because of Wahlund's selectivity, focus only on certain technical details.

How did it come about that an attempt to build a Newcomen engine in Sweden was made as early as in the 1720s, and why did it fail? These are the main questions considered in this study, the aim of which is to examine what was specific and what was of general validity in this case of failure of technology transfer. But the study may also fulfil another purpose. The huge bronze cylinder of the Dannemora engine will serve as our mirror, and in its convex surface we may find reflected an image of early eighteenth-century Sweden.

DEFINITIONS

Four fundamental concepts employed in the following study—technology, engineer, science and scientist—are defined on two levels. They are used first in the sense defined here on the basis of present-day usage, and secondly as they were defined and used in the era being studied. The distinction between the eighteenth-century and the modern definition is important to maintain.

Technology is defined as those activities, directed towards the satisfaction of human wants, which produce change in the material world.[14] This definition is not so vague as it may perhaps sound, and has in fact to be as broad as this in order to accommodate all the phenomena that we call "technology" today. The distinction between human "wants" and more limited human "needs" is crucial, for we do not use technology only to satisfy our essential material requirements. Helmer Dahl has written that historically technology has had basically to fulfil three different functions in society:

- A *productive* function
- A *military* function
- A *symbolic* function[15]

The interesting thing about Dahl's suggestion is that it places the productive function of technology on a par with its military and symbolic functions. Elsewhere, of course, it is common today to analyse technology only in terms of its productive function, although the other two functions have in the past been of at least equal importance. Technology has been put to military use from time immemorial in order to dominate, conquer or annihilate other societies, groups or individuals. It has been used as a symbol to express, advance or maintain social conditions and cultural values. The art of casting large cylinders in bronze, for example, was applied in the eighteenth century in the manufacture of cannons (military function), church bells (symbolic function) and steam engine cylinders (productive function).

It has been said that "technology embodies the values of the society that creates it".[16] This view of technology is relevant to the question of technology transfer. If a technology is moved to another society with a different set of values, its success will depend on the degree to which these values coincide. The failure to introduce the Newcomen engine into Sweden in the early eighteenth century may, in this perspective, be seen as a result of a conflict of values. A study of this conflict might thus not only illuminate the nature of technology transfer, but also tell us something about Swedish society in the early eighteenth century.

There was no direct equivalent of the term "technology" in Swedish in the early eighteenth century.[17] The nearest approximation was the Swedish word *"mechanik"* or the Latin *mechanica*. The concept represented by this word was seen as having a practical and a theoretical component: in Swedish *"practik"* (Latin *praxis*) and *"theorie"* (Latin *theoria*) respectively.[18]

The word *engineer* is used to denote a trained person who is responsible for the development and/or maintenance of a technological system. It should be noted that the word "engineer" has been used both as a description and as a title, and that its meaning has changed considerably over the years.[19] "Engineer" taken as a unit of current sociological analysis refers to a person who has gained the title "engineer" by graduating from a technical school or technical university. This is a professional group that appeared during the nineteenth century, but for a long time it only accounted for a fraction of all the individuals engaged in capacities that we, today, describe as "engineering". It goes without saying that there have always been engineers in the descriptive sense of the word, but it is important to stress this difference between description and title. Thus, a number of persons will be referred to in the following pages as engineers although they held other titles: for example, *"Konstmästare"* at the mines, *"Capitaine-Mechanicus"* in the Fortifications Corps, *"Konst-Byggmästare"* for millwrights, and the universal title of *"Mechanicus"*.

In the early eighteenth century, *"ingenieur"* (the Swedish version of the word "engineer") was used principally as a title of officers in the Fortifications Corps,[20] but it had also begun to be applied occasionally to describe civilian engineers.[21] Not until the latter part of the century did it also become an unofficial title of consultant engineers in the emerging mechanical workshops. Those referred to as engineers in this and the following chapters thus make up a much larger group than those known as or possessing the title of *"ingenieur"* at the beginning of the eighteenth century.

Science is defined as systematized knowledge of the physical world derived from observation, study, and experimentation carried out in order to determine its nature, principles, and general laws.[22] During the eighteenth century the Swedish equivalent of the term was *"wetenskap"*, but usage varied and no less than four different meanings of *"wetenskap"* may be discerned in the early part of the century.[23] The first is a survival of a seventeenth-century usage, signifying knowledge in general. Secondly, the term was also used more specifically to denote specialist knowledge or professional understanding of a practical and utilitarian nature, e.g. in agriculture and technology. Thirdly, during the first part of the eighteenth century the term *"wetenskap"* also had a temporary meaning, corresponding to the content of the modern "science" in French and English and signifying systematized knowledge of nature. Finally, the term had begun also to be used in the sense in which it is used in modern Swedish, to denote strictly systematic knowledge in general. The context usually reveals which of these senses was intended, and in the following chapters "*wetenskap*" has been translated as "knowledge" when used in one of the first two senses and as "science" when used in the third or fourth sense.

The term *scientist* is used to denote a person devoting himself to scientific activity in the third or fourth sense defined above. There was no contemporary definition of the word "scientist"—the term does not appear until the nineteenth century.[24]

METHOD

This dissertation has been prepared in the Department of History of Ideas and Science at the University of Uppsala. This implies that the choice of subject matter and method has been influenced by the "Nordström tradition" in the discipline.[25] The classic dissertation subject in this tradition has been the introduction of a new idea or current of thought to Sweden from abroad, and I depart from this only in having chosen to study a new *technology*. This has also made it natural to focus on technology as knowledge, and to regard this development as a part of intellectual history in general.

My background is in engineering, and even though I have not had any benefit here from my own field of study (applied physics) it has made it easier to concentrate on the essentials as far as the technical aspect of this study is concerned. I began my training as a historian of technology under Torsten Althin (1897–1982) at the Royal Institute of Technology in Stockholm.[26] This was what is nowadays called an internalist training, but this is not—contrary to what is often claimed by those who lack it—a bad thing in the background of a historian of technology, who must always strike a balance between the concrete technical and the general historical perspective.[27]

Torsten Althin had many friends among the historians of technology of other countries, and he provided introductions generously. Through him, I came into contact with the Newcomen Society in London and with British colleagues.[28] This has been important since my study to a large extent deals with British technology in the eighteenth century. Althin had been inspired in the United States to begin teaching the history of technology, and he was a close friend of all those who are now known as the "Founding Fathers" of the Society for the History of Technology (SHOT).[29] Most of the ideas and problems that I take up here were inspired by discussions in SHOT and its journal, *Technology and Culture*.[30] History of technology has not yet crystallized as an academic discipline in Sweden. It is only during recent years that the subject has established itself,[31] and theory and methodology have begun to be discussed.[32] I have benefitted from these discussions, but I have also felt free to find my own way.[33]

The point of departure has been the basic characteristics of the technology in question. The Newcomen engine was the first practical working heat engine, i.e. it converted thermal energy (heat) into mechanical energy (work). Since natural resources of thermal and mechanical energy were limited, it seems clear that the advantages and disadvantages of the Newcomen engine must have been assessed in terms of these competing needs. Part I is a study of this issue in Sweden as a whole during the eighteenth century: first in outline as it can be interpreted from the number of different mechanical models housed in the Royal Chamber of Models (Ch. 2), and then in greater detail as illustrated by the proposals in the literature that were designed to save thermal and mechanical energy resources (Chs. 3 and 4). Part II is a study of how knowledge of the Newcomen engine was transferred to Sweden, and how information concerning the new technology was evaluated by the Board of Mines (Chs. 5–9). The conflict between the need for thermal and for mechanical energy resources is illustrated by a more particular case, namely the assessment of the new technology by the Board of Mines in 1725. Part III is a study of the first attempt to build and operate a Newcomen engine in Sweden. This was Mårten Triewald's engine at the Dannemora Mines, which was built in 1726–1728 and operated, intermittently, until 1734 (Chs. 10–13). A dispute between the mineowners and Triewald led to a lawsuit in which the new technology was brought to trial, and the arguments of the parties are considered in Ch. 14. The concluding chapter (Ch. 15) is a discussion of the critical factors in the transfer of technology, some of which were specific to this particular failure in technology transfer, and some of which were general difficulties always attending the movement of new technologies across national boundaries.

I am the first to acknowledge that the space devoted to the different sections is inconsistent, i.e. that the number of pages is not always commensurate with the significance of the conclusions. But this merely reflects the state of research in Sweden. We lack general surveys of the history of technology, and there are only a handful of problem-oriented monographs. As far as Mårten Triewald's background and the reasons for his proposal to build a steam engine at the Dannemora Mines in 1726 are concerned, for example, I was obliged to resort to exhaustive studies of the archives in order to avoid the traditional image of the individual entrepreneur as a hero. As for the need for mechanical and thermal energy, it was necessary to extend the time perspective to cover the whole eighteenth century in order to be able to see the Dannemora engine in a wider context. If therefore the proportions strike the reader as misplaced, it is only because I have tried to go beyond what Steven Shapin has dismissed as "footnote contextualism".[34]

THE AUTHOR'S STANDPOINT

As the work has taken shape, I have become increasingly aware of how much my own value judgments have affected the presentation, and it is therefore only proper to declare these value judgments (as far as I am conscious of them) openly to the reader. This study is not only an attempt to understand and, to the best of my ability, explain the failure of the efforts to introduce steam power technology in Sweden. It is also an attempt to argue a case on five levels.

(1) On one level it is an argument against the accepted view of Triewald's steam engine at the Dannemora Mines as a technical failure. This verdict is based on an inadequate study of the sources, and has been quoted uncritically in the literature. The course of events was more complex than that.

(2) On another level it is also an argument against much of what has previously been written about the history of Swedish technology. This example sets out to demonstrate something of the richness of the archival material available but still unexploited for research into Swedish technological history. It also shows that we should be wary of uncritically quoting even such acknowledged Swedish authorities as Carl Sahlin or Torsten Althin before we have taken the trouble to go to the sources.

(3) On a third level it is an argument against the traditional picture given by the literature of the importance of science to technical progress in Sweden in the eighteenth century. Many historians have accepted at face value the pronouncements of the eighteenth century about the importance of "combining theory and practice", or have approached the problem with a view of the link between science and technology that is drawn from our own period. What is in fact required is a historical research that starts from the peculiarity of technology as a field of knowledge and social process, and that studies each period on its own terms.

(4) The work on this dissertation was done during a period when the expression "Technology and Society" was the catchphrase of the day. As an expression of the opinion that technology should be studied from a social perspective, I approve of it, but I have been unable to escape a feeling of irritation at the indiscriminate way in which this label has been used to lend legitimacy to the most varied, and often superficial, attempts to study technology. I have often wondered what assumptions, if any, underlie the unexpressed incongruity in this pair of concepts, as I have at the cause-and-effect relationship that the sequence in which they are named seems to imply. Technology is far too often treated as an entity whose characteristics do not appear to require closer specification, but which can nevertheless be used as an explanatory factor in all manner of circumstances. Nathan Rosenberg has recently described the common conception of technology as a "black box",[35] but this is

almost too optimistic. Often (to extend his metaphor) neither its colour nor its shape are discussed. What I argue against on a fourth level is, as I perceive it, a simplified conception of the phenomenon called technology.

(5) On a fifth level this study is an argument against the modern conception of technology as simply applied science. Of all definitions of technology, this is the most superficial and also, because it is so widely held, the most dangerous. It is a view that has considerable social and economic consequences, because it provides the basis—often unstated or unconscious—for crucial policy decisions affecting education and research. It is something of a cultural myth, because it satisfies the definition of a myth as "an emotionally charged notion with no sound basis in reality". That this belief is an emotionally loaded one is confirmed by the indignation that even its simplest questioning arouses in many. This is not to say that science is unimportant to modern technological development—obviously not—but it is absurd for one of the fundamental elements of social progress to be controlled by a simple belief that there is a direct and straightforward causal connection between science and technology, and for so little interest to be devoted to studying the nature and growth of technological knowledge on its own terms.

For example: "research and development" is today not only a frequent entry in the accounts of industrial enterprises but also a common concept in economics and social science (its abbreviation, "R & D", in particular, seems to exercise a charismatic attraction on scholars).[36] It is the kind of concept that is so commonly used that no one dares ask exactly what it stands for; but this, too, implies the assumption of a simple, one-way causal relationship between science and technology and the taking for granted of scientific advances as prerequisites for technological change. Recent figures for Swedish industry show that as much as 88 % of "research and development" costs are actually spent on product and process development, and only 3 % on basic scientific research.[37] The fact that the two, epistemologically distinct, activities are often lumped together under the entry "R & D" in the statements of accounts does not mean that they should be studied as a single concept, which says nothing of their ratio, individual significance or causal relationship.

As I am a historian I have chosen to question this myth in a time other than our own. But I believe that the problems and conclusions of this study may also be of relevance today.

PART I
The Need for Energy

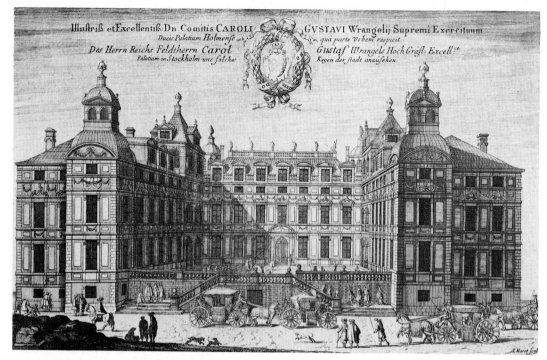

Fig. 2.1. The old Royal Palace (*Wrangelska palatset*) on Riddarholmen in Stockholm. The Royal Chamber of Models was located on the top floor between 1757 and 1802. The roof of the palace has been rebuilt since then, and this seventeenth-century engraving gives a better impression of what the palace looked like in the eighteenth century than a modern photograph. Engraving by Jean Marot in Eric Dahlberg, *Suecia antiqua et hodierna* (Stockholm, 1716). (Photo Royal Library, Stockholm)

2. Wooden Images of Technology

A visit to the Royal Chamber of Models (*Kungl. Modellkammaren*) was almost *de rigueur* for any foreign visitor to Stockholm at the end of the eighteenth century.[1] It was one of Stockholm's major public attractions. The German traveller Johann Wilhelm Schmidt, who visited Stockholm in 1799, wrote: "No matter how often I have seen this chamber of models before, I am always glad to visit it again."[2] The English traveller and mineralogist Edward Daniel Clarke (1769–1822), who visited Stockholm in the same year, described it as "one of the most pleasing of the public exhibitions of their capital".[3] His description is representative of the enthusiasm for the Royal Chamber of Models that several foreign visitors expressed in their travel journals:

As a repository of models of all kinds of mechanical contrivances, it is the most complete collection that is known. We went several times to view it; and would gladly have brought to *England* specimens of the many useful inventions there shewn. In this chamber, it is not only the number of the models that strikes the spectator, but their great beauty and the exquisite perfection of the workmanship, added to the neatness with which they are arranged and displayed. Every thing necessary to illustrate the art of agriculture in *Sweden* may here be studied;—the models of all the ploughs used in all the provinces from *Smoland* to *Lapland*; machines for chopping straw, for cutting turf to cover houses, for sawing timber, for tearing up the roots of trees in the forests, and for draining land; stoves for warming apartments, and for drying all sorts of fruit; machines for threshing corn; corn-racks; wind-mills; pumps; all sorts of mining apparatus; fishing tackle; nets; fire-ladders; beds and chairs for the sick; in short models for almost every mechanical aid requisite for the comforts and necessaries of life, within doors or without. — There can be no doubt but that patents would be required for some of them, if they were known in England: and possibly patents may have been granted for inventions that were borrowed from the models in this chamber.[4]

The Royal Chamber of Models was established in 1756, when several separate collections of mechanical models were brought together to form one single collection.[5] As an idea, these collections were only partly descendants of the Renaissance cabinets of artificial curiosities, although these had contained such items as scientific instruments and mechanical devices.[6] Derek J. de Solla Price has said that the main purpose served by the early scientific instruments was as embodiments of scientific theories: "They were not even for practical pedagogy but to symbolize the possession of the theory."[7] Most of these instruments were never used for serious measurement, but were intended to serve as symbols of the social position and education of their owners. The mechanical devices in the cabinets of artificial curiosities were likewise mainly restricted to perpetual motion machines and automata, and they were objects to ponder upon rather than prototypes for technological change.[8] No, the collections of mechanical models owed more to the new educational philosophy of object teaching, which had its origin in the empiricism of Bacon.[9] This educational philosophy was expressed in practical proposals for technical education during the seventeenth century. The Moravian educational reformer Johann Amos Comenius (1592–1670) wrote:

Mechanics do not begin by drumming rules into their apprentices. They take them into the workshop and bid them to look at the work that has been produced, and then, when they wish them to imitate this (for man is an imitating animal) they place tools in their hands and show them how they should be held and used.[10]

In 1648 Descartes had proposed the foundation of establishments for the perfecting of arts "which should be open to the public: various large rooms for artisans, each to be devoted to one craft, and to each would be joined a room filled with all the mechanical instruments necessary or useful to the art taught".[11] This proposal was a part of Descartes' attack on the contemporary emphasis on training in rhetoric, and Descartes argued that a teacher should instruct students to observe and analyze the world about them before they turned to abstract ideas.[12] Frederick B. Artz has written that "the new proposals for education 'through the senses', and of proceeding from the known to the unknown and from the concrete to the general were still in the process of formulation [during the seventeenth century] and were largely without practical programs for their realization. They continued to remain mostly on paper until the nineteenth century".[13] But they were not totally confined to paper: the new educational philosophy was embodied in wood by the establishment of collections of models.

Sweden was influenced early and directly by the new educational philosophy. Comenius was invited to Sweden in 1642 to reform its educational system, and spent some years in the 1640s in Swedish service, writing textbooks in Latin.[14] Olof Rudbeck the Elder

(1630–1702) was one of the first in Sweden to be influenced by Descartes.[15] As a professor at the University of Uppsala he created a mechanical laboratory in the 1660s filled with models where he taught students agriculture, surveying, mechanical and hydraulic engineering, building technology, naval architecture, fortifications and pyrotechnics.[16] This was the first institution for technical education in Sweden, and existed for nearly forty years until it was destroyed by fire in 1702. But the idea of object teaching in technology was to serve a utilitarian purpose throughout the eighteenth century when several other collections of mechanical models were established. Polhem's *Laboratorium mechanicum* (see Chapter 4) is well-known,[17] but it should be made clear that this was only one of several similar collections. By 1750, the Board of Mines, the Board of Commerce, the War Office, the Fortifications Corps, the Royal Swedish Academy of Sciences and the Swedish Ironmasters' Association all had such collections. There was also a collection in the new Royal Palace, which consisted mainly of models of Polhem's inventions, and a private collection established by the architect Carl Johan Cronstedt (1709–1777), who had been a pupil of Polhem.[18] In 1754 it was proposed that these collections be brought together to form a public, centralized collection of mechanical models to serve the utilitarian purpose better.

This suggestion was made by Carl Knutberg (1712–1780), *Capitaine-Mechanicus* in the Fortifications Corps, in his inaugural address to the Royal Swedish Academy of Sciences in 1754, *Tal om nyttan af ett Laboratorium Mechanicum* (On the utility of a mechanical laboratory).[19] Knutberg had studied for ten years at the University of Uppsala in 1729–1738, and had been engaged by the War Office in 1738. In 1742–1744 he was sent to Europe to report on innovations in military technology, and visited Germany, the Netherlands, France and England. He spent the winter of 1743–1744 in Paris, where he visited public libraries and attended "the professors' lectures in mechanics".[20] This means that he must have become aware of the technocratic function of the Académie Royale des Sciences. J. B. Gough has written in a biography of Réaumur:

> The French Academy, unlike the English Royal Society, was an integral part of the French bureaucratic system. This governmental role of the Academy became more and more pronounced throughout the eighteenth century as academicians assumed administrative control of French technology as consultants, inspectors, and even directors of industry. Réaumur was one of the earliest and most enthusiastic supporters of this technocratic function of the Academy, and perhaps it was for this reason that he was given charge of writing the vast industrial encyclopedia that Colbert had projected.[21]

This was the famous *Descriptions des arts et métiers*, published in 1761–1788, which Colbert in 1675 had requested the Academy to prepare.[22] The Academy had also been charged with the task of

Fig. 2.2. The collection of mechanical models of l'Académie Royale des Sciences in the Louvre in 1698. Engraving by S. le Clerc in Ernest Maindron, *L'Academie des Sciences* (Paris, 1888). (Photo Deutsches Museum, München)

examining and reporting on new machines of all kinds. A collection of models of inventions had been established in the late seventeenth century (Fig. 2.2), and although the collection was not open to the public, the Academy began publishing descriptions of the inventions with engravings in 1735.[23] It seems probable that it was in Paris that Carl Knutberg conceived his idea of a centralized collection of models in Sweden, since he suggested that it should come under the auspices of the Royal Swedish Academy of Sciences, and that it should be used by the Academy for the testing of machines and models submitted to it.[24] But the Swedish Academy bore a greater resemblance to the Royal Society on which it had been modelled, and lacked the strong links with the bureaucracy and government which characterized the Académie des Sciences.[25] When the Royal Chamber of Models was established by royal charter in 1756, it was made subordinate to the Board of Mines.[26] Although Knutberg's administrative suggestion had been rejected by the Swedish Government, his basic idea of a public, central-ized collection of models had been appreciated. He urged its usefulness as follows:

At Academies mechanics is now learned from books, but I believe that the best way to learn it and the best aid to the memory is when it is learned at the mechanical workshops themselves, or, since this cannot be done without extensive travel and heavy costs, most easily and conveniently in chambers of models. Botany is best learned on the ground or in gardens; anatomy is most clearly learned by the dissection of animal and human bodies; medicine quickest in infirmaries and hospitals; chemistry most soundly in chemical laboratories and so on and mechanics best of all in a *Laboratorio Mechanico* or a Chamber of Models, where beside the actual models, the different movements and their laws, as well as the assembly of machines etc., can most clearly be described and shown.[27]

The utility of such a chamber of models would not only be in the training of apprentices. Knutberg thought that it would also enable posterity to see what had been done in earlier times and that this might inspire the improvement of machines not yet perfected. This is, he wrote, what has happened with the fire engine ("*Eld-machin*"), which is so useful in England and elsewhere.[28] It is interesting that Knutberg chose the Newcomen engine as his example in this plea for continuity in technological development. He had realized the progressive, cumulative nature of technology, and suggested a chamber of models at a time when few constructions were recorded graphically and when most machines were built of perishable wood. These two concerns—the training of apprentices and the promotion of technological development—were his main reasons for establishing a chamber of models, which thus had the same functions as any institution of higher education: teaching and research.

The Royal Chamber of Models was located on the top floor of the old Royal Palace on Riddarholmen (Fig. 2.1). It consisted of a large hall for the exhibition of the models, and three adjoining rooms—a storeroom, a workshop and a dwelling for a modeller. Most of the various collections of mechanical models were brought here, repaired, arranged and catalogued. The collection grew as new models were made and others were handed over from other institutions. In 1779 it consisted of 212 models, and in 1801 of more than 350.[29]

All foreign visitors were struck by the size of the collection in terms of the number and the variety of models. Clarke called it the most complete collection known, and Johann Georg Eck (1777–1848), a German who viewed it in 1801, wrote: "This collection is indisputably among the greatest of its kind".[30] Giuseppe Acerbi (1773–1846), an Italian traveller who visited Sweden in 1798/99, declared it "the most complete collection of the kind that I have ever seen or heard of".[31] And Schmidt reported:

> Upon entering the hall set aside for it, one is amazed not only by the quantity (their number must run to almost 500) of models belonging to the fields of domestic economy and agriculture, mining and manufacturing, but even more by the expediency, sometimes simply achieved, sometimes by more complex means, of the models.[32]

The mechanical ingenuity of the models elicited a variety of admiring comments from foreign visitors. Schmidt saw the models as a proof of the wonderful inventiveness of man. Here, in the Royal Chamber of Models, a visitor could apprehend "how with the aid of his intelligence

man has known how to use the simple laws of nature to noble ends".[33]
Something akin to the fascination that the collection held for travellers
from abroad may be experienced today when touring any large technical museum, but there was one essential difference between the Royal
Chamber of Models and a technical museum. The latter exhibits
existing or long-abandoned technologies, and the models are made on
the pattern of established technologies, whereas the models in the Royal
Chamber of Models were prototypes of technology. They were images
of technological aspirations in Sweden during the eighteenth century.

This difference appears from the observations made by some foreign
visitors on the originality of the models. Clarke wrote in the passage
quoted above that he "would gladly have brought to England specimens of the many useful inventions there shewn", and speculated that
patents may possibly have been granted in England for inventions that
were actually borrowed from the Royal Chamber of Models. Acerbi
had similar reflections:

> My inspection influenced me to remark, that many mechanical inventions
> and improvements, which are produced to the English nation as new, may be
> found to have originated in Germany, and to have been previously known in
> Sweden. This should put the people of England upon their guard not to betray
> their ignorance in giving approbation and patronage to things that are bor-
> rowed from other nations, and held out to them as inventions. That favoured
> country possesses so much original genius, and has been the fountain of so
> much excellence, that it is vain, foolish and superfluous in its inhabitants to
> plume themselves on mechanical novelties first brought to light in other
> nations.[34]

The Royal Chamber of Models filled a *pedagogic function* as it was open
to anyone—Swedes and foreigners alike. Acerbi had paid the doorkeep-
er of the Chamber a gratuity to be allowed the privilege of seeing it
alone "without being interrupted by a crowd of spectators".[35] This
crowd also included the ironmasters, i.e. owners of ironworks. Carl
Daniel Burén (1744–1838), the owner of Boxholm Ironworks and a
typical example of a successful ironmaster of the late eighteenth century, wrote in his diary for November 7, 1792: "Today I have visited
with great pleasure the Chamber of Models in Stockholm, which
foreigners admire as quite rare of its kind."[36] The educational function
was appreciated by the South American patriot General Francisco de
Miranda (1750–1816), who did not neglect the opportunity to see "Sala
de modelos" during his visit to Stockholm in 1787:

> It is a large hall, which contains a collection of models of various machines,
> about 400 or 500 in number, which serve to give everyone who comes here and
> is willing to study them, an understanding of the construction—an admirable
> and useful thing. One sees here a series, which is called the alphabet of
> Polhem, a famous mechanical engineer here, because it shows the individual
> movements in great detail by a really alphabetical method. Only 100 *riksdaler* is
> available to keep this useful institution running.[37]

Another traveller who found this a useful institution was the French-man De Bougrenet de Latocnaye, in 1798: "The chamber of mechanical models is a well-sited attraction and of real utility".[38] It was a public institution, open to everyone, and useful because the models revealed the construction of the machines in an instructive manner—a technical school without entrance requirements. Its public character is evident from the catalogue published in 1779 by its Director, Jonas Norberg (1711–1783), *Inventarium öfver de machiner och modeller, som finnas vid Kungl. Modell-Kammaren i Stockholm, belägen uti gamla Kongshuset på K. Riddareholmen* (Inventory of the machines and models which are kept in the Royal Chamber of Models in Stockholm, situated in the old Royal Palace on Riddarholmen).[39] Note that the address was given in the title! The catalogue was inexpensively produced "so that the poorer as well as the wealthier may be able to afford to buy it".[40] This was clearly a public institution, intended to be not only known but also visited. Norberg assured the reader in his preface that:

Should anyone reading this be interested in any of the inventions, and desire to become more thoroughly informed, would the same kindly announce himself in the Royal Chamber of Models, where I am present every weekday, and will take pleasure in being of agreeable service to the Public.[41]

The most remarkable thing about the Royal Chamber of Models was the very existence of a public institution, established by royal charter and supported by the State, to function as a centralized collection of all mechanical models. *"Bien placé"*, as de Latocnaye wrote, in the heart of the city in the old Royal Palace. It provided tangible proof of the great importance attached by the State to technological development and education. It was a royal institution of technology in Stockholm a hundred years before the Royal Institute of Technology.

Teaching was not by lesson but by example. The Royal Chamber of Models constituted a collection of examples in the shape of three-dimensional objects, for every model represented a solution to a technical problem. The visiting student could take with him the visual memory of these solutions and try to recreate them in three dimensions and full size. In his article "The Mind's Eye: Nonverbal Thought in Technology" in 1977, Eugene S. Ferguson asserted that "thinking with pictures" is an essential strand in the intellectual history of technological development:

Many features and qualities of the objects that a technologist thinks about cannot be reduced to unambiguous verbal descriptions; they are dealt with in his mind by a visual, nonverbal process. His mind's eye is a well-developed organ that not only reviews the contents of his visual memory but also forms such new or modified images as his thoughts require.[42]

Ferguson shows that technical information has been transmitted and exchanged in pictorial form among technologists throughout history.

Fig. 2.3. A large number of the original models from the Royal Chamber of Models have been preserved and are today exhibited at the National Museum of Science and Technology in Stockholm. This photograph of the original arrangement of the exhibits in the museum in 1947 conveys an impression of what the Royal Chamber of Models may have looked like in the eighteenth century. (Photo National Museum of Science and Technology, Stockholm)

Technologists have converted their nonverbal knowledge either directly into objects or into drawings that have enabled others to build what was in their minds. This intellectual component of technology, Ferguson argues, has generally been unnoticed because its origin lies in art and not in science.[43] The existence of the Royal Chamber of Models points to the importance of this nonverbal component in eighteenth century technology. A technology that was disseminated by a nonverbal process was in itself by and large nonverbal.

The examples in a textbook are grouped in chapters according to the subject matter, i.e. to the different types of problem discussed. The exhibits in the Royal Chamber of Models were likewise arranged in groups in the hall according to the subject matter, i.e. to the technical problems they were designed to solve. Clarke had noted "the neatness with which they are arranged and displayed", but it was more than neatness. The French visitor Louis de Boisgelin observed in 1791 that they were "arranged in proper order",[44] and Acerbi also commented that they were "arranged in a very proper and elegant manner".[45] The subjects covered in the Royal Chamber of Models may be seen from the catalogue issued in 1779.[46] It listed 212 models in six main sections:

Section	Number of models
1. Agricultural models	43
2. Fireplaces designed to conserve wood	29
3. Mining machinery and other mills powered by water	30
4. Machines for factories and handicraft	33
5. Sluices, bridges, dams and other hydraulic works	35
6. Miscellaneous	42
	212

Since this collection comprised most of the previous individual collections of models we may be justified in regarding it as *a reflection of the general interest in technology during the eighteenth century* (up to 1779). The subjects and their order of priority in the catalogue support this assumption. The mechanization of agriculture was the topic of liveliest technological debate in the second half of the eighteenth century, when this catalogue was published, and agricultural models were therefore listed first.[47] The second section consisted of fireplaces designed to conserve wood, and although it is well known that Swedish attempts to improve tile stoves during the eighteenth century were a consequence of fear of depletion of the forests, it is nevertheless surprising to find this section ranked so high and represented by such a large number of models. The mining machines in the third section are less surprising, since the production of bar iron was Sweden's major export industry and a high level of technology in this area was consequently a matter of national importance. The fourth section was indicative of the State's persistent, but unavailing, attempts to create factories (mainly textile mills) and to mechanize handicraft production during the Era of Liberty (1718–1772).[48]

The detailed descriptions in the catalogue of the intended use of each model allow an alternative classification under more general headings. Three major technical problems seem to have been discussed during the eighteenth century: (1) mechanization of manual work, (2) conservation of wood fuel and (3) increasing efficiency in the utilization of mechanical power. An attempt to classify the models under these headings gives the following result:

Technical objective	Number of models
1. Mechanization of manual work	58
2. Conservation of wood fuel	49
3. Increasing efficiency in the utilization of mechanical power	39
4. Miscellaneous	65
	211

"Miscellaneous" is the largest group, as it is in any attempt to characterize interest in technology in the eighteenth century, when the

general optimism of the time was expressed in technology in a fascination with a great diversity of inventions.[49] The first of the major
problems, *mechanization of manual work*, mainly concerned agriculture
and the textile industry. The machines were in both cases to be driven
by muscle power—that of animals in agriculture and that of humans in
the textile industry. There was felt to be no need for alternative sources
of power in these areas during the eighteenth century, and the technological interest focussed mainly on designing muscle-powered machines
that imitated the work of a man with a tool, but with a higher
productivity.[50]

The other two major technical problems of the eighteenth century
that were reflected in the Royal Chamber of Models were conservation
of wood fuel and increasing the efficiency with which mechanical power
was utilized, or in short: *the need for energy in the forms of heat and work*. It
should, however, be pointed out that heat and work were perceived as
totally separate phenomena in the eighteenth century.[51] Developments
in what we call the field of energy technology therefore occurred in two
quite separate areas. A distinction will nevertheless be made between
the need for *thermal energy* and the need for *mechanical energy*, since the
true significance of the steam engine is that it converts thermal energy
(heat) into mechanical energy (work). It promised to meet the need for
mechanical energy that could not be provided by the traditional sources
of mechanical energy (water, muscles and wind). But on the other hand
it posed a threat to natural resources of thermal energy, and, since
Sweden lacked fossil fuel, to supplies of wood fuel. In other words, the
steam engine brought two of the major areas of technological development in eighteenth-century Sweden into conflict. The steam engine
came to Sweden as something of a paradox; both as a solution and at
the same time as a threat.

The observant reader will already have noted that the total number
of models in the last table was only 211, not 212. The missing item was
a "Model of a Fire and Air Engine to draw up water in the English
manner".[52] It was a model of a Newcomen engine on a scale of 1 to 8,
made in 1772, and classified in Norberg's catalogue under the sub-
section hoisting machines in the mining industry.[53] Only one of a total
of 212 models in the Royal Chamber of Models was of a steam engine,
and this suggests a summary explanation for the small number of steam
engines in Sweden during the eighteenth century. The steam engine
was of no help in mechanizing manual work, because the machines in
agriculture and the textile industry were to be powered by the traditional sources of muscle power (58 models). There was a great interest in
fulfilling the need for mechanical energy (39 models), and the steam
engine was thought to be useful in this respect in the mining industry (1
model). But the steam engine gave rise to a conflict between the need
for mechanical energy and the much stronger interest in conserving
wood fuel (49 models).

This is all a study of the Royal Chamber of Models reveals about the introduction of steam power technology into Sweden, and it illuminates only certain aspects of the story. It does, however, suggest the importance of understanding the two major areas of technological activity that came into conflict with the introduction of the steam engine. Chapter 3 will examine the need for thermal energy and the debate on technical improvements aimed at conserving the natural resources of wood fuel. Chapter 4 will examine the need for mechanical energy, and the debate on technical improvements aimed at increasing efficiency in the utilization of mechanical energy resources.

3. Thermal Energy

INTRODUCTION

Before the industrial revolution, when fossil fuels began to be used for the production of iron and the running of steam engines, economic expansion had of necessity to be based on timber. Virtually all aspects of material culture were entirely dependent on the forests, and the latter constituted a natural resource that appeared to be diminishing rapidly. It was a common belief in eighteenth-century Sweden that there would soon be a shortage of timber. The possibility of such a scarcity was seen as a threat to the country's leading export industry, the production of bar iron. This fear of an imminent shortage of timber led to a great interest in technical improvements aimed at reducing the number of trees felled.[1]

This chapter will describe the main features of the debate in eighteenth-century Sweden on technical improvements aimed at reducing the number of trees felled, as it is reflected in the contemporary literature. An attempt will be made to ascertain which questions attracted the most attention, which factors led to this interest and which social groups were involved in the discussion.

NATURAL RESOURCES AND TECHNOLOGY

A country's natural resources can never be quantified in absolute figures. The natural resources that are to be found in the geographical environment are exploited with the aid of technology, and it is the relationship between the geographical environment and the level of technology which defines a country's natural resources. Abbott Payson Usher has written:

> For purposes of economic and social activity, the geographic environment is not the totality of physical features, but only that part of the complex which we

can conceivably use, immediately or ultimately. This effective geographic environment is determined by our skills in using it; it is, therefore, related to the development of technology. The environment is enlarged by new knowledge and new skills.[2]

The relationship between the geographical environment and the level of technology emerges clearly if we distinguish between *potential resources* and *available resources* and regard technology as the link between them. *Potential resources* are those natural resources which meet a need in a society and which have had their location discovered and their quantity assessed. Needs change in response to general political, economic, social and technological developments. The discovery of new supplies increases the potential resources. The size of the potential resources also changes as knowledge of the natural resources in question grows and their quantity can be more accurately assessed. *Available resources* are determined by the technology that is used to exploit the potential resources, and the technology must be not only capable of achieving its intended result but also economic to employ. Nathan Rosenberg has written that "The economic usefulness of [natural] resources is subject to continual redefinition as a result of both economic changes and alterations in the stock of technological knowledge".[3] The difference between potential and available resources is thus determined by the degree to which the available technology can be efficiently and profitably used, and the level of technology in a society is decisive in defining its potential and available natural resources. However, such a definition is a relative one and only applies to a given society at a given point in time. There is consequently both a temporal and a cultural dimension in any definition of a country's natural resources. The level of technology must be studied not only in terms of technical efficiency but also in a historical and cultural perspective, because, as Rosenberg has written:

the production and use of technological knowledge must be seen against the backdrop of specific societies with different cultural heritages and values, different human capital and intellectual equipment, and confronting an environment with a very specific collection of resources.[4]

THE ENERGY NEEDS OF INDUSTRY

The production of bar iron was the most important industry in Sweden and accounted for about 70 % of the country's exports during the eighteenth century.[5] About 50 000 tons of bar iron were produced every year, and every stage in the process of production required very considerable quantities of timber.[6] It is difficult to calculate the total amount of charcoal used in blast furnaces and forges. Gösta Wieslander (1936) and Gunnar Arpi (1951) each gave a different average for the different amounts of charcoal consumed in order to produce each ton of pig iron or bar iron.[7] Leif Mattsson and Einar Stridsberg (1979) have calculated that the total consumption of charcoal by the iron industry in the second half of the eighteenth century was about 3 million m^3 of timber a year.[8] In addition, there was the actual extraction of iron ore in the mines by fire setting, which also consumed large quantities of timber, perhaps as much as 1 million m^3 a year.[9]

The importance of copper as an export product declined during the eighteenth century, and it was replaced by two industries that drew their raw materials from the forest: tar boiling and sawmills. The volume of tar exports fluctuated greatly, but on average amounted to 80 000 barrels a year and constituted 8–10 % of Sweden's total exports. The export of timber increased throughout the eighteenth century and averaged 150 000 *tolfter* (a *tolft* was 12 boards) a year, representing about 5 % of total exports. A number of other industries were also wholly dependent on the forests for fuel or raw materials: potash production, tanneries, glassworks, saltpetre works, train-oil works, lime works etc.[10]

It was not only industry that needed the forests. Eli F. Heckscher has underlined the extent to which virtually all aspects of material culture were dependent on them: for wood for heating houses; dry kindling wood; timber for making houses, fences, ships, carts, barrels, agricultural implements; burn-beating; the drying of grain and malt; etc.[11] It is impossible to calculate the total consumption of timber to meet industrial and domestic needs during the eighteenth century, but an extrapolation backwards in time from calculations for the nineteenth century suggests that it was of the order of 15–18 million m^3 a year.[12]

Heckscher concluded, when attempting to assess the proportion of timber consumption accounted for by different activities in the eighteenth century, that "domestic consumption constituted the quite preponderant part of timber consumption in earlier times and that such industrial demand for timber as there was came above all from the iron industry".[13] He estimated domestic consumption at 10–12 million m^3 a year and industrial consumption at 5–6 million. However, as we shall see, the making of such estimates was alien to eighteenth-century thinking. The development of technology was governed by other considerations.

POTENTIAL RESOURCES OF THERMAL ENERGY

The forests were practically Sweden's only thermal energy resource, and there was general anxiety during the eighteenth century that the country would be afflicted by a dearth of timber, a *"skogsödande"* as it was called at the time. It was believed that the forests were being laid waste by excessive felling, and "that many large areas of the realm are in danger of soon becoming desolate because of the shortage of timber and that the mines and towns in many parts of the country are threatened, for the same reason, with a ruin that cannot long be delayed if an early remedy is not found".[14] In the absence of such a remedy, "the fatherland will in the course of time be reduced to a miserable condition".[15]

The question of whether there was an imminent risk of a shortage of timber can be viewed from a national, a regional or a local perspective. The total supply of timber, viewed from a *national* perspective, was, of course, more than sufficient to meet the total demand in Sweden at the time. However, the potential resources that were constituted by the huge expanses of forest in the north could not be unlocked with the transportation technology of the eighteenth century.[16] This was impossible in these remote and deserted areas, where there were neither roads nor labour to fell timber and to transport it several hundred kilometres southwards to the industries of the mining region (*Bergslagen*). The potential natural resources of thermal energy were therefore never assessed on a national basis. If the situation is considered from a *regional* perspective, it seems that the mining region was never threatened with a shortage of timber during the eighteenth century. Arpi writes that the amount of timber consumed appears to have been roughly the same as the new growth, if a broad, regional view is taken. If in contrast, however, a *local* perspective is adopted, a shortage of timber was sometimes apparent in certain districts. There was on occasion insufficient timber for certain ironworks, which resulted in temporary loss of production.[17]

The notion that the country was threatened with a shortage of timber arose from the belief that too many trees were being felled. The forest is a renewable natural resource, and any assessment of its potential size must take into account both new growth and consumption. It was believed during the eighteenth century that the forest was being impoverished because consumption was greater than new growth, but the calculations of both were incomplete and erroneous. It was in fact only after the great national survey of the country's forests in the 1920s that a clear picture of the country's timber resources and of new growth and consumption was obtained. However, the question whether or not there really was an imminent risk of a timber shortage is an uninteresting one for our purposes. What matters is that the belief that such a shortage was imminent was generally accepted, and that this belief exercised a powerful influence on men's minds in the eighteenth century. This chapter will be concerned with how this belief, and also the real problems that sometimes existed at the local level, affected interest in technological development.

To be sure, the forest was the main source of thermal energy, but it was thought that there were good prospects of discovering alternative energy resources.[18] For example, it did not seem inconceivable that the country might contain unsuspected quantities of petroleum, and indeed some deposits had been found in Dalarna.[19] It was therefore only a question of increasing knowledge of the potential resources. The inventories of Sweden's natural resources that were made during the eighteenth century, of which Linnaeus's journeys are the foremost example, may be seen partly as an expression of these hopes. Peat was experimented with as a fuel, and attempts were made to use it in blast furnaces and forges.[20] Alum slate, which was used primarily as a raw material in the production of alum, was also sometimes employed as a fuel.[21] However, the greatest interest was devoted to coal. The country's only deposit of coal, in north-west Skåne, had been mined on a small scale since the seventeenth century, but it was believed that there might be other deposits elsewhere in Sweden.[22] It was felt that it was important to increase the fund of knowledge of coal mining while waiting for new seams to be discovered so that, when they were, lack of technical knowledge would not impede their rapid exploitation. A great deal of interest was also devoted to increasing the extent of the traditional, potential resources. Progress in this area was achieved mainly by reducing burn-beating, being more prudent in the felling of trees, planting new types of tree, combating insects that damaged the forests etc.

TECHNOLOGICAL DEVELOPMENT IN THE FIELD OF THERMAL ENERGY

As we have seen, the greatest industrial need for timber was to be found in Sweden's three great export industries: the production of bar iron (which included mining, pig iron production and refining the iron), tar boiling and saw mills. What technological developments took place in these areas with the object of reducing the difference between potential and available resources? And what technological improvements were made in order to reduce the need for timber for household consumption?

INDUSTRY

Fire setting (*Tillmakning*) was one of the major items in industry's consumption of timber, and it has been estimated that about 1 million m^3 of wood was used each year in the mines for the extraction of iron ore. The amount of ore obtained through fire setting could vary by as much as a factor of 5 within a mine, and this difference arose mainly from the quality of the wood employed. It was known from experience that the species of wood used, its age and moisture and the length of the

Fig. 3.1. Fire setting (*Tillmakning*) in a mine. The heat from the fire made the stone brittle and easy to excavate, or even split it by increasing the internal stresses. It is a common misconception, often repeated even in scholarly works, that the heated rock was quickly cooled off by watering, but although this might be the way to break pottery in the kitchen, it was not the way that ore was excavated in the Swedish mines. Detail of a painting from a map of the Great Coppermine in Falun by Hans Raine in 1683. (Photo Stora Kopparbergs Bergslags AB)

logs were factors that influenced the result,[23] but no systematic comparative studies were undertaken with the object of improving the technology employed. In spite of the great importance of this technology, knowledge of it was astonishingly primitive. The length of the logs used, for example, was determined by the tradition that the peasants delivered logs of two Swedish feet to the mines, and it was in any case difficult enough to persuade them to keep even to this agreed length.[24] Throughout the eighteenth century fire setting was in all essentials carried out in the traditional way,[25] and there was virtually no technical literature on the subject. For example, Rinman only devoted *one* page to this topic in his *Bergwerks lexicon* (Mining dictionary) of 1 250 pages.[26] Lindroth's explanation for this lack of interest in attempting to improve the technology is that fire setting had been practised for several hundred years, and that it was therefore assumed that improvement was hardly possible.[27] In other words, it was believed that an established and traditional technology was also an effective one.

However, an alternative technology did exist, namely the use of gunpowder. The method had been introduced into Sweden from mines on the Continent as early as the 1620s, but was very slow to gain acceptance in Swedish mines. There appear above all to have been greatly exaggerated notions about the risks involved, and a reluctance to cause explosions for fear of the rock being disturbed by "the thunder" and caving in. This misapprehension was general, from the Board of Mines to the miner at the rock face.[28] The lack of knowledge of the consequences of the alternative technology thus prevented the acquisition of the experience that might have provided that knowledge. This is something of a truism, but it is nonetheless a state of affairs that has influenced technological development in all ages. One of the advantages claimed for the alternative method was that it required less wood, but nevertheless it long remained unclear whether it actually offered any economic benefit.[29] In this respect, the results achieved did not provide unambiguous evidence, and as late as 1789 Rinman expressed the belief that fire setting was only half as expensive as the use of gunpowder.[30] According to Lindroth, those involved in mining at Stora Kopparberg in the eighteenth century did not even consider whether the use of gunpowder was cheaper than fire setting. "The use of gunpowder had been introduced, because in many cases it was more effective; economic considerations played a small role during the eighteenth century."[31]

In 1769 a new method of using gunpowder was tested.[32] The main innovations concerned the shape of the drill, which also had an improved steel tip, and heavier sledge hammers. At a test arranged by the Board of Mines the new method proved twice as efficient as the one using the old drills. The Board of Mines, which had previously regarded the use of gunpowder with "benevolent, albeit somewhat awestruck interest",[33] actively promoted its use in Swedish mines from the 1770s onwards. However, when gunpowder was used, it was employed in combination with fire setting, and the latter remained a common method until far into the nineteenth century.

Another large quantity of wood was used to make charcoal. There were two alternative methods of charcoal burning: the wood was either stacked horizontally in piles called *liggmilor* or vertically in piles called *resmilor*. It was unclear throughout the eighteenth century which of the two methods extracted more charcoal from a given quantity of wood,[34] and there was virtually no technical literature on the subject. In 1746 Magnus Wallner (1714–1773) published his *Kolare konsten uti Swerige, korterligen beskrifwen* (A brief account of charcoal burning in Sweden), a dissertation for Celsius at Uppsala in 1741, which Wallner translated from Latin into Swedish and published at his own expense.[35] As the title suggests, it was a brief description of methods and tools, supplemented by the author's own ideas, statements by charcoal burners he had met and quotations from foreign literature. The only other technical literature on this subject was a handful of articles and pamphlets concerning the relative merits of *resmilor* and *liggmilor*. The conclusions presented in these works were incomplete in that the quantity of wood used in each method was not given; nor was it clear which units of measurement were employed for the charcoal.[36] Any attempt to give quantities in defined units and to carry out systematic experiments under comparable conditions was lacking.

Studies of this kind were only made at the beginning of the nineteenth century and at the expense of the Swedish Ironmasters' Association, which was a private association of the independent ironworks. In 1814 Carl David af Uhr (1770–1849) published the results of a series of experiments in charcoal burning that had been carried out between 1811 and 1813.[37] A detailed comparison of the results achieved by the different methods was presented in tabular form in this work and this laid the foundations for the development of charcoal-burning technology. The results of these studies that were of practical utility were summarized in a handbook for charcoal burners which appeared in several editions.[38] Throughout the whole of the eighteenth century little interest was thus shown in attempts to improve charcoal-burning technology, despite the fact that, as we have seen, the supply of charcoal to the iron industry amounted to the largest item in industry's consumption of timber, about 3 million m^3 a year.

Charcoal burning was exclusively an occupation of the common people, and was undertaken by peasants and crofters who were under a tenant's obligation to deliver charcoal to the ironworks. Work in the forest was linked with the changing of the seasons and the tilling of the soil. In the autumn, after the harvest had been gathered, the wood was burnt to charcoal in the forest where it had been felled, and in the winter, when the snow made the trackless forest passable, the charcoal was carried to the ironworks by sledge.[39] The methods of charcoal burning had been formed by local conditions and established by tradition. This had resulted in many local variations, which were reflected in the names of the different types of pile. Methods were therefore characterized by total decentralization. Production was in the hands of individuals: tens of thousands of peasants and crofters, each working on his own, deep in the forests.[40] They could not be supervised or influenced, and their system was not receptive to centralized technological developments.

The common people burnt wood to obtain charcoal in the same way as they always had done in this area. They were blissfully unaware of any dissertation presented to Professor Celsius by Wallner in Uppsala. In 1760 the Swedish Ironmasters' Association decided, after pondering over whether the charcoal burners could read, to buy up Wallner's book and publish it in an enlarged edition, but the plan was abandoned when it transpired that as many as 216 copies of the 1746 edition remained unsold.[41]

The methods employed in the blast furnaces of the eighteenth century had been established in the sixteenth century and it was not until the nineteenth century that technological changes were introduced.[42] In all essentials, the technology remained unchanged throughout the eighteenth century, though certain developments did lead to some saving in charcoal consumption. There were about 400 blast furnaces in Sweden during the eighteenth century.[43] They fell into two different categories. Some, *"brukspatronshyttorna"*, belonged to the ironworks, while others, *"bergsmanshyttorna"*, were owned and worked jointly by independent miners. In their social organization, the miners' blast furnaces represented a combination of industrial activity and peasant craft. As in the case of charcoal burning, work in these blast furnaces was seasonal and linked to the tilling of the soil and the rhythm of the year. During the period of the autumn and spring floods the joint-owners took turns to use the blast furnace.[44] In comparison with the ironmasters' blast furnaces, those of the miners were small and badly built. The miners often used ore of a lower quality, and the quality of the pig iron produced was often uneven, since each of them worked the furnace by himself.[45] The ironworks obtained about half of their pig iron from the miners' blast furnaces and the remainder from their own.[46] In 1751 the Swedish Ironmasters' Association, acting at the prompting of the Board of Mines, established *Övermasmästareämbetet*, a supervisory bureau, in order to check the quality of the pig iron produced by the miners and ensure that a high standard was maintained.[47] *Övermasmästareämbetet* had both powers of inspection and an advisory function. Åke Kromnow has written that it was, as an attempt at state control of the economy, a pure expression of the mercantilist thinking of the period.[48]

During the latter part of the eighteenth century the officials of *Övermasmästareämbetet* struggled with intractable miners in an attempt to achieve improvements in the miners' blast furnaces.[49] The activities of the officials led to the accumulation of considerable practical experience, and some of them published this new knowledge in voluminous monographs such as *Försök till järnets historia* (An essay on the history of iron) in 1782 and *Bergwerks lexicon* in 1788–89, both by Sven Rinman, and *Handledning uti svenska masmästeriet* (A guide to Swedish pig iron production) in 1791 by Johan Carl Garney.[50] These works are today regarded as classics in the history of the Swedish iron industry, but this does not mean that they reflected the real situation at the time or that after their publication the miners built and worked their blast furnaces with one finger on an open page of a well-thumbed copy of Rinman or Garney.[51] The books demonstrate rather the experience that had been accumulated within *Övermasmästareämbetet* and show that the level of practical knowledge of blast-furnace technology was very high among

Fig. 3.2. Tar boiling in Österbotten. Engraving by J. H. Seeliger from A. Dahlsteen 1749. (Photo Royal Library, Stockholm)

the leading Swedish experts during the latter part of the eighteenth century. The establishment of *Övermasmästareämbetet* was an aspect of the state's efforts to maintain the quality of Swedish iron that were inspired by fear of competition from Russian iron on the international market. It was felt that no quantitative advances should be made at the expense of the quality of the product,[52] and the reduction of charcoal consumption was regarded as a secondary issue. Nonetheless, the efforts of *Övermasmästareämbetet* did produce such a reduction. It was reported in 1763 that the improvements which it had achieved within the blast furnaces "had led in many places to its now being possible to produce a *skeppspund* (a Swedish unit of weight equalling 170 kg) of pig iron with only 9–12 barrels of charcoal instead of the 15–18, or even 24, barrels that were required previously", i.e. a reduction of about 40 %.[53] The total size of this reduction nationally is an open question, and the decline in charcoal consumption, whatever its size, was not primarily the result of attempts to achieve quantitative economies but a by-product of technological developments aimed at achieving qualitative improvements.

The refinement of pig iron into bar iron at forges was carried out by two different methods in the eighteenth century: the "German process" (*Tysksmidet*), which accounted for about 90 % of the bar iron exported, and the "Walloon process" (*Vallonsmidet*), which accounted for about 10 %. These two processes did not change during the eighteenth century, and the discussions of their respective merits concerned above all the question of quality.[54] Concern for the forest as a thermal energy resource did not affect technological development and was expressed in other, purely organizational forms.

In the seventeenth century the state's policy towards the iron industry had been to reserve the forests of Bergslagen to meet the needs of the mines and blast furnaces by relocating the forges, the second stage in the process of producing iron, in areas that were better supplied with forests. In the 1740s the state succeeded in securing a restriction of the total production of the forge attached to each individual ironworks. The reason given for this measure was the danger of a timber shortage, but the real objective was to safeguard the position of Sweden's foremost export industry on the world market.[55]

The question of quality was always the most important aspect in Swedish considerations about foreign competition, above all from Russia, on the British market. The dominant idea, which has survived to our own day, was that whatever changes might affect supply and demand in foreign countries, it would nonetheless remain possible to sell Swedish iron abroad so long as the principle of "always delivering a good product" was adhered to.[56] In the view of *Övermasmästareämbetet*, the implication of this for the forges was that the primary aim was to use charcoal of the right quality, and not to reduce the quantity of charcoal consumed.[57]

Tar was one of Sweden's largest export products and accounted for 8–10 % of the country's total exports. It is hard to estimate the timber consumption involved in this industry, since the tar was largely produced through the dry distillation of the resinous duramen of pine stumps.[58] However, the needs of the tar industry were regarded as a major cause of forest depletion, and it was calculated, for example, that

Fig. 3.3. Water-powered sawmill with a single blade. Painting on a window-glass depicting the farmer Rasmus Paulson Westning in Västinge, Gotland, 1674. (Photo Nordiska museet, Stockholm)

close on 60 000 trees a year were used for tar production alone in one parish in Österbotten during the 1750s.[59] Early in the eighteenth century the state finally abandoned the plans it had entertained the previous century of establishing a tar monopoly; tar production remained purely a peasant occupation, with its economic organization unchanged. The peasants distilled the tar in the forest and sold it to merchants in the towns, who in their turn delivered it to merchants in Stockholm.[60] As in the case of charcoal burning, the methods employed varied from one locality to another and could not be supervised or influenced by the state; and, as in the case of fire setting, it seems to have been assumed that an established and traditional technology was an efficient technology.[61] In tar boiling, as in charcoal burning and fire setting, no technological development occurred.[62]

At the beginning of the eighteenth century the sawing of timber was a peasant occupation, undertaken in small and primitive mills, using frame saws driven by waterwheels (Fig. 3.3).[63] Here, too, the work was seasonal, with the partners in the mill working the saw in turn at the time of the spring and autumn floods.[64] The primitive nature of the sawmills lay not in their mechanical construction but rather in the quality of the home-forged saw blades used.[65] The blades used by the peasants were very thick and this led to a considerable loss of timber as sawdust.[66] It was, in other words, a technology that gave a poor yield from the potential resources.

In the course of the seventeenth century the Dutch had developed sawmills that employed fine-bladed and multi-bladed saws. The thinner blades reduced the loss of timber involved and produced smoother boards, while the use of several blades within the same frame made it possible to saw greater quantities and also thinner dimensions. Dutch saw blades began to be imported into Sweden in the 1730s when a number of fine-bladed and multi-bladed sawmills were established in Finland and on the west coast of Sweden. This was a technological change that was in the interests of the state, and in 1739 certain special privileges were granted to the fine-bladed sawmills.[67] In his speech to the Royal Swedish Academy of Sciences in 1748, which was subsequently printed under the title *Tal om skogarnes nytjande och vård* (On the exploitation and conservation of the forest), Ulric Rudenschöld (1704–1765) pointed out that as much timber could be produced from 6 818 2/12 *tolfter* of logs when using fine-bladed saws as from 10 000 *tolfter* of logs when using the traditional thick blades.[68] This meant, he continued, that the use of the traditional method of sawing cost the realm the equivalent of 6½ barrels of gold a year. This was an argument which his contemporaries could appreciate.

The new technology achieved its breakthrough in Sweden between 1740 and 1760,[69] and it was described at the end of the 1760s by Carl Knutberg, *Capitaine-Mechanicus* in the Fortifications Corps (see Chapter 2), in the *Proceedings of the Royal Swedish Academy of Sciences*.[70] Knutberg gave a detailed description, accompanied by drawings, of a two-framed, fine-bladed and multi-bladed sawmill (Fig. 3.4). His account was based on practical experience of various sawmills, and it is clear from his article that there was sound knowledge of the new technology by this time. The introduction of fine-bladed saws involved a change in the organization of the sawmill industry. By comparison with the small thick-bladed sawmills of the peasantry, mills incorporating the new technology entailed higher costs and required a form of economic organization more reminiscent of that of the iron industry. In the course of the eighteenth century the sawmill industry therefore passed from the peasantry into the hands of other groups in society.[71]

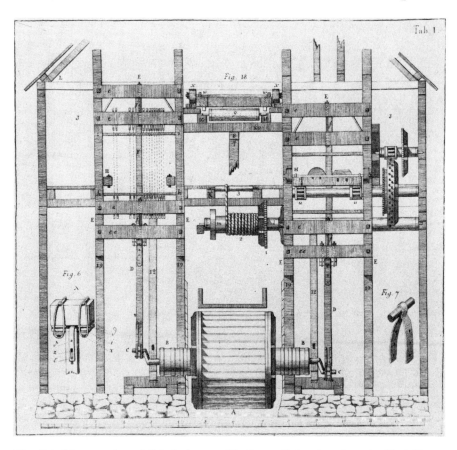

Fig. 3.4. Fine- and multibladed sawmill. One of the drawings in Carl Knutberg's description in the *Proceedings of the Royal Swedish Academy of Sciences* in 1769. (Photo Bengt Vängstam)

Concern for the forests as a source of thermal energy was not the only consideration behind the technological change from thick to fine-bladed saws. In 1741 Christopher Polhem had pointed out the advantages to be derived from the introduction of fine-bladed saws, which could produce "many and smooth boards from each log".[72] In other words, they provided not only increased yields but also better quality. Whereas the thick saws used by the peasantry produced boards that were, in Polhem's words, "shaggy like bears",[73] the fine-bladed saws produced boards that were easy to plane. They thus turned out a more finished product. It was no longer necessary to export half-finished boards which were then recut in the importing country.[74] The fine-bladed sawmills therefore represented a technological change that was entirely in accord with the tenets of mercantilism, since it provided for a better exploitation of the country's own resources combined with maximum processing within the country itself. As much value was attached to the

Fig. 3.5. Model of a fine- and multibladed sawmill. This showed an improvement of Knutberg's construction (Fig. 3.4) by Jonas Norberg, Director of the Royal Chamber of Models. The model was on display in the Royal Chamber of Models, and listed in Norberg's catalogue in 1779 as no. 99. This model, like most of the other original exhibits in the Royal Chamber of Models, was moved from one institution to another as higher technical education developed in Sweden during the nineteenth century. In 1925 the Royal Institute of Technology deposited the remains of the collection at the newly founded National Museum of Science and Technology, where this model was renovated in 1927 and where it is kept today (TM No. 1207). (Photo National Museum of Science and Technology, Stockholm)

qualitative improvement in the export product as to the quantitative benefits which the fine-bladed saws provided.

DOMESTIC CONSUMPTION

Heckscher has demonstrated how difficult it is to estimate the amount of timber used for domestic consumption during the eighteenth century, but it is quite clear that it accounted for the major part, perhaps as much as two-thirds of the total.[75] Little was known about the number of trees felled for different purposes in different parts of the country. A few local studies yielded such information as, for example, that fencing alone needed 2 828 4/7 trees each year in one parish or that building construction required 81 532 trees each year in another.[76] But the precision of these figures was misleading: they were arrived at by making a rough estimate and then multiplying it by a known factor (in this case the total acreage and the number of farmsteads in the parish), but without rounding off the result. This was a kind of rough calculation that was common in the eighteenth century. The total number of trees felled for domestic consumption could therefore never be accurately calculated. Instead, the supposedly devastating effects on the forests were alleged with a mixture of sweeping assertion and rough estimate:

The shortage of timber, especially in some provinces of the realm, which is growing with every passing day, must give rise to serious concern and consideration of the means by which it might be remedied. We inhabit a country whose climate creates a demand for a much greater quantity of firewood than is found in most other European countries. It is hardly possible to calculate the amount of timber our ironworks require or indeed the quantity consumed in the production of distilled liquor, which has reached such excessive levels at the present time. However, it is not difficult to make an approximate calculation, if we bear in mind that a household in a town, made up of 8–10 people, needs at least 10 cubic cords (*famnar*) of wood merely to heat the necessary rooms of the house, the stove, the oven and for beer-brewing.[77]

The strong conviction that the domestic consumption threatened the forests could not therefore be supported with figures, but this was hardly necessary anyway. The local examples which could be adduced served to strengthen this general conviction. 2 828 4/7 trees merely for fencing! The very amount was astonishing, but it was the precision of the figure that was fascinating: the four numerals and the four-sevenths not only indicated a quantitative value but also served to confirm a qualitative judgement. No other area of technological development intending to preserve the forests attracted such great interest as efforts to reduce domestic consumption. Many of the proposals made may seem trivial, but the ultimate motive was always a concern to conserve timber supplies to meet the needs of industry. As we shall see, it was primarily engineers who devoted themselves to these questions and to an astonishingly high degree.

Most trees felled for domestic consumption were used to provide timber for building and firewood for heating. *Timber* was used for the construction of buildings, carts, fences, tools etc. It was the most common building material, because the technology of the eighteenth century was based on wood in all essentials. Efforts to reduce the need for timber therefore focussed partly on methods of preserving existing timber buildings and partly on finding alternative building materials. The timber houses of eighteenth century towns were often ravaged by fire, and their reconstruction provides a palpable illustration of the great quantity of timber required to build an eighteenth century town. Measures to prevent fires therefore included the protective impregnation of timber with various fire-resistant substances.[78] The slow decay of timber due to rot was less dramatic but equally devastating in its

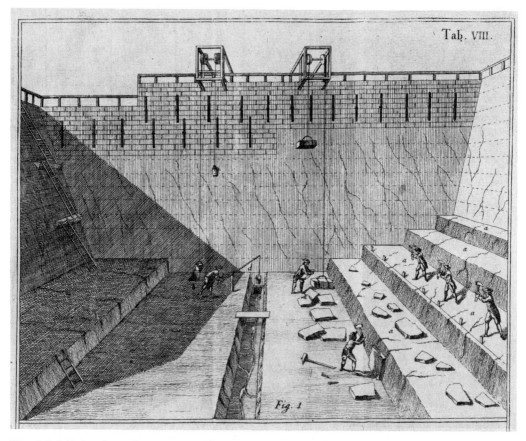

Fig. 3.6. Mining in a slate quarry. Illustration to an article by Samuel Gustaf Hermelin in the *Proceedings of the Royal Swedish Academy of Sciences* 1771, 271–294, on the mining and cutting of slate and its advantages for roofing. Hermelin based his account on a description published some years before in the French *Descriptions des arts et métiers* (see Fig. 3.7). (Photo Bengt Vängstam)

effects, and plaster was used experimentally to combat this problem. Special efforts were made to find alternative building materials, particularly ones that were incombustible. The use of stone bricks, slag bricks, clay and soil was suggested for the construction of buildings, and interest was also shown in different types of mortar, limestone and quicklime. The use of straw, tiles, iron sheeting and slate as roofing alternatives to split logs and birch-bark was recommended. The normal wooden fences were regarded as particularly damaging to the forests, since they were made from young trees, the guarantee of future forest resources, and stone walls, banks of earth, and hedges were therefore suggested in place of fencing.

Firewood was used as fuel for heating buildings, stoves and ovens and for drying malt and grain in malt houses and dry kilns etc. Tile stoves received particular attention, and various systems of bricked flues were

Fig. 3.7. The pattern for Hermelin's illustration (Fig. 3.6) was taken from this engraving in "Art de tirer des carrières la pierre d'ardoise" by Fougeroux de Bondaroy in *Descriptions des arts et métiers* (Paris, 1762), and this, in turn, was based upon an account found in the posthumous papers of Réaumur. Figs. 3.6–7 illustrate how technical innovations abroad diffused to Sweden when they supplied a need. Note that Hermelin had simplified the drawing, and reduced the number of workmen from sixteen to six. The Swedish *Proceedings* were less extravagant than the French *Descriptions*, and the purpose of Hermelin's engraving was to convey the main features of a new technology that was thought to be potentially highly useful in Sweden. (Photo Bengt Vängstam)

tested. It is well known that in 1767 Carl Johan Cronstedt was commissioned by the state to reduce the fuel consumption of tile stoves, but his proposals were only part of a long, continuous development.[79] In 1739, for example, Anders Johan Nordenberg (1696–1763), a captain in the Fortifications Corps, wrote on the subject of tile stoves in the first volume of the *Proceedings of the Royal Swedish Academy of Sciences* (Fig. 3.9).[80] He had improved the flues in the tile stoves in his own home in Stockholm, and wanted to recommend the method he had used to others:

Fig. 3.8. Fence of willows ("living fence") suggested by Johan Julius Salberg in the *Proceedings of the Royal Swedish Academy of Sciences* in 1740, 326–328. Engraving by Jean Eric Rehn. (Photo Bengt Vängstam)

The adoption of this method in the many thousands of tile stoves that can be found in this great city could reduce the consumption of wood by at least one cord a year on average for each tile stove (taking both large and small stoves into account). In this way, many thousands of cords would be saved, each household would benefit financially and, more particularly, our forests, which have suffered severely in recent years, would be preserved for the future.[81]

This is a typical expression of the eighteenth century's optimistic view of technological development and of its ability to preserve the forests: a single experiment is taken as the basis for a rapid rough calculation, and the author assumes that all good citizens will hurry to act upon his proposals. The interest in improving tile stoves was remarkable in view of the fact that such stoves were still not particularly common. The rural population constituted about 75 % of the country's inhabitants in the eighteenth century and they, and indeed many others, still used simple, open fireplaces without dampers.[82] Minor improvements to tile stoves, which were still largely restricted to the homes of the well-to-do, were therefore of relatively small importance to the country's total domestic consumption of timber. However, this was not the kind of estimating that determined technological development in the eighteenth century. If it had been, then Cronstedt would have been given the task of developing a simple and inexpensive damper, an

Fig. 3.9. Improved tile stove suggested by Anders Johan Nordenberg in 1739. The engraving by Carl Bergquist in the *Proceedings of the Royal Swedish Academy of Sciences* also illustrated an article on the poisonous blue helmet flower (*Aconitum* L.). Technology and Linnaean botany kept each other company in the proceedings of the new Academy. (Photo Bengt Vängstam)

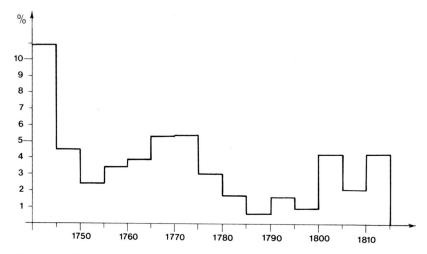

Fig. 3.10. The proportion of articles in the *Proceedings of the Royal Swedish Academy of Sciences* between 1739 and 1815 which deals with technological changes aimed at reducing the difference between potential and available thermal energy resources. Five-year average as a percentage.

improvement that would certainly have done more to reduce the domestic consumption of timber.[83]

INTEREST IN TECHNOLOGICAL DEVELOPMENT

Lindroth has written that the Royal Swedish Academy of Sciences was borne forward by the spirit of the time and itself bore forward that spirit.[84] It was something of a national institution whose purpose was to promote the country's material development. Its objective was to enlarge and disseminate within Sweden knowledge of mathematics, natural history, economics, trade, applied art and manufacturing. The degree of interest showed by the Academy in thermal energy technology, as it was reflected in its published Proceedings, can therefore be regarded as representative of the interest in technology in Sweden as a whole. In the history of the Royal Swedish Academy of Sciences which Lindroth published in 1967, he employed certain simple statistical calculations to show how the composition of the Academy and the relative degree of interest in different topics recorded in its Proceedings changed in the course of the eighteenth century.[85] The same method will be used in this chapter to examine the relative interest in different aspects of thermal energy technology and to ascertain which groups most concerned themselves with this subject.

The most important task of the Royal Swedish Academy of Sciences during the eighteenth century was the publication of its Proceedings. The president of the Academy was duty bound to publish one issue of its Proceedings every quarter, and the Proceedings were edited by the secretary. The contents of each issue were decided by the members of the Academy at their meetings and Lindroth has written that these gatherings may in reality be regarded as a sort of editorial meeting. Manuscripts submitted for publication were read out, discussed and remitted to experts for their consideration. Manuscripts were often returned to their authors for revision or parts of them were excised before printing.[86] All this constituted normal editorial work, and when a new issue was being compiled the members of the Academy must have had to decide between manuscripts which had just been received and which contained interesting new material, manuscripts which had been returned after examination by experts and manuscripts which had previously been accepted for publication but which had been set aside for inclusion, when necessary, in an issue that would otherwise have been too thin. The contents of the published Proceedings therefore came to reflect the relative importance which the Academy as an institution attached to different topics.

A perusal of the Proceedings between 1739 and 1815 shows that in total 105 of its articles dealt with technological changes designed to reduce the difference between potential and available resources of thermal energy. In some cases, the line drawn between the articles included and those excluded in this perusal is an arbitrary one (for example, certain articles about the nature of thunder ought perhaps to have been included), but these doubtful cases do not affect the overall picture. The change in the percentage of articles devoted to such topics during this period (Fig. 3.10) reveals broadly the same tendency as Lindroth found for articles dealing with the larger subject area of economics and practical matters as a whole, namely that they accounted for a very large part of the contents of the Proceedings during the first decade of the Academy's existence, but then declined in number and reached their lowest proportion in the 1780s.[87] The figure thus mirrors the enthusiasm for economics and practical matters that characterized Sweden during the Era of Liberty and also the weakening of this enthusiasm during the Gustavian period. The level of interest in such a limited and comparatively trivial subject as thermal energy technology is thus able to reflect the spirit of the time.

Of these 105 articles, 9 were devoted to mining, 4 to forestry, 32 to the heating of buildings and 60 to building technology. The energy problems of industry were thus the subject of only 12 % of the articles, while attempts to reduce domestic consumption of timber accounted for as much as 88 %. This confirms the picture that emerged above, namely that technological development within industry related primarily to *qualitative* improvements in export products, whereas there was a great deal of interest in various proposals for securing a *quantitative* reduction in the domestic consumption of timber. This attitude was wholly in accord with the mercantilist view that a country ought to refine and export its raw materials to the greatest possible extent. If raw material exports were to be maximized, it was important for them not to be "wasted" on domestic consumption, while quality was essential when attempting to ensure that they underwent as much refinement as possible before export. Mercantilism thus strongly influenced the interest shown in technological development within the field of thermal energy resources.

Nonetheless, it is remarkable that so few articles were devoted to the energy problems of industry. The consumption of wood in fire setting, charcoal burning and tar production was so large that even small improvements in the methods employed would have resulted in very considerable savings. One reason for this state of affairs was discussed earlier: technology was linked to an economic organization that was not receptive to centralized technological development.

However, there was also another and less easily defined reason for this lack of articles on the energy problems of industry. The authors of the articles which appeared in the Academy's Proceedings primarily took an interest in the technology which played a part in their own lives and which they experienced every day, palpable technology as it were. Every winter's day they saw how wood was consumed in the fires of their tile stoves. As they sat at their desks writing their articles with fingers that were numb with cold, the flames rattled the shutters on their tile stoves. They were distressed by all the timber that was destroyed when the timber-built towns they inhabited were swept by raging fires. They were amazed by the amount of timber required for roofing a house, and when they travelled along the roads of the countryside they saw mile after mile of fencing made from the trunks of young trees. On the other hand the extraction of iron ore down in the mines and charcoal burning and tar production in the depths of the forests were remote activities whose enormous consumption of wood remained invisible to them. The technology involved there was anonymous and their writings concerned technology that was visible and close at hand. In the Proceedings of the Academy, the authors wrote about matters that affected them personally, like tile stoves, stone houses and towns. They wrote mainly about their own environment and not about the open hearths and timber cottages of the common people or about work in the forests or mines. In their choice of problems to consider, technological development was circumscribed by social factors.

In his study of the composition of the Academy, Lindroth divided its members into 10 groups according to "social position and/or *vitae genus*".[88] This is a classification which reflects the social mobility in Sweden during the eighteenth century, which witnessed the growth of a class of state officials who did not own land as their predecessors had done. Lindroth's means of classification has also been adopted here and applied to the authors of the 105 articles about thermal energy technology.

	Industrial sector	Domestic sector	Total
1. Aristocrats, senior bureaucrats and senior officers	1	1	2
2. Other bureaucrats		2	2
3. Industrialists, financiers, ironmasters and landowners	1	6	7
4. University teachers		11	11
5. Physicians, surgeons and apothecaries		9	9
6. Officials of the Board of Mines	8	21	29
7. Architects, engineers and precision instrument makers	1	14	15
8. Officers in the Fortifications and Artillery Corps	2	14	16
9. Clergymen		4	4
10. Others		10	10
	13	92	105

Fig. 3.11. The authors of the 105 articles dealing with thermal energy resources, grouped by social position or profession. The ten categories are taken from Sten Lindroth, *Kungl. Svenska Vetenskapsakademiens historia*, Vol. 1: 1 (Stockholm, 1967), 27–35.

Fig. 3.11 shows how the 105 articles were divided between these 10 groups and also the number of articles from each group on industrial and domestic consumption respectively. It is clear from the figure that almost all the discussions of the energy technology of industry emanated from officials of the Board of Mines (Group 6). They also produced the greatest number of articles on the problem of the domestic consumption of timber. Their daily contact with industry made them acutely conscious of the importance of saving wood, but they were obliged to direct their efforts towards the domestic sector, because it was generally accepted that no economy measures should be allowed to put the quality of Swedish iron at risk. As a result, it was the officials of the Board of Mines who evinced the greatest interest in stoves that consumed less wood, the use of plaster on timber houses, buildings made from stone and brick, roofs made from slate and the like. The two next most frequent writers were those in groups 7 and 8, which were made up of architects, engineers and officers in the Fortifications Corps. The latter played an important role in technological development in general during the eighteenth century. We have already mentioned Carl Knutberg and Anders Johan Nordenberg, and Mårten Triewald was also an officer in the Fortifications Corps. Groups 6, 7 and 8 were made up of what we would today call engineers and together they accounted for more than half of the 105 articles. A more surprising fact is that university professors (Group 4), the trained scientists who were the driving force in the scholarly life of the Academy,[89] only contributed articles on the domestic sector. In comparison with the officials of the Board of Mines they wrote considerably fewer articles, even though they represented a much larger proportion of the Academy's members. Among the chemists, Pehr Adrian Gadd (1727–1797) of Åbo wrote about mortar in 1770 and slate roofs in 1780, while Torbern Bergman (1735–1784) of Uppsala wrote about the manufacture of roof tiles in 1771.[90] In 1760 Johan Gottschalk Wallerius (1709–1785), Bergman's predecessor in the chair at Uppsala, published an essay on different kinds of lime, which contained "what one ought to know about *which sort of lime is best for bricklaying*" (Wallerius's italics).[91] Wallerius had carried out a number of comparative experiments not only with normal limestone but also with the shells of snails, mussels and even eggs, and had reached the conclusion that it was preferable to use limestone rather than mussel shell when making mortar.

If a bricklayer of the period had happened to learn of these conclusions, he would have roared with laughter. What use were such experiments to him? Even if egg and mussel shell had proved to be better than limestone, it would have made no difference to him, since it would have been quite impossible to obtain enough shell to build a whole house. However, it is not very likely that any bricklayer heard about the conclusions Wallerius drew from his scholarly labours. While Wallerius was carrying out his experiments with egg shells, he was also supervising the construction of Uppsala University's new chemistry laboratory,[92] and it is probable that the bricklayers working on this building used the same mixture of lime, sand and water as had been used in medieval churches 500–600 years earlier and as is used today. Wallerius's experiments were certainly a first and necessary step towards a scientific study of mortar, but they also tell us something about the gap between everyday reality and the eighteenth century's lofty talk about "the practical application of the natural sciences, which has so often helped us to overcome our deficiencies and to meet our needs".[93]

Historians of today regard technology and science as two quite distinct fields.[94] Historically, the relationships between science and technology have been of three different kinds: (1) from antiquity to the middle of the eighteenth century, the only relationship was that technology served to provide a source of examples for scientists, and when the latter decided to study what technicians could already do, the result was often an increased understanding of nature; (2) in the middle of the eighteenth century, technology began to take its methods from the work of scientists, who were carrying out systematic experiments whose results were quantitatively determined in fixed units by scientific instruments; (3) during the nineteenth century scientific discoveries began for the first time to result in new technology, for example in the field of electricity and the chemical industry. The connection is more compli cated today and is often described by using an interactive model which contains all three different kinds of relationship.[95] However, in the eighteenth century types (1) and, to some extent, (2) represented the only relationships between technology and science. Thomas S. Kuhn has written about the first kind of connection, i.e. when technology serves to provide examples for science, that:

> With few exceptions, none of much significance, the scientists who turned to technology for their problems succeeded merely in validating and explaining, not in improving, techniques developed earlier and without the aid of science.[96]

This applies not only to Wallerius and his egg shells, an example of the first kind of relationship, but also to practically all scientific contributions to technological development during the eighteenth century. The systematic charcoal-burning experiments which af Uhr carried out at the request of the Swedish Ironmasters' Association at the beginning of the nineteenth century are an example of the second kind of relationship, i.e. of how technology began to take its methods from science. However, eighteenth-century Sweden offers no examples of the third kind of relationship, at least not in the field of thermal energy resources, i.e. of advances in science which led to technological development during this period (with the introduction of the lightning conductor as the only possible exception).

The link between science and technology is characterized by a time lag which complicates the relationship between them. Scientific discoveries are not immediately applied to technology. In certain areas of high technology in the present century there has, to be sure, only been a delay of about ten years, but usually it is much longer. In earlier times, it was longer still, and it was precisely in this respect that miscalculations were made during the eighteenth century. It was generally believed that the third kind of relationship existed, i.e. that scientific discoveries led to technological development, and that the link was a direct one with only an inconsiderable period of delay. This belief has provided the justification for science ever since Bacon's day, even during periods when the only kind of relationship was in reality the first.

However, there were individuals who realized that the connection between science and technology was influenced by a time factor and that science had to be brought to a certain level before it could contribute to technological development. In his inaugural speech in the Royal Swedish Academy of Sciences in 1764, Torbern Bergman stated that:

> The sciences ought to help to use nature to our advantage and to satisfy our needs, and may expect to receive in return the honour, support and reward that are their due. However, before they can serve us effectively, they must

reach a certain height or level of perfection, and they should be protected and helped to do this because of their promise for the future.[97]

The foundations of the scientific knowledge which influenced techno-logical development in the nineteenth century were laid by scientific research as a whole and not only by experiments aimed at solving specific technological problems. Fig. 3.11 therefore understates the contribution made by the scientists who constituted group 4. Techno-logy and science are two distinct branches of knowledge and cannot be included in the same table as two comparable entities, though to do so is a mistake that was also made in the eighteenth century. Nonetheless, we can draw the conclusion that the scientists of the period displayed very slight interest in contributing to the technological development of thermal energy resources.

SUMMARY

One consequence of the general belief during the eighteenth century that Sweden was threatened by a shortage of timber was a strong interest in technological development aimed at increasing the country's available thermal energy resources. If we take the contents of the *Proceedings of the Royal Swedish Academy of Sciences* as a measure of interest in this topic, we can conclude that interest was very strong during the first years of the Academy's existence. In the period 1739–1745 about 11 % of all the articles in the Academy's proceedings dealt with thermal energy resources. The level of interest subsequently declined and reached its nadir in the 1780s. This development reflects the general trend during the Era of Liberty and the Gustavian period.

The main features of this interest in technological development can be summarized in a number of dichotomies. The numerous ways in which the forest, which constituted practically the only thermal energy resource in Sweden, was used fell into two categories: *industrial* and *domestic consumption.* In the industrial sector, attention focussed above all on securing *qualitative* improvements to export products, and measures to conserve the forest had to take second place to quality. Efforts to achieve *quantitative* reductions in timber consumption therefore concen-trated on the domestic sector. Only 12 % of the articles dealing with thermal energy resources in the Academy's proceedings between 1739 and 1815 were concerned with the conservation of timber within the industrial sector, while 88 % dealt with the reduction of domestic consumption.

The remarkably low level of interest in attempting to increase the efficiency of the methods used in fire setting, charcoal burning and tar production can also be explained in part by the fact that these were small-scale undertakings carried out by the common people. They all employed a *decentralized* technology, rooted in a social pattern and a local tradition, which was not receptive to influence from without. In contrast, the state exercised a very high degree of control over bar iron production, which was a *centralized* and large-scale industry. An intermediate position was held by the miners' blast furnaces and the sawmills, which combined some of the characteristics of both peasant occupation and industrial activity. In the course of the eighteenth century, the state succeeded in obtaining control of the miners' blast furnaces and the sawmills, with the primary intention of improving the quality of export products, though an incidental result was a reduction in timber consumption.

The high level of interest in reducing domestic demand for wood fuel and timber can also perhaps be explained in part by the fact that the domestic sector employed a *palpable* technology that was close at hand and visible, while the consumption of timber in the mines for fire setting and in the forests for charcoal burning and tar production was remote and invisible, an *anonymous* technology. The men who produced the articles which appeared in the Academy's proceedings wrote about their own environment, and there was therefore a social limitation on technological development in their choice of problems.

Scientists showed no interest in attempting to make a direct contribution to technological development in the industrial sector and comparatively little interest in the domestic sector. They were convinced that there was a link between advances in science and technological progress. There was a contrast between *theory* and *practice*, since although it was believed that science could be applied to technology, in reality technology served as a source of phenomena for scientific investigation.

It was above all officials of the Board of Mines who wrote about the energy problems of industry. They were also the largest single group among those who concerned themselves with improvements within the domestic sector. Military and civilian engineers also manifested great interest in reducing the domestic consumption of timber in the heating and construction of buildings. Officials of the Board of Mines, civilian engineers and officers in the Fortifications Corps together accounted for more than half of all the articles on thermal energy technology that were published.

4. Mechanical Energy

> Waterpower doesn't need food
> and wages.[1]
>
> *Polhem*

THE NEED FOR MECHANICAL ENERGY IN INDUSTRY

The Swedish industries of the eighteenth century that were dependent on mechanical energy included mines, blast furnaces, tilt hammers and all other metal industries, saw mills, paper mills, gunpowder factories, brickworks, oil mills, glassworks etc. What industry required was *high power* (work per unit of time) and *continuous operation*, and the extent to which these requirements could be met by the traditional sources of power varied.

Muscle power (human and animal) was the most important source of energy in the eighteenth century. It could provide continuous operation, but only at a relatively low output. A man working for a 10-hour day could only produce approx. 0.1 hp continuously, whereas a horse in a good harness could produce roughly six times as much.[2] The power supplied by two persons turning a windlass or a crank, or walking in a treadmill (Fig. 4.1), might often be sufficient, but human muscle power became inadequate as soon as higher output was demanded. The horse whims used at the mines, for example, were usually driven by two horses and could thus develop approx. 1.2 hp (Fig. 4.2).[3] For continuous operation six horses were needed, because they had to work three shifts. Some forty men would have been needed to do the same amount of work. The economics of human and animal muscle power in the eighteenth-century Swedish context have not been subjected to comparative study, but animals were always preferred when power of this magnitude was required.[4] From the Great Coppermine in Falun there is information that in 1764 horse whims were preferred to windlasses when the depth of a new mine exceeded 10 fathoms.[5] It should be remembered that the horses not only needed forage and grazing, but also had to be managed at work. There was little interest during the eighteenth century in trying to improve the efficiency of horse whims and treadmills,[6] which were probably regarded in very much the same

Fig. 4.1. Full-size model (3.6 m. in diameter) of a treadwheel from 1520 preserved in the loft of Storkyrkan in Stockholm. The wheels that were used in Swedish industry during the eighteenth century were of the same kind. This replica was built at the National Museum of Science and Technology in 1980 by the author and students at the Royal Institute of Technology, and it is today exhibited at the mining museum in Ludvika. (Photo by the author, 1980)

Fig. 4.2. Horse whim used to hoist ore out of a mine. Detail of a painting from a map of the Great Coppermine in Falun by Hans Raine in 1683. (Photo Stora Kopparbergs Bergslags AB)

way as fire setting and tar boiling, i.e. as established and traditional technologies and *ipso facto* efficient. Muscle power was in any case quite insufficient to provide the high output needed for blast furnaces, tilt hammers, saw mills, mine pumps etc.

The *wind*, on the other hand, could supply much greater power. An eighteenth-century windmill developed approx. 5–10 hp.[7] Even at the lower value, this meant that a windmill produced more power than about 25 horses or 150 men (both of which would have to work in three shifts). Windpower was freely available all over the country, although varying with topographical conditions. However, it was totally dependent on unpredictable changes in the weather. Days of gale could be followed by weeks of calm. This was of little importance in a flour mill since the miller could store the grain in the mill and grind it whenever there was a wind. But continuous operation could never be guaranteed by using windpower, which was therefore of no use for industrial production. The problem is illustrated by Carl Knutberg's proposal —recorded in the *Proceedings of the Royal Swedish Academy of Sciences* for 1751—to combine a windmill with a horse whim to guarantee operation even on days of calm.[8]

Waterpower was the only traditional source of power that promised both high output and continuous operation. Muscle power could fulfil only the second of these requirements and windpower only the first. A waterwheel produced about the same power as a windmill (approx. 5–7 hp),[9] and the only problem was to secure a water supply large enough for continuous operation. The floods of spring and autumn might vary, and a year of high precipitation might be followed by several years of relative drought. But the running water of streams and rivers was nevertheless a fairly reliable source of power. By damming watercourses, operation could be sustained over a period longer than that of the seasonal floods, and the daily variations in precipitation could be evened out. Waterpower allowed continuous operation over a predictable period of time, and this made it superior to windpower.

Before the advent of electric power transmission systems in the 1880s, waterpower was firmly rooted in the geographical environment. A waterfall stayed where it was, and this truism shaped the human geography of Sweden for hundreds of years. It had governed the location of industry since medieval times when the first simple blast furnaces and sawmills were built. Around them communities grew up, and the original reason for their location is reflected in the many Swedish place names that end in "*fors*" (rapids), e.g. Bergfors, Billingsfors, Bjurfors, Bofors, Borgfors, Brattfors, Bråfors, Bäckafors ... Industries that needed high power and continuous operation were always located along rivers or streams.

The Scandinavian peninsula has the highest average precipitation in Europe. Sweden also has a large number of lakes connected by streams and rivers, and this gave the country a generous supply of waterpower. High-grade iron ore, large forests for charcoal burning and numerous small waterfalls were the three advantages enjoyed by Swedish industry over the rest of Europe, and the reasons for the dominance of the Swedish iron industry on the European market in the seventeenth and eighteenth centuries. But the waterfalls in the industrialized areas were fully utilized, and there was virtually no unexploited capacity. The newer industries were not only competing with one another for the right to use the water, but also with the flour mills, which usually claimed time-honoured priority. Attempts to dam lakes invariably led to lawsuits with the landowners who used the land for farming and grazing. The economics of this struggle for energy remains to be fully investigated, but it is clear that it was essential for the potential natural resources of waterpower to be used as efficiently as possible. To reduce the difference between potential and available waterpower resources, the water mills had to be designed to maximize output. This was mainly a matter of designing the waterwheels, a question to which we shall shortly return, but there was one area of industry in which this problem was particularly acute.

If the location of a waterfall was immovable, so also was that of a mine, which meant that a mine could not easily be supplied with waterpower. In large ones like the Great Coppermine and Sala Silvermine, both of which had been worked on an industrial scale long before the eighteenth century, the problem had been solved by damming distant lakes and leading the water to the mine through extensive systems of canals.[10] But where no water was available, or where the mine was situated at a higher altitude than the water, muscle power had to be used. The importance of treadmills in Swedish mines and other industries has probably been underestimated, but Harald Carlborg showed in 1967 that treadmills were common in Swedish mines from the early sixteenth century until the late nineteenth century.[11]

Stangenkunst technology (*Stånggångar*) was introduced to Sweden from the mines of Germany in the 1620s.[12] This was a technology that made it possible to supply mines with waterpower even over a distance or on high ground. The reciprocal motion of the connected rods was used to power the mine pumps, and large *Stangenkunst* systems were built at Swedish mines in the seventeenth century.[13] Christopher Polhem's contribution at the turn of the century was to convert the reciprocal motion of the rods to rotary motion so that it could also be used to power the hoists.[14] *Stangenkunst* was an established technology in Sweden in the eighteenth century, but it had its limitations. The loss of power due to friction was approximately 20 % per kilometre, but this meant that a waterwheel developing 5 hp still gave more power at a distance of 3 kilometres than two horse whims at the head of the mine shaft.[15] Transmission distances of up to about 2 kilometres were considered more economical than horse whims, and this appears usually to have been the maximum length.[16] But sometimes not even a combination of *Stangenkunst* and other methods was able to keep a mine free from water.[17] This was, as we shall see in Chapter 11, the situation in the Dannemora Mines that led the owners to sign a contract in 1726 with Mårten Triewald for the construction of a steam engine.

We may sum up by saying that industry needed sources of mechanical power that provided high output and continuous operation. Only waterpower could fulfil both of these requirements, but the supply of water was limited. Waterwheels that gave maximum power were thus a necessity. In the field of mechanical energy the design of waterwheels was therefore the most-debated technological problem of the eighteenth century, and we will devote the rest of this chapter to a study of the development of hydrodynamics. In *A History of Technology and Invention* Maurice Daumas and Paul Gille wrote on the development of waterwheels as a source of power during the eighteenth century:

The use of waterpower acquired all the more importance during the eighteenth century when an expansion of industry occurred before the steam engine was capable of providing a ready source of power [...] this explains why during this period many engineers and mathematicians devoted much effort to establishing a satisfactory theory of waterwheels and to improving their output. The first goal was not achieved in the eighteenth century, for until the beginning of the nineteenth none of the theoretical views proposed were satisfactory [...] The evolution of these theoretical ideas had no direct effect [...] on the construction and installation of the wheels, which continued to be done according to a time-honored tradition that evolved very slowly until approximately the end of the eighteenth century.[18]

This was also true in Sweden, but the urgency of improving the utilization of the potential energy resources was perhaps perceived as even more acute in a country so heavily dependent on its mining industry. At all events, the problem certainly attracted considerable interest during the eighteenth century and engaged some of the nation's best engineers and mathematicians. A common historical generalization today is that they tried to "unite theory and practice".[19] It will be shown, however, that the development was far more complex. Events in Sweden differed from the general pattern abroad in one important respect, and that was the influence of the remarkable hydrodynamic experiments performed by Christopher Polhem during the first years of the century.

POLHEM'S HYDRODYNAMIC EXPERIMENTS, 1702–1705

In 1696, Christopher Polhem proposed to the Board of Mines the establishment of a mechanical laboratory, where he intended to teach students mechanics and experimentation with models, *instrumenta experimentalia*, that would facilitate the construction of more efficient machinery for mining and other industries.[20] The Government granted the funds in 1697, and in 1700 the *Laboratorium mechanicum* was established as an equivalent to the *Laboratorium chemicum* in the organization of the Board of Mines. The mechanical laboratory was at first located in Falun, where Polhem had his official post as Technical Director (*Konstmästare*) of the Great Coppermine. The practical work was entrusted to Samuel Buschenfelt (1666–1706), who as Mine Surveyor (*Markscheider*) in Falun was also in the service of the Board of Mines.[21]

Buschenfelt's first assignment was to construct an apparatus that Polhem had designed for hydrodynamic experiments (Fig. 4.3).[22] This remarkable machine is well-known to Swedish historians of technology, but it has only recently attracted attention among scholars abroad thanks to Terry S. Reynolds' book *Stronger Than a Hundred Men: A History of the Vertical Water Wheel*, published in 1983.[23] Buschenfelt built the apparatus in 1701–1702, and experiments were performed in 1702–1704. Polhem, who preferred to spend as little time as possible in Falun, directed activities from his ironworks in Stjernsund. Buschenfelt was assisted by Göran Wallerius (1683–1744), who had been appointed Stipendiary in Mechanics by the Board of Mines in 1703.[24]

Sten Lindroth has written that Polhem's experimental apparatus represented a remarkable and original achievement, but that he may have been influenced by the French physicist Edmé Mariotte (ca. 1620–1684), who had made experimental studies of the effect of water

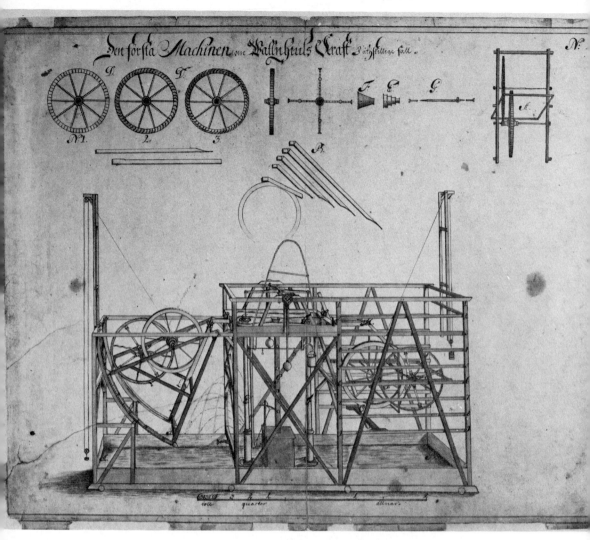

Fig. 4.3. Christopher Polhem's apparatus for hydrodynamic experimentation, built in 1701–1702. Drawing in Göran Wallerius' report to the Board of Mines in 1705. The apparatus was 3.6 m. long and 2.6 m. high. (Photo Bengt Vängstam)

and wind in mills.[25] It is not known to what extent Polhem was actually influenced by Mariotte's work, but it is clear that he adhered to the same empirical tradition. Polhem was, like Mariotte, more concerned with the articulation and application of experimentally de-termined generalizations than with their reduction to more fundamen-tal principles, and he, too, relied on common sense to guide his reason-ing.[26] But Mariotte, a scientist in seventeenth-century France, lived in a different society from that of Polhem, the *Director mechanicus* at the Great Coppermine and Stjernsund Ironworks. Whereas Mariotte's hydrodynamic engineering aimed at improving the fountains of palaces,[27] Polhem wanted to apply his results to mining machinery.

The value of priority claims in the history of technology is a dubious matter. British historians of technology have long claimed that John Smeaton in 1752–1753 was the first to test models, the earliest application in any field of engineering of scale model analysis. Whether Polhem had been influenced by Mariotte or other physicists is an open question, but it must at least be said that Christopher Polhem tested models and performed scale-model analysis in 1702–1705—fifty years before John Smeaton.

It is not necessary to describe the experimental apparatus in detail.[28] The intention was to optimize the output of waterwheels for five different parameters. The first was the *type of blade*, and three waterwheels with different types were used, all 18 Swedish inches in diameter. The second was the *ratio between the diameter of the waterwheel and the crank*, which was varied by using a winding-drum on the axle of the waterwheel with four different diameters (1, 2, 3 and 4 Swedish inches). The third was the *vertical drop*, which was varied by placing the waterwheel at eight different levels, the difference between each step being one-quarter of the diameter of the wheel. The fourth was the *inclination of the headrace*, which was moved in steps of 5 degrees between 5 and 90 degrees. The fifth was the *weight raised* by the waterwheel, which was varied by loading the wheel with multiples of 2.5 Swedish ounces.

An experiment with this apparatus involved first setting all the parameters at the specific values that were to be examined, and then opening the sluice gate to the headrace for one minute. The speed of the waterwheel was measured by counting the number of revolutions the wheel made during this time, and this figure could be determined accurately to one decimal place, as the waterwheel had ten spokes. Time was measured by a pendulum clock that marked seconds and minutes. We can calculate the total number of possible experiments by multiplying the number of values for the different parameters:

Parameter	*Number of values*
Blades	3 types
Ratio between the diameter of the waterwheel and the crank	4 diameters of the winding-drum
Vertical drop	8 heights
Inclination of the headrace	18 steps between 5 and 90 degrees
Weight raised	approx. 15 multiples of 2.5 ounces

This means that the total number of possible experiments was in the region of 25 000, and this figure is consistent with Buschenfelt's own estimate that they had performed between 20 000 and 30 000 experiments.[29] It is easy to imagine why they found the work "toilsome and plodding".[30] When the experiments were finally completed in 1704, Wallerius went to Stjernsund, where Polhem showed him how to

	1. Hjulet		2. Hjulet		3. Hjulet		4. Hjulet	
0	370	----	370	----	370	----	370	----
1	35.6	35.6	34.3	68.6	33.3	99.9	31.2	124.8
2	33.4	66.8	32.1	128.4	30.2	181.2	27.8	222.4
3	31.7	95.1	29.2	175.2	26.2	235.8	22.5	277.0
4	30.1	120.4	26.4	211.2	22.6	271.2	19.5	312.0
5	29.0	145.0	23.4	234.0	19.0	285.0	16.4	328.0
6	28.0	168.0	20.9	250.8	16.2	291.6	14.2	342.2
7	27.0	189.0	18.9	264.6	13.8	289.8	12.0	336.0
8	25.9	207.2	17.5	280.0	11.7	280.8	9.8	313.6
9	24.7	222.3	15.7	282.6	9.7	261.9	7.8	280.8
10	23.3	233.0	14.0	280.0	7.5	225.0	6.0	240.0
11	22.1	243.1	12.7	279.4	8.2	171.6	4.5	198.0
12	20.5	246.0	11.4	273.6	3.1	111.6		
13	19.5	253.5	10.1	262.6				
14	18.5	259.0	8.8	243.6				
15	17.4	261.0	7.2	216.0				
16	16.4	262.4	5.7	182.4				
17	15.5	263.5	4.2	142.8				
18	14.7	264.6						
19	14.0	266.0						
20	13.4	268.0						
21	12.8	268.8						
22	12.1	266.2						
23	11.3	259.9						
24	10.6	254.4						
25	9.9	247.5						
26	9.2	239.2						
27	8.3	224.1						
28	7.4	207.2						
29	6.4	185.6						
30	5.5	165.0						
31	4.5	139.5						
32	3.6	115.2						
33								

Fig. 4.4. One of the 64 tables in Wallerius' report in 1705 on the hydrodynamic experiments. The heading states "overshot wheel at the second level and 30 degrees inclination". This meant that three parameters in the experiment were fixed. The four column heads are for the four different diameters of the winding-drum. The column on the extreme left gives the different weights that were raised. The first column under each winding-drum diameter gives the number of revolutions the wheel made during one minute. The second column the calculated output. (Photo Bengt Vängstam)

present and interpret the results.[31] Wallerius wrote a report on the project in 1705, which was sent to the Board of Mines together with drawings and tables.[32] There were some additional experiments to be carried out before Polhem was prepared to formulate any general rules for the design of water mills.[33]

But when Polhem himself resumed work with the experimental apparatus in 1710, he discovered that two quantities had been measured inaccurately.[34] First the length of the pendulum in the clock had not been correct, which meant that the values for the speed of the waterwheel were incorrect—and hence also the values for the output. Polhem tried to reduce all these figures to their proper value with a correction coefficient, but found the work "so difficult and tedious, that no patience would have been enough for it".[35] More seriously, the protractor used to measure the inclination of the trough gave different readings when the waterwheel was placed at different levels—and this made the results incommensurable and the whole series of experiments non-reproducible. Polhem told Wallerius all this in a letter in November 1710, and concluded: "In fact between ourselves, this work is as useful as a fifth wheel on a carriage."[36]

Two years of work and 25 000 experiments had been in vain because the measuring instruments were inaccurate! "If only I had more than just sufficient to live on I would have it all done again for my own curiosity", Polhem told Wallerius.[37] However, Christopher Polhem, always hard-headed in financial matters, was far from being impoverished. His scientific curiosity was obviously for sale, but in this case he had already received his fees and the Board of Mines its report. Polhem may have feared that the Board would insist that he repeat the experiments at his own expense if it became known that the data in the report were useless. He told Wallerius not to mention the incorrectness of the experiments to anyone.[38]

Polhem's hydrodynamic experiments in 1702–1705 are a very early example of science-influenced technology. When technology began to show the influence of science in the eighteenth century, it was in the adoption of scientific methods: systematic experiments where quantities were measured in fixed units with precision instruments.[39] Polhem's experiments were scientific on two of these counts, since they were systematic and the quantities were measured in the established units of inches, ounces, seconds and degrees. He failed, however, to meet the third requirement—measurement with precision instruments. Polhem was, however, as Friedrich Neumeyer pointed out 1942, aware of the problem of scale.[40] In a letter to Emanuel Swedenborg in 1716, Polhem wrote: "no moving machine keeps the same proportions on a large as on a small scale, even if all parts are made identical and proportional".[41]

Polhem presented this enormous volume of quantitative data in diagrammatic form, and this method is perhaps the most noteworthy feature of his hydrodynamic experiments.[42] The output was calculated

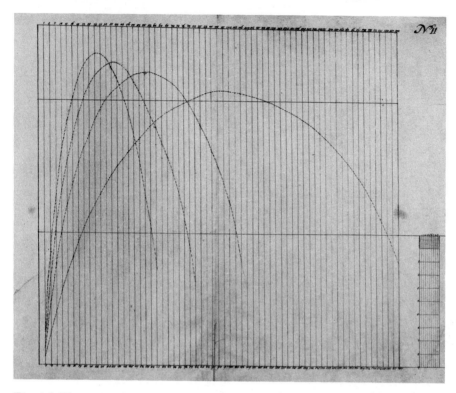

Fig. 4.5. The numerical results of the experiments were interpreted graphically by plotting the output of the waterwheels against the weight that was raised in a co-ordinate diagram. A table such as that in Fig. 4.4 gave four parabolic curves as shown in this figure: one for each diameter of the winding-drum. (Photo Bengt Vängstam)

by multiplying the weight lifted by the speed of the waterwheel, and plotted against the weight in a co-ordinate diagram. For each type of waterwheel, set at a specific level and with a specific inclination of the headrace, this resulted in four parabolic curves: one for each winding-drum diameter (Fig. 4.5). It was then easy to see which of the four diameters gave the highest output and at what weight. This was a geometrical method of finding the maximum output as a function of the weight.

The three different types of waterwheel were plotted on the same diagram, and this made it possible to see which one had the highest output (Fig. 4.6). By this method, all the quantitative data recorded in the experiments were expressed graphically. The design of a water mill could then be optimized for three parameters (blade, winding-drum and weight) simply by placing a ruler parallel to the abscissa in the diagram and moving it upwards.

There were, as has been mentioned, some additional experiments to be conducted before Polhem was willing to draw up any general rules

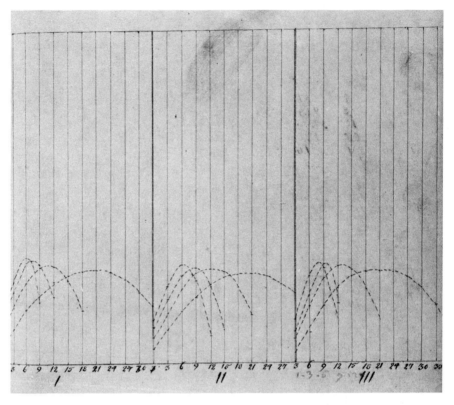

Fig. 4.6. The three different types of waterwheels were plotted in the same diagram for each level and inclination. This made it possible to determine which type of waterwheel gave the highest output at this setting of the parameters. There were 21 such diagrams in Wallerius' report. (Photo Bengt Vängstam)

for the design of water mills, but the Board of Mines was pressing him for results.[43] Polhem therefore settled for an interim report, which was sent to the Board of Mines, and this remained the only report on his hydrodynamic experiments, since he discovered the primary data to be inaccurate when resuming the work in 1710.[44] No general conclusions for the design of water mills were ever drawn from these experiments, but they nevertheless had a significant influence on later developments. Polhem's work had shown beyond any doubt that for each water mill there was an optimum speed that would give maximum output.[45] The parabolic curves showed visually the existence of these maxima, and that the output could be increased considerably if only the speed of the waterwheel was properly adjusted.

Thus, the Board of Mines was convinced in 1705 that waterpower technology had not yet reached the limit of its possible development, and that it promised to deliver a considerably improved performance. In other words, better technology could reduce the difference between

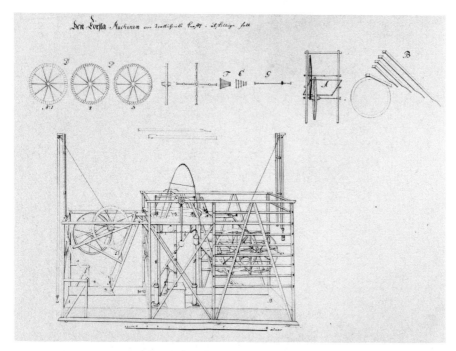

Fig. 4.7. Drawing of Polhem's experimental apparatus by Carl Johan Cronstedt in 1729. National Museum of Science and Technology, Stockholm, MS 7404, Cronstedtska planschsamlingen. (Photo National Museum of Science and Technology, Stockholm)

potential and available waterpower resources. This opinion remained largely confined to the Board of Mines for the next thirty years, until Mårten Triewald published the second volume of his lectures on experimental physics, *Föreläsningar öfwer nya naturkunnigheten*, in 1736. Triewald reproduced Wallerius' report almost in full, together with the sixty-two tables of experimental results, one engraving of the experimental apparatus (Fig. 4.8) and three illustrations of the diagrammatic method of optimization.[46] The faith of the Board of Mines in waterpower now became more widely shared and helped to shape the development of hydrodynamics in Sweden during the remainder of the eighteenth century. Triewald had concluded in his book that there was no doubt of "the use and advantages that these experiments could bring about when actually applied in practice".[47]

TWO THEORETICAL METHODS

In 1742 the Royal Swedish Academy of Sciences decided to publish a book by Pehr Elvius (1710–1749), *Mathematisk tractat om effecter af vatndrifter* (A mathematical treatise on the effect of water mills).[48] This was

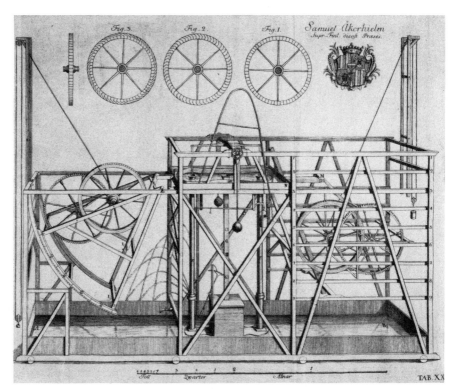

Fig. 4.8. Engraved copy of Wallerius' drawing of Polhem's experimental apparatus as illustration in Mårten Triewald, *Föreläsningar öfwer nya naturkunnigheten*, Vol. 2 (Stockholm, 1736). Engraving probably by Carl Bergquist. (Photo Bengt Vängstam)

the first time the Academy acted as the publisher of a monograph, but the result was discouraging.[49] Elvius, who was to become Secretary of the Academy in 1744, had studied mathematics and astronomy in Uppsala.[50] He had also been a pupil of Polhem, and his main interest was mechanical engineering. In his book, he attempted to solve the same problem as Polhem: how should the speed of a waterwheel be chosen between the two extremes of running freely with the speed of the current when not loaded, and standing still when the load equalled the effort applied by the water?

in the former the speed is at its highest, but in the latter the load, and in both cases there is no output. The difficulty consists therefore in how, among the numerous outputs in between, to find the highest.[51]

But, unlike Polhem, Elvius intended to solve the problem by a deductive method. Contemporary mathematicians, he wrote, had been enlightened on the laws of dynamics by the work of Galileo, Huygens and Newton, who had also given an efficient method of studying these laws in the differential calculus:

And it is owing to this, in particular, that Natural Science has been brought to its present perfection. Encouraged by success, I have therefore tried to advance *Mechanics* in the matter of increasing knowledge of machinery for producing useful power.[52]

The result was a highly mathematical treatise of 260 pages, divided into numerous propositions, corollaries, lemmas and theorems. Elvius had stated in the preace that his aim had been to produce a useful manual, while at the same time admitting that he was unable to formulate any general rules for the construction of water mills.[53] He claimed, however, that the maximum output from an undershot water-wheel was obtained when its peripheral velocity was one-third of the speed of the current.[54] This had already been asserted in 1704 by the French physicist Antoine Parent (1666–1716), who had also been using differential calculus and was likewise convinced that the practical application of mathematics had a very real utility.[55] Parent's rule was accepted as correct for almost half a century until challenged by the work of John Smeaton and Charles de Borda.[56] Smeaton established experimentally in 1752–53 that the optimum velocity ratio between wheel and current was at least $2:5$ and under certain conditions approached $1:2$.[57] In 1767, de Borda demonstrated that the power developed was proportional to the velocity of the water and not, as Parent had assumed, to its square, and that maximum output was obtained when the velocity of the wheel was equal to half the speed of the current.[58] It is interesting to note that in all his writings throughout the first half of the eighteenth century Polhem maintained his opinion that the speed of an undershot waterwheel should be half the speed of the current.[59]

Elvius' book was referred to with reverence by Swedish authors dealing with waterpower later in the eighteenth century, but it is clear that few had read it, fewer still had understood it, and perhaps none had found it useful.[60] Only 500 copies of the book were printed, but there was little demand for it although it had been advertised in the *Proceedings of the Royal Swedish Academy of Sciences*.[61] In 1784, there were still 229 copies left,[62] i.e. almost half of the edition remained unsold more than forty years after it had been published. Polhem mentioned Elvius' book the year it was published in an article on "The combination of theory and practice in mechanics" in the Proceedings, and gave it patronizing praise:

and although this book is really written for the learned, who are already familiar with the modern mathematics, which by its discoverer the learned Leibnitz is called *calculus differentialis* and by Newton, *fluxio curvarum*, so does yet Mr Elvius show his profound knowledge of such puzzling matters, that he gives hope of becoming a good *Mechanicus* with time, as well in *Practice* as now to begin with in *Theory*. For his use and that of others I wish to present the following ...[63]

In the article, Polhem gave several general rules, based on his own practical experience, for designing a water mill for maximum output. He warned that if these rules were "not carefully observed, one may easily strain at a gnat and swallow a camel, as the proverb goes".[64] Evidently Polhem thought that Elvius had swallowed a camel in his attempt to solve the problem by using the "modern mathematics".

Elvius replied half a year later in the Proceedings of the Academy with an article entitled "The theory of water mills compared with experiments".[65] He wrote that the reason why he had not compared his theory with experimental evidence before was that he had lacked sufficient data for a complete comparison.[66] The data he referred to were Polhem's hydrodynamic experiments forty years earlier, and Elvius pointed out that Polhem himself had admitted that these experiments were incomplete. "Nevertheless, I would like to present the comparison I have been able to make [...] from which one will find an agreement, which could scarcely be wished greater."[67] From the tables of Polhem's experiments, published in Triewald's lectures in 1736, Elvius had chosen two tables as "the most sufficient and most complete",[68] and showing a seemingly striking resemblance between theory and experiment. This was in fact only a *relative* agreement since Elvius had used some of the figures from the tables to calculate his own values. Moreover, these were also the results that Polhem had reached by using inaccurate measuring instruments, and they therefore lacked an *absolute* agreement with the experiments on the scale model. Elvius' comparison was in fact a specious similarity between two levels of abstraction, neither of which bore much resemblance to physical reality.

Thus by the mid-1740s there had emerged two theoretical methods of hydrodynamic research in Sweden, both of them being pursued in the hope of discovering general rules for the most efficient design of water mills. One method, first attempted by Polhem in 1702–1705 and given publicity by Triewald's lectures in 1736, was *inductive*: to formulate general rules from systematic experiments by means of parametric variation and optimization. The other method, introduced by Pehr Elvius in 1742, was *deductive*: to formulate general rules from the fundamental laws of motion by means of mathematical analysis and in particular by differential calculus.

A modest early attempt to merge the two methods was made in 1752 by the architect Carl Henrik König (1726–1804) in *Inledning til mecaniken och bygnings-konsten* (Introduction to mechanics and the art of building).[69] In the preface to his book he wrote:

An abstract theory without application is less pleasing, I have therefore not proceeded further with the theory, than is obviously useful and is indispensable to practice, and since it is necessary to have knowledge of the latter, I have included several examples of this.[70]

He devoted one chapter to waterpower, in which he presented a few simple calculations and some practical examples.[71] But it was, as the title indicated, no more than an introduction and it had little to offer either theorists or practitioners.

Olof Åkerrehn (or Åkerrén, 1754–1812) was one of the more important engineers during the early industrialization of Sweden, and perhaps the country's first consulting engineer in the modern sense.[72] He had studied mathematics, astronomy and physics at the universities of Åbo and Uppsala for ten years, 1774–1784. In 1784, he published a dissertation, *De lege Kepleriana, comparandi distantias planetarum medias a sole*, in which he tried to use Newton's theory to prove Kepler's laws of planetary motion.[73] He was appointed Auscultator in the Board of Mines in 1786, but had begun his work as a consulting engineer in 1784.[74] Over the next thirty years he designed 49 tilt hammers with 83 hearths, 51 nail hammers, 7 sheet hammers and also rolling mills, blast arrangements for furnaces, mining machinery, threshing machines, and 89 flour mills. He also directed major technical improvements at Wedevåg Ironworks, Nyköping Brassworks, the vinegar factory at Gripsholm and the white lead works at Tyresö, and he built the Kinda Canal in 1802–1810.[75]

What Åkerrehn called his "unbridled passion" was his desire to combine theory and practice in his work as a consulting engineer.[76] He made a declaration of his intentions in 1788 in his book *Utkast til en practisk afhandling om vattenverk, grundad på physiske och mechaniske lagar* (Draft of a practical treatise on water mills based on physical and mechanical laws).[77] In this book, he criticized the traditional practice of waterpower utilization:

> That an Art has long been known does not prove that it has reached perfection [...] There are no greater bunglers in this matter, than those who have to refer for what must be done to the custom, dimensions and form of Father and Grandfather.[78]

A handy farm hand can usually be found, Åkerrehn said, to build what is needed cheaply and adequately. But "only to know how to handle the axe and the plane is not enough, where Nature's own Laws demand to be consulted, and where they never, without damage and loss, can be neglected or misrepresented".[79] No, a millwright (*"Konst-Bygmästare"*) is needed who knows natural philosophy, and particularly mathematics.

> A millwright without sound knowledge of the said Sciences cannot avoid perplexity and failure, which can never be remedied by experience alone.[80]

Thus it is that "*Theory* has to be united with *Practice*".[81] This could have been written by Polhem fifty years earlier, but it was now an expression of an opinion more generally embraced: the new type of engineer had to have a theoretical education. It is not surprising to find Åkerrehn as a spokesman for this sentiment, since he himself had ten years of academic study behind him. As he tried to make his living as a consulting engineer, he must have competed with local craftsmen for commissions: craftsmen who had never heard of Kepler's laws nor of Newton. Åkerrehn's own social position may have influenced his views on the relationship between theory and practice.

But unlike Polhem, Åkerrehn wanted to unite theory and practice in the deductive method. In the introduction to his book, he reviewed what had been undertaken in the way of theoretical studies of water-power.[82] Some articles had been published in the *Proceedings of the Royal Swedish Academy of Sciences*, but he thought that few readers had access to such an expensive work. Then there was Elvius' book, but Åkerrehn considered it mathematically too advanced. "Almost no one who is not as skilled in higher Mathematics as its author was himself can derive any use from the book."[83] Åkerrehn knew the works of Bélidor, Kraft and Bossut,[84] but thought them not only rare in Sweden and difficult to understand, but useless to those most concerned since the subject was only treated mathematically and the results were not applied in practice. But how well did Åkerrehn himself succeed in his efforts to "unite theory with practice"?

His book contained only four chapters and dealt with how to supply water to water mills through dams, sluice gates, headraces etc. It was excellent as far as it went, but intended only as the first section of a larger work. This was to be entitled *Samlingar i konstbygnadsvetenskapen* (Collected papers in civil and mechanical engineering), and was planned as a major work illustrated by 80 or 90 engravings.[85] But it was never published, although as late as in 1811, a year before his death, Åkerrehn still expressed hopes of finishing the work he had begun almost thirty years earlier.[86] Åkerrehn thus failed to achieve his ambition of combining theory and practice in the deductive method.

A year after the publication of Åkerrehn's book, in 1789, Sven Rinman published the second volume of his *Bergwerks lexicon* (Mining dictionary).[87] Rinman had by then worked for twenty-five years in the Swedish Ironmasters' Association as Supervisor of blast furnaces (*Övermasmästare*), and his job had been to bring about practical improvements based upon the most efficient processes and constructions known.[88] Under the heading "Waterwheel" he described the various types that were used for different purposes, and gave some characteristic dimensions for the essential parts.[89] The article concluded:

> Concerning how waterwheels of the kind and power common here in the realm are built, no dissertation has yet, to my knowledge, been published.[90]

He was soon to publish such a book himself. Note that Rinman was not talking about how waterwheels *ought to be* constructed, but about how they *were* commonly constructed. His method was practical, or based upon what Åkerrehn dismissed as the practice of father and grandfather. But Rinman was even more widely read in the internation-

Fig. 4.9. The title page of Vol. 2 of *Afhandling rörande mechaniquen* by Sven Rinman. The decorative design was discussed by Torsten Althin in *Technology and Culture* 10 (1969), 185 f. (Photo Bengt Vängstam)

al literature than Åkerrehn, and he referred to the works of Bossut, Kraft, Poda, Delius, Leupold and others.[91]

TREATISE ON MECHANICS: VOLUMES 2 AND 1

In 1784 the Swedish Ironmasters' Association decided to publish a handbook on mechanical engineering for the mining industry.[92] It was to contain two parts: one theoretical part on mechanics, hydrostatics and hydrodynamics, and one practical part for the instruction of mill-wrights who were to carry out the construction work. The first volume was to contain the theoretical part, and the second volume the practical part—all "to preserve the proper order".[93] There seems to have been general agreement that theory was superordinate, or at least prerequisite, to practice. But the first volume published was Volume 2 (1794), and another six years were to elapse before Volume 1 was published (1800). This delay made for an interesting incongruity between the two volumes that is not obvious when they stand on a bookshelf. In the library, the two identically bound volumes stand together, with the first volume to the left of the second. Here, at least, it seems as if theory and practice were united "in the proper order" at the end of the eighteenth century.

Volume 2 of *Afhandling rörande mechaniquen, med tillämpning i synnerhet til bruk och bergwerk* (Treatise on mechanics, with particular application to mills and mines) was written by Sven Rinman (Fig. 4.9).[94] It contained 574 pages, and had a separate appendix of 53 folio engravings. There were chapters on tilt hammers, sheet hammers, forging hammers, rolling mills and slitting mills, wire-drawing and boring machines. All these machines were to be powered by water, and the text and the engravings showed the most efficient way to construct them. Improved utilization of waterpower was thus the main theme of the book and there were separate chapters on dams and headraces, while the third chapter concentrated on waterwheels.[95]

Rinman made several references to the still unpublished Volume 1, e.g. "as may be seen in the previous theoretical volume", and "more on this in the first volume". Whether Rinman really believed in this, or whether he only paid lip service to the necessity of theory, is an open question. But the fact is that his book gave all the information needed to design and build waterwheels for all kinds of mining and metalworking machinery. For example, his general rules for the design of a tilt hammer deal with the subject most thoroughly,[96] but he ends by saying: "More on this may be found under Hydraulics, in the previous Theoretical Volume."[97] But who needed more? Rinman gave a detailed specification of the dimensions and design of waterwheels for various purposes, supplemented with scale drawings. He also recommended suitable materials for the various parts, saying, for example:

Wheel spokes are best chosen from pure and good marsh spruce, as being light as well as both strong and hard, when felled at the right time, before the sap rises, cut and dried, in the shade, or under cover, so that no damaging cracks are caused by quick drying in the sun. Where there are very thick and clean-boled spruces, it is best if they are split along the heartwood, since such splits do not crack again.[98]

It is clear that Sven Rinman had expert practical knowledge of his subject. One can almost hear the sound of the carpenter's axe and smell

Hammarverkens namn.	Hela fallets höjd.	Hammarskaftets längd.	Hammarens tyngd.	Lyftarmens längd ifrån centrum.	Sprängvattnets höjd.	Qvadrat-tums öpning.	Hammarflag i minuten.
1:o. *Vid Underfall.*	aln. t.	aln. t.	lisp.	aln. t.	aln. t.		
Braås i Småland.	3: 9.	3:18.	32.	1: 9.	2: 6.	370.	80.
Bäcks hammare i Avefta.	3:18.	4:18.	41.	1:18.	2: 8.	528.	82.
Wifors i Geftrikeland.	5:17.	4.	44.	1:12.	3:12.	384.	78.
Skagersholms gamla hammaren.	4:12.	4.	36.	1:12.	2: 6.	816.	70.
Dito förbättrad.	lika.	3:18.	36.	1:12.	2:18.	336.	84.
2:o. *Vid Öfverfall.*							
Strömbacka.	7: 6.	4.	38.	1:12.	1:12.	263.	90.
Galtftröm.	7.	3:12.	36.	1:12.	1:18.	336.	107.
Skagersholm.	7.	3:18.	56.	1:12.	1:12.	264.	90.
Dormfjö, nya hammaren.	6.	3:12.	38.	1: 9.	1.	336.	90.
Tallbo vid Götheborg.	7.	3:12.	44.	1:12.	1:12.	230.	91.

Fig. 4.10. Table of the design and output of ten tilt hammers, five with undershot and five with overshot waterwheels, located in all parts of Sweden. The table showed the advantage of overshot waterwheels in terms of water consumption, and the reduction in water consumption that could be gained through improvements based upon practical experience. From Sven Rinman, *Afhandling rörande mechaniquen*, Vol. 2 (Stockholm, 1794), 107. (Photo Bengt Vängstam)

the resin! After reading Rinman's book, even today's reader feels ready to go out and build a water mill. And for good measure Rinman added a section of practical hints on the building of waterwheels, and listed the necessary tools.[99]

On the design of waterwheels, Rinman stated that it had been found by experience that an undershot waterwheel gave the greatest power when it ran not more than half as fast as when running with no load.[100] This was also, as we have seen, what Polhem had maintained throughout the first half of the eighteenth century.[101]

Rinman's method is illustrated by the section comparing the output of various tilt hammers.[102] He gave a table of ten tilt hammers, five with undershot and five with overshot waterwheels, from all parts of Sweden (Fig. 4.10). For each hammer, he listed six parameters and the output measured in strokes per minute. Two of the parameters were varied by Nature: the vertical drop and the area of the open sluice gate, the latter being dependent on the available water resources. The other four parameters were varied by local practice, three concerning the design of the hammer (weight, total length and lever arm) and the final one being the depth of the pent-up water in the headrace. Rinman compared them in order to obtain:

further information on the advantages and disadvantages of various waterfalls, as well as on the saving that can be gained by the correct construction of the waterwheels, both overshot and undershot.[103]

The table showed how the old Skagersholm tilt hammer (no. 4 in the table) had been improved (no. 5) by reducing its length from 4 Swedish ells to 3 ells 18 inches, and by increasing the height of the pent-up water in the headrace from 2 Swedish ells 6 inches to 2 Swedish ells 18 inches.[104] The area of the sluice gate had then been reduced from 816 to 336 square inches, representing a reduction in water consumption of almost 60 per cent, and yet the hammer made an additional 14 strokes per minute! This was the sort of saving in waterpower resources that could be achieved with the application of practical experience.

He then compared the water consumption of the undershot wheel for the improved Skagersholm hammer (no. 5) with that of the overshot wheel at Dormsjö (no. 9).[105] The water consumption of the undershot wheel was equal to that of the overshot wheel, yet the hammer of the latter made 6 more strokes per minute.

These observations seem to confirm that with an overshot wheel, even a low one, one may nevertheless obtain a great saving in water.[106]

This was an opinion that had gained currency during the second half of the eighteenth century, mainly due to the work of de Parcieux and Smeaton,[107] and Rinman's conclusion confirms the merit of his practical method. His *Afhandling rörande mechaniquen* was one of the outstanding technical works published in Sweden during the eighteenth century. It made an important contribution to the revival of the mining industry that took place in Sweden during the early nineteenth century, but that will not be examined here. For our purposes, we are more interested in what Rinman's book tells us about waterwheel design *before* it was published in 1794.

Since this was the first book of its kind, we may conclude that in eighteenth-century Sweden the designing and building of water mills was an *empirical* art. What Rinman did was to document the state of the art in words and pictures, revealing to us the extent and nature of the empirical knowledge of the millwright. Rinman took the normal skills of a master builder for granted, and did not include them in his book:

As for other matters pertaining to the construction of waterwheels, which properly speaking concern manual skills, assistants, tools, measurements, scaffolding, organization of work etc., it befits every millwright to understand them.[108]

His book included only the specific knowledge needed for water mills. Rinman described about 50 different kinds of mill, and for each one he listed 35–55 technical terms for the different parts. This meant that a millwright had to know the meaning of more than 2000 technical terms. He had—not unlike a student of anatomy—to know the positions of the parts concerned, the way they were connected and their function in the construction as a whole. But he had furthermore to know their dimensions and the best material for each, and to be able to construct, assemble and repair them. He had also to have some basic knowledge of mathematics, such as for example how to use π to calculate the circumference of a circle of a given diameter. Even if he did not use our concepts of force, output and efficiency, he had at least to possess some sort of an intuitive understanding of them. Most of his knowledge took the form of nonverbal, mental pictures of three-dimensional constructions, and most of his skills resided in his hands. This was the kind of knowledge that could be handed down from one generation to another by practical experience. It was acquired by working as an apprentice under an older millwright for a number of years, and by participating under his instruction in the building of a large number of mills in different local conditions. Louis C. Hunter has written of the millwright's craft:

Millwrighting as the art of designing and building water mills was perhaps the most demanding craft of its time. Its practice by the millwright required familiarity not only with the building crafts of the carpenter, the joiner, and the mason but with blacksmithing, the wheelwright's trade, and the emerging craft of the machinist. Since the running of lines and the fixing of levels with fair accuracy were indispensable for the proper location and layout of mill and hydraulic facilities, the millwright had to be something of a surveyor as well. He was the key technician of the preindustrial and early industrial age, the craftsman predecessor of the civil and mechanical engineer.[109]

The professional competence and importance of the millwrights were overshadowed by the emergence of professional consulting engineers at the beginning of the nineteenth century. The latter, with their higher education and social position, were better able to speak for themselves. In Åkerrehn's judgement on the millwright in 1788, we have already seen an early example of how the new, formally educated engineers tried to discredit the millwrights as a part of their effort to establish a professional identity. The millwright of the eighteenth century was described by William Fairbairn in his *Treatise on Mills and Millwork* in 1861.[110] Although this concerned England, it was probably also true of millwrights in Sweden:

The millwright of former days [...] was the engineer of the district in which he lived, a kind of jack-of-all-trades, who could with equal facility work at the lathe, the anvil, or the carpenter's bench. In country districts, far removed from towns, he had to exercise all these professions, and he thus gained the character of an ingenious, roving, rollicking blade, able to turn his hand to anything, and, like other wandering tribes in days of old, went about the country from mill to mill, with the old song of "kettle to mend" reapplied to the more important fractures of machinery.[111]

We will return to the question of the social position of the millwright in Chapter 9, and in this context it is enough to say that the empirical knowledge of the millwright was far more extensive than that of any ordinary craftsman. The millwrights were the polytechnicians before the days of the polytechnics.

The first volume of *Afhandling rörande mechaniquen* was finally published in 1800, and it was written by Erik Nordwall (1753–1835, ennobled as Nordewall in 1816).[112] He had studied at the University of Uppsala and had worked under Sven Rinman in the Swedish Ironmasters' Association.[113] His book set out to develop:

a theory of the movement of water, and its effect on waterwheels; which would be more reliable for practical use and generally more available to the millwright than the one we had before.[114]

This was an obvious allusion to Elvius' book. Nordwall was critical of the deductive method, and thought that mathematical reasoning "often led a practitioner away from the light, instead of enlightening him".[115] Like Polhem, he wanted to perform experiments with scale models of waterwheels. Nordwall had his experimental apparatus ready by 1792, but to the annoyance of his employer, the Swedish Ironmasters' Association, the completion of the work was delayed year after year and he kept excusing himself on the grounds that additional experiments were necessary.[116] His situation was very similar to that encountered some ninety years earlier by Polhem, who had also been urged by his superior, the Board of Mines, to deliver his report although he felt that much remained to be done. Due to the large number of parameters in a water mill, the inductive method led them both to discover a need for more and more experiments until their principals said that enough was enough.

In his book, Nordwall stuck faithfully to his basic idea that theory should be based upon "evidence, confirmed by, or founded on incontrovertible experience".[117] He had diligently examined the change in output of a large variety of waterwheels when he varied parameters such as the number, angle and shape of the blades; the position, angle and height of the headrace; the diameter and width of the waterwheel etc. But Nordwall optimized each parameter under different conditions, and his experiments therefore resulted only in hundreds of incommensurable conclusions. "The conclusions that can be drawn from these experiments are almost as numerous as the experiments", was one verdict on Nordwall's work in 1822.[118] Another, more charitable, passed in 1817, was that his book was "more a priceless stock of experiments than a fully developed theory".[119] But a practitioner who wanted to know how to design a water mill for a specific purpose could turn the pages in vain. It would be a great coincidence if he found a situation examined by Nordwall that corresponded to his own set of circumstances.[120]

THE FALUN EXPERIMENTS

It was Nordwall's belief in the value of quantitative data in great bulk that had delayed the publication of his book. Afterwards, he continued to maintain that additional experiments would ultimately lead to a general theory of waterpower.[121] This belief was shared by others, and an experimental apparatus, twice the size of Nordwall's, was built in 1804 at the Great Coppermine in Falun.[122] It was intended for investigating the efficiency of winding gear at the mine in Falun, and experiments were conducted in 1804–1805 that seemed to confirm Nordwall's results.[123] The famous metallurgist Eric Thomas Svedenstierna (1765–1825) and others supported the idea that the Swedish Ironmasters' Association should finance a lengthy new series of experiments with the apparatus in Falun.[124]

This step was decided upon in 1811, and the intention was that Nordwall and Jonas Gustaf Möllenhoff (?–1812) should direct the experiments.[125] The practical work was to be carried out by Pehr Lagerhjelm (1787–1856), with the assistance of Jacob Henrik af Forselles (1785–1855).[126] When Möllenhoff died the following year, he was replaced by Zacharias Nordmark (1751–1828), professor in experimental physics at Uppsala.[127] But Lagerhjelm soon took charge of the project and abandoned the original plan devised by Nordwall and Möllenhoff. He directed the experiments himself, with the assistance of af Forselles and Georg Samuel Kallstenius (1786–1863).[128] The result was not the additional information that Nordwall had hoped for, but a completely new series of experiments and a highly critical revision of Nordwall's work.

These experiments, carried out in 1811–1815, became widely known as the "Falun experiments".[129] The publication of the results in two volumes in 1818 and 1822 represents a turning point in the history of hydrodynamics in Sweden and marks the beginning of scientific technology in the modern sense in the country. There was, for example, a precision in the measurements quite lacking in earlier experiments by Polhem and Nordwall. *Linear measurements* were taken on "a precisely graduated decimal scale, comprising two Swedish feet", made by Johan Gustaf Hasselström (1742–1812), manufacturer of mathematical instruments to the Royal Swedish Academy of Sciences.[130] *Weights* were measured with "a set of weights, calibrated against the Swedish original standard, which is kept in the Archives of the Royal Treasury", and a balance made by Gabriel Collin (1761–1825), manufacturer of optical instruments to the Academy.[131] *Time* was kept by a clock borrowed from the Astronomical Observatory of the University of Uppsala.[132] Another feature of this work that distinguished it from earlier publications was the thorough and critical review of the international literature in hydrodynamics. The first fifty pages of the second volume were devoted to comments on the works of Smeaton, Euler, Borda, Bossut, Banks, Langsdorf and others.[133] The review of Elvius and Nordwall was particularly critical, almost derisive, in tone.[134]

Elvius' greatest mistake, said Lagerhjelm, had been to assume that all the water remained in the buckets until they reached their lowest position.[135] This approximation simplified calculation, since the water could be assumed to be uniformly distributed around the wheel, but it led to absurd results. Lagerhjelm stated that Euler, Borda and Bossut had all worked with the same erroneous approximation until Langsdorf "declared this formula fundamentally wrong" in *Lehrbuch der Hydraulik* (1794).[136] The pronouncement on *A Treatise on Mills*, by John Banks (1795), was: "A remarkable example of the inaccuracy of crude experience, since the Observer does not know how to differentiate the accidental from the essential."[137] John Smeaton's famous work *Experimental Enquiries Concerning the Natural Power of Wind and Water* (1796) was more favourably reviewed, but even so Lagerhjelm concluded:

> This author, who only and solely reports the actual experiments, seems to take it for granted that experience is the only teacher in these matters.[138]

Lagerhjelm was obviously mindful of the two ways of reasoning, and he did not favour either the inductive or the deductive method in his critique. In the preface to the second volume, Lagerhjelm gave an interesting epistemological programme for the Falun experiments. On the theory of a new type of breast wheel, where the water only acted through its weight, he wrote that it was "completely independent of experience".[139] The design of the waterwheel was, according to Lagerhjelm, determined by fifteen parameters, of which six were constants and the remaining nine could be determined through equations.[140] The theory of the design of such a waterwheel was therefore "evident *a priori* (if the expression may be permitted)".[141] But for other types of waterwheel, there was a more complex relationship between theory and experience:

> Since the water acts either only and solely by its impact, or by impact and weight together; so is the treatment of the subject dependent on experience; although in no way as if experience alone, i.e. experiments, were demanded for the investigation of the subject. Knowledge is universal, and is thereby distinguished from abstraction from a given experience, in that the latter is only

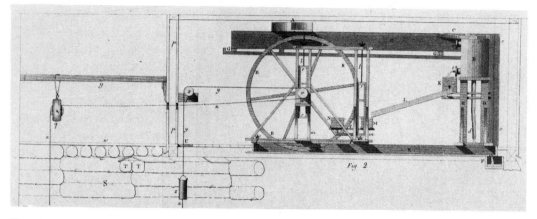

Fig. 4.11. The machine used in the Falun experiments in hydrodynamics in 1811–1815. The output of different waterwheels was examined by hoisting weights up the mineshaft (left). Water was supplied by the system that powered the mining machines at the Great Coppermine. The machine was almost 6 m. long and 2.5 m. high. Illustration from Pehr Lagerhjelm et al., *Hydrauliska försök, anställda vid Fahlu grufva, åren 1811–1815*, Vol. 2 (Stockholm, 1822). (Photo Bengt Vängstam)

valid under the circumstances and within the boundaries, that essentially attended that class of phenomenon, which one experienced. This universality (form) gathers knowledge, from its speculative root; but truth, its consistency with real facts (content), is gathered from experience. The resolution of the conflict between these opposite extremes within this subject lies in every true Natural Philosopher's own method.[142]

Thus did Lagerhjelm philosophize as he stood beside his huge machine at the head of the mine shaft in Falun, with the water splashing against the blades of the whirling waterwheel (Fig. 4.11). The concepts and ideas in Lagerhjelm's epistemological programme bear a resemblance to Immanuel Kant's theory of knowledge, and when we consider Lagerhjelm's background this becomes less surprising than it at first may seem.

Pehr Lagerhjelm came from a noble family with long military traditions, but at the age of seventeen he had chosen to follow a different career in industry.[143] He studied at the University of Uppsala, where he passed the mining examination (*Bergsexamen*) in 1807 at the age of twenty. This university degree, established in 1750, was a prerequisite for officials in the service of the Board of Mines and the subjects were physics, mechanics, geometry, chemistry and law.[144] After graduating, Lagerhjelm was duly appointed Auscultator in the Board of Mines and, in 1808, under-secretary in the Swedish Ironmasters' Association. He spent three years, 1808–1810, in Stockholm, where his relatively undemanding duties left him time to devote to his other interests.

He became a pupil of the famous chemist Jöns Jacob Berzelius

(1779–1848), and worked in his laboratory for about two years.[145] He assisted Berzelius in calculating the percentage composition of nearly 2 000 chemical compounds, their work being published as a supplement to the third volume of Berzelius' *Lärbok i kemien* (Textbook in chemistry).[146] Lagerhjelm was obviously an able mathematician, who was not daunted by large volumes of statistics. During the tedious work of calculation there was time to talk about natural philosophy, and Berzelius wrote in 1812 of these discussions that Lagerhjelm "always keenly defended the German manner during the conversations at home in my laboratory".[147] Lagerhjelm had also pursued his literary interests in Stockholm, and he belonged to the circle from which the Romantic movement was emerging in Sweden at that time.[148] Indeed, his three years in Stockholm coincided with the formation of the movement, and he was a friend of the author Lorenzo Hammarsköld (1785–1827), one of its foremost spokesmen.[149] The natural philosophy of the Romantics had its roots in German philosophy, "the German manner" which Lagerhjelm defended so keenly in the discussions in Berzelius' laboratory.

When Lagerhjelm claimed that the theory of the new type of water-wheel was evident *a priori*, he meant with Kant and his followers that it could be deduced by mathematical demonstration. He also used the terminology of the Kantian school when he spoke of "phenomenon", "form" and "content". Kant had attempted to synthesize the empiricism of Hume and the rationalism of Leibniz, and Pehr Lagerhjelm tried—on a much smaller scale to be sure—to synthesize the empiricism of Polhem and the rationalism of Elvius. He pointed out that it was not possible to reach any conclusions of general validity inductively, because "abstraction from a given experience [...] is only valid under the circumstances and within the boundaries, that essentially attended that class of phenomenon, which one experienced". This implied that the inductive method, which Polhem and Nordwall had pursued with so many thousands of experiments, was epistemologically impossible. But neither was the deductive method, the "speculative root" of knowledge in which Elvius and Åkerrehn had placed their confidence, sufficient on its own. Truth was not in fact found until one could reach a synthesis between "form" and "content". In this manner Pehr Lagerhjelm succeeded in reconciling the claims of reason and experience,[150] and the two theoretical schools in hydrodynamics were merged at the beginning of the nineteenth century under the influence of the Romantic movement.

DISCUSSION

In the previous chapter some simple statistical calculations were made in order to illustrate the fluctuations in interest in thermal energy

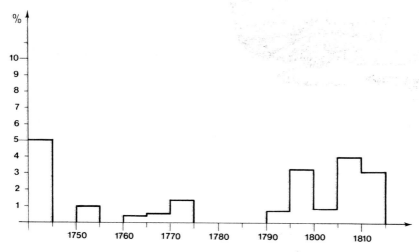

Fig. 4.12. The proportion of articles in the *Proceedings of the Royal Swedish Academy of Sciences* between 1739 and 1815 which deals with technological changes aimed at reducing the difference between potential and available mechanical energy resources. Five-year average as a percentage.

technology. The basic assumption was that the degree of interest within the Royal Swedish Academy of Sciences, as reflected by its Proceedings, could be regarded as representative of the debate in Sweden as a whole. A similar study of the Proceedings between 1739 and 1815 shows that 35 of the articles dealt with suggestions for technological changes designed to reduce the difference between potential and available re-sources of mechanical energy.[151] Of these articles, 24 were on water-power, 6 on muscle power and 2 on windpower.[152] This is a confirma-tion of the statement at the beginning of this chapter that the greatest interest in technological change in this area was devoted to waterpower. The reason for this was that waterpower was the only source of energy that promised both high power and continuous operation. Christopher Polhem's hydrodynamic experiments in 1702–1705 had shown that waterpower technology had not reached the limit of its potential devel-opment, and that it carried the promise of considerably greater efficien-cy. For each type of water mill there existed an optimum speed for the waterwheel to give maximum power. The only difficulty was to find the rules for determining this speed, and this was the problem to which solutions were sought throughout the eighteenth century.

This study of the development in hydrodynamics has revealed a fact not previously observed: namely that the insistence during that period on the necessity of uniting theory and practice referred to two theoreti-cal methods of establishing general rules for the efficient design of water mills. One method, first introduced by Pehr Elvius in 1742, was *deduc-tive*. It had its roots in Newtonian physics and the new differential calculus, and was used to try to derive general rules from the fundamental laws of motion by means of mathematical analysis. But before the 1740s the *inductive* method was unchallenged, and this involved formulating general rules from systematic experiments by means of parametric variation and optimization. This had first been attempted by Christopher Polhem in his hydrodynamic experiments in 1702–1705. Great expectations were attached to both these methods during the eighteenth century, and immense amounts of work invested in what remained unavailing projects until the 1820s. The deductive method, on the one hand, only resulted in formulas that became far too complicated as the physical reality was expressed in mathematical terms. The inductive method, on the other hand, only resulted in large amounts of quantitative data that were not only unreliable due to inaccurate measuring techniques but also too extensive to be systematized in a few simple rules. A waterwheel, peacefully turning in a stream, proved to be far more complicated than the heavenly clockwork.

The change in the percentage of articles devoted to mechanical energy technology in the *Proceedings of the Royal Swedish Academy of Sciences* during the period is presented in Fig. 4.12, which may be compared with Fig. 3.10 (thermal energy). It is evident that this diagram reflects the same pattern, although less markedly so. This becomes clearer when we compare the two curves in one diagram when both curves have been smoothed out (Fig. 4.13). A comparison shows that interest in thermal energy technology was at least twice as great as in mechanical energy technology during the period from 1739 to ca. 1790. This supports the conclusion drawn from the study of the Royal Chamber of Models in Chapter 2. It may be recalled that the catalogue of models in 1779 showed that there were 49 models relating to the problem of conserving wood fuel but only 39 relating to improved utilization of mechanical energy resources. We may thus infer that *there was in general a stronger interest in the conservation of wood fuel than in the improvement of prime movers.*

A classification of the seventeen authors of the thirty-five articles on mechanical energy by occupational category gives very much the same results as in the previous chapter. Those who took the greatest interest in mechanical energy technology were the officials of the Board of Mines (six authors) and engineers, architects and officers in the Fortifications Corps (eight authors). Fig. 4.13 shows that there was a relatively lively interest in energy technology during the first years (1739–1745) of the Academy. This may be explained by the fact that its first members were to a large extent recruited among the officials of the Board of Mines.[153] It should also be remembered that Wallerius' report on Polhem's hydrodynamic experiments was sent to the Board of Mines in 1705, but not made public until it was published in Triewald's lectures in 1736. The conviction that waterpower technology could be made considerably more efficient thus remained largely confined to the Board of Mines during the period with which we are concerned.

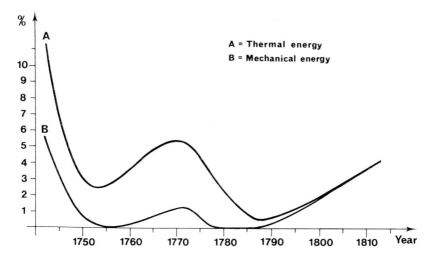

Fig. 4.13. The proportion of articles in the Proceedings of the Academy between 1739 and 1815 which deals with (A) thermal energy and (B) mechanical energy. This diagram is based on Figs. 3.10 and 4.12 where the curves have been smoothed out.

Furthermore, in no other industry was the problem of supplying mechanical energy as pressing as in the mines, and this was the responsibility of the Board. In the previous chapter it was shown that the Board's officials were those who wrote most about ways to conserve natural thermal energy resources, in industry as well as in the domestic sector. We may thus conclude that *the greatest interest in energy technology was shown by the officials of the Board of Mines.*

It should be noted that only one of these 35 articles on mechanical energy dealt with steam power technology. This was an article in 1809 on the measurement of steam pressure by Abraham Niclas Edelcrantz (1754–1821), well-known as the introducer of the James Watt steam engine in Sweden.[154] This means, in fact, that there was only *one* article on steam power technology out of the 2 596 articles that were published in the *Proceedings of the Royal Swedish Academy of Sciences* during the period 1739–1815.

In Chapter 2 it was pointed out that heat and work were not perceived as different forms of the same phenomenon in the eighteenth century. Developments in what we call energy technology therefore occurred in two quite distinct areas. It has been shown that these were two of the major areas of technological activity in Sweden and that it was primarily the officials of the Board of Mines that took an interest in these matters. The introduction of the steam engine led to a conflict between these two areas, and the Board of Mines became the scene of a clash between two incompatible interests: the conservation of wood fuel and the improvement of prime movers.

PART II

Diffusion of Knowledge

Fig. 5.1. The facade of the building ("*Gamla Myntet*") in Stockholm in which the Board of Mines was housed from 1674 until 1845. The building was on the same site as the present "*Gamla Kanslihuset*". The view is from the north-west side of the Royal Palace (*Lejonbacken*), looking west. Detail of a painting by Johan Sevenbom in 1761. (Photo Stockholm City Museum)

5. The Social Context: The Board of Mines

In the early seventeenth century, the central administration of Sweden was reorganized on a foreign model in a system of collegia, or administrative boards.[1] The Swedish collegia were government agencies in which a matter was dealt with by several persons at once (as distinguished from modern systems under which a decision is made by one civil servant on behalf of his department). A collegium consisted of a president and a number of assistants of equal rank, known as assessors. The decisions of a government agency were reached by these members, acting in concert. The original collegia (*Rikskollegierna*) were the Svea Court of Appeal (*Svea Hovrätt*), the War Office (*Krigskollegium*), the Admiralty (*Amiralitetskollegium*), the Chancery (*Kanslikollegium*) and the Treasury (*Kammarkollegium*). To these were added, later in the seventeenth century, the Board of Commerce (*Kommerskollegium*) and the Board of Mines (*Bergskollegium*). Each collegium was in contact with the Government and acted as a kind of government department. The duties of the collegia were to act as bodies to which appeal could be made against the decisions of subordinate authorities, and to carry out investigations for and supply information to the Government on matters within their spheres of activity.

The Board of Mines was the collegium which exercised ultimate authority over the mining industry. Sten Lindroth has written that the advent of the Board of Mines "ushered in a new era in Swedish mining history. Henceforth the Government was to have mining under effective control".[2] However, the significance of the part it played in the country's most important industry during the seventeenth and eighteenth centuries—the great age of Swedish mining—cannot be evaluated in detail, because its history has never been written. It never celebrated a centenary by publishing a history, nor has it attracted the attention of modern historians. This is an unfortunate lacuna in our historiography,

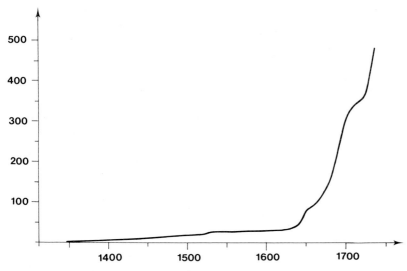

Fig. 5.2. The growth in the number of mining statutes (*Bergsordningar*), 1347–1735. The sharp increase in the 1630s coincides with the establishment of the Board of Mines.

as the history of the private ironworks is comparatively well documented. This imbalance gives an inadequate picture of the influence of the state on private industry. The only historical account which exists is Johan Axel Almquist's excellent but summary history of the Board's administration and officials, published in 1909.[3] This is a valuable outline, which was produced in conjunction with the transfer of the archives of the Board of Mines to the National Record Office, but it does not deal with the policy, activities and importance of the Board. The following is therefore a modest attempt to describe the Board of Mines during one period of its history, the decade 1715–1725.

Powers were granted and instructions issued to the Board of Mines in 1637.[4] Its influence extended over all mining in Sweden, and its task was to make sure that "all may proceed rightly and properly".[5] In a later formulation (1736) it was stated that the Board of Mines had "all the mines in the realm, with their appurtenances, under its control and charge".[6] What this entailed was put in more concrete terms in the regulations, statutes, privileges and decisions issued by the Government and the Board of Mines, known as the "mining statutes" (*Bergsordningar*). The oldest official mining statute which has survived was issued as early as 1347,[7] but it was not until the start of the seventeenth century that the number of statutes began to grow rapidly. Fig. 5.2 shows the increase in the number of mining statutes in the period 1347–1735,[8] and the leap in the rate of growth coincides with the creation of the Board of Mines in the 1630s. It was the extension of operations in the Great Coppermine in Falun in the early seventeenth century which gave rise to the need for greater government control.[9] After the 1630s, ordinances were issued in rapid succession, and during the period 1637–1715 about four mining statutes per year were issued.

The mining statutes covered all aspects of the mining industry: forests, mines, blast furnaces, tilt hammers, manufacturing industry and sales. They consisted of everything from general ordinances relating to the industry as a whole to the most detailed instructions to cover specific events. The statutes dealt not only with production, but also with the financial and judicial administration of the mining districts. They regulated the relationship between different groups in the industry and also between the mining industry and the rest of the community. The statutes also included disciplinary measures and prescribed, for example, punishment by flogging and running the gauntlet for crimes such as negligence at work or the stealing of iron.[10] The mining industry formed a state within the state, and only in cases of homicide did the law of the land have precedence.[11]

The duty of the Board of Mines was principally *executive*, in that its field officials in the mining districts saw to it that the statutes were observed. As the Board itself issued many of these statutes or proposed and drafted those issued by the Government, it also had *legislative* power. The second duty of the Board of Mines was *judicial*, in that it had to sit in judgment on cases in which there was appeal from lower courts in the mining districts (*Bergstings- och gruvrätter*).

There were several organizational changes in the Board of Mines during the seventeenth century, but the Board acquired a more permanent form at the beginning of the eighteenth century. The organizational structure established in 1720 remained unaltered for thirty years,[12] and may be seen in Fig. 5.3. The executive members consisted of the President, two Mine Councillors and five Assessors.[13] Subordinate to these members were the central administration in Stockholm (left column in Fig. 5.3), the local administration of mining districts in the provinces (centre column) and the apprentices, known as Auscultators, who served in the Board of Mines without salary (right column). Altogether the officials in the Board of Mines numbered about ninety.[14]

The *central administration* in Stockholm consisted of nine sections, with about twenty officials. There were, in addition to sections usual in any administration, four technical sections: "*Proberkammaren*" (Chamber of Assaying), "*Ingenjörsstaten*" (Land and Mine Surveying), *Laboratorium chemicum* and *Laboratorium mechanicum*. The last-named of these has already been mentioned in Chapter 4 in connection with the experiments in hydrodynamics conducted by Christopher Polhem in 1702–1705. Polhem had been appointed Director of the laboratory in 1697 and became Director of Mining Mechanics (*Bergsmekaniken*) in 1716. He held this position until his death in 1751, but the work of the laboratory was neglected.[15] Polhem was far too busy with other mat-

Bergskollegium

1 President
2 Bergsråd
5 Assessorer

Kansliet

1 Sekreterare
1 Förste kanslist
2 Relationsskrivare

Fiskalämbetet

1 Advokatfiskal
1 Kanslist

Kammarkontoret

1 Kamrerare
1 Kammarskrivare

Notariekontoret

2 Notarier

Arkivet

1 Aktuarie

Proberkammaren

1 Proberare

Ingenjörsstaten

1 Markscheider
1 Undermarkscheider
1 Lantmätare

Laboratorium chemicum

1 Riksvärdie
1 Laborant
2 Stipendiater i kemin

Laboratorium mechanicum

1 Direktör i bergsmekaniken
2 Stipendiater i mekaniken

1:a Bergmästaredömet

1 Bergmästare
1 Geschworner
1 Bergsfiskal
1 Bokhållare
1 Konstmästare
1 Bergsvärdie

2:a Bergmästaredömet

1 Bergmästare
1 Geschworner
1 Bergsfiskal
2 Kronogruvfogdar
1 Bokhållare
1 Inspektor
1 Konstmästare
1 Kronoafradsinspektor

3:e Bergmästaredömet

1 Bergmästare
2 Bergsfogdar
2 Konstmästare
1 Proberare

4:e Bergmästaredömet

1 Bergmästare
4 Bergsfogdar

5:e Bergmästaredömet

1 Bergmästare
1 Geschworner
4 Bergsfogdar
1 Konstmästare

6:e Bergmästaredömet

1 Bergmästare
1 Bergsfogde

7:e Bergmästaredömet

1 Bergmästare
1 Gruvfogde
2 Bergsfogdar

8:e Bergmästaredömet

1 Bergmästare
4 Bergsfogdar

9:e Bergmästaredömet

1 Bergmästare
2 Bergsfogdar

10:e Bergmästaredömet

1 Bergmästare

11:e Bergmästaredömet

1 Bergmästare

Ämnen i bergsväsendet

Ca 15 Auskultanter

Fig. 5.3. The organization of the Board of Mines in 1715–1725. Subordinate to the executive board were the central administration in Stockholm (left column), the local administration, divided into eleven mining districts (centre column), and the apprentices (right column). The total number of officials was about ninety.

ters and spent most of his time at his ironworks in Stjernsund. The two Stipendiaries in Mechanics (*Stipendiater i mekaniken*) during the period 1715–1725 were Polhem's son, Gabriel Polhem (1700–1772), and Anders Gabriel Duhre (ca. 1680–1739). Gabriel Polhem was his father's dutiful assistant at Stjernsund,[16] and Duhre was an imaginative conceiver of projects who had his hands full with his attempts to realize his many inspirations.[17] Mechanical engineering was therefore in poor shape in the Board of Mines, and the reasons were Christopher Polhem's preoccupation with other interests and his dominant position. He had little interest in the routine work of the Board, and his eminence as the "Archimedes of the North" probably made it impossible for the Board to reorganize the mechanical laboratory during his lifetime. It was only after his death that mechanical technology became efficiently organized with the establishment of the Royal Chamber of Models under the Board of Mines (see Chapter 2).

The development of the mechanical laboratory has a parallel in that of *Laboratorium chemicum* a few decades earlier.[18] This, too, had been established with royal support in the late seventeenth century, and great expectations attached to its Director, Urban Hiärne (1641–1724). But by the early eighteenth century Hiärne was beginning increasingly to be regarded as too advanced in years and out of date by contemporary chemists. In the second decade of the century work at the laboratory appears to have been discontinued, and Magnus von Bromell (1679–1731), who succeeded him in 1719, also accomplished little.[19] It was not until after Hiärne's death that *Laboratorium chemicum* could be revived as a functioning laboratory within the Board of Mines.[20] Both Hiärne and Polhem stand out as representatives of a bygone era in the Board of Mines of 1715–1725. They had not risen to the higher levels of the Board through the normal channels, having both been appointed by royal command (Hiärne's appointment had infuriated the Board).[21] The fact that the laboratories were under their personal control meant that the focus of operations was influenced by their personal interests, and that as they aged, activity in the laboratories declined. The relationship between science, technology and organization during this transitional period may merit a separate study, but it is safe to say that mechanical engineering was poorly organized during the period 1715–1725.

The *local administration* of the Board of Mines consisted of eleven mining districts (*Bergmästaredömen*), which covered all those parts of Sweden in which there was mining.[22] The local administration was highly flexible, changing with the changing structure of the industry. Posts were established or withdrawn and sub-districts were created or amalgamated as needed. Each mining district was headed by a Mine Inspector (*Bergmästare*), who was the highest local representative of the Board.[23] The number of officials in each district varied according to the district's size and importance, and the influence of German mining

Fig. 5.4. The average official in the Board of Mines in the period 1715–1736 is the central character in this study. Portrait of an unknown man by an unknown artist, dated ca. 1730. (Photo Swedish Portrait Archives)

organization is reflected in their titles: *Bergshauptman, Markscheider* and *Geschworner*. The largest number of officials was in Mining Districts Nos. 1 and 2, which included the Great Coppermine in Falun and Sala Silvermine. The smallest number was in Mining Districts Nos. 10 and 11, which consisted of Lapland and Finland. There were about fifty senior officials in local administration.

The *Auscultators* were the third main group in the Board of Mines.[24] These were participants in an apprenticeship system for young men wishing to gain experience and knowledge of the mining industry in order to qualify for employment with the Board. An appointment as Auscultator gave the right to attend the sessions of the Board, and the Auscultators worked either in the central administration in Stockholm or in the mining districts. Only about half of all applicants were accepted,[25] which indicates that a certain level of competence and experience was required, even on the lowest step of the Board's ladder. On average, three or four Auscultators were admitted each year between 1715 and 1725, and the total number at any one time was about fifteen.

A comparative study of the biographical details of the Assessors during the period 1720–1815 enables us to build up a picture of the average Board of Mines official in the eighteenth century.[26] He studied at Uppsala University until he was 20 years old, after which he made a successful application for an appointment as an Auscultator in the Board of Mines. After a couple of years' experience, either in the central offices in Stockholm or in the mining districts, he obtained employment at the age of 23 in some minor capacity in the central offices. The next year he went abroad for a year or two at the expense of the Board of Mines to study mining in Europe. During the decades which followed he again served either in Stockholm or in the provinces. He was promoted slowly, and in those cases where he served in one of the mining districts he became a Mine Inspector at the age of 35. Not until he was 43 did he become an Assessor, i.e. one of the eight executive members with a seat and a vote on the Board. By then he had had 23 years' experience of the collegium and of the mining industry. As an Assessor he attended meetings every weekday almost all the year round, and the meetings included both morning and afternoon sessions. At this point he also became a member of the Royal Swedish Academy of Sciences, and if he contributed to the Proceedings of the Academy, it was probably by writing, as we saw in Chapter 3, about various proposals for reducing domestic consumption of wood fuel in order to conserve the forests for the needs of the mining industry. At the age of 52 he became a Mine Councillor, and he died in office about ten years later, having served the Board of Mines for over forty years.

This description of a typical official of the period also shows that nobody could become an Assessor without having acquired extensive experience by working for the Board in various positions for many years. Competence was guaranteed by the organizational pyramid of the collegium, which ensured that there were many capable applicants for every vacancy. The system of continuous admission of Auscultators guaranteed that there was always a broad base of candidates suitable for promotion in the hierarchy.[27] The Board of Mines itself thus ensured the competence of its members. Before the Era of Liberty (1718–1772), however, there were a number of occasions on which the Government appointed Assessors Extra Ordinary against the wishes

and even without the knowledge of the Board.[28] There was a conflict
here between the Government and the Board of Mines which manifested itself in those official matters in which the Board tried to assert its
position as an independent department with the right to appoint its
own members. The Constitution Act of 1720 stipulated however that
the members of the Board should all be:

such, who by experience have made themselves suitable for Judicial Office and
have also gained all the necessary and thorough knowledge of all that concerns
the mining industry with regard to production and to the financial administration of the mining districts.[29]

This did not, however, mean that a higher level of competence was
required than previously,[30] but it was a rule which guaranteed that
nobody without the necessary competence could be appointed a member of the Board. We shall see in Chapter 9 how Emanuel Swedenborg
was subjected to this particular rule in the period of transition between
the period of the Absolute Monarchy and the Era of Liberty. He had
been personally appointed Assessor Extra Ordinary by King Karl XII
in 1716, and he had to work hard for several years in the early 1720s
before he was accepted by the Board as a worthy member in 1724. This
was to be of some consequence to the introduction of steam power
technology into Sweden.

The period 1715–1725 was a time of upheaval in several respects. For
one thing, changes were now made to the composition of the collegium.[31] Of the executive members of the collegium in 1720, half were
people who conform to the picture of what was to be the typical Board
of Mines official for most of the eighteenth century: Johan Angerstein
(1672–1720), Adam Leijel (1669–1744), David Leijel (1660–1727),
Anders Svab the Elder (1681–1731), Johan Bergenstierna (1668–1748)
and Lars Benzelstierna (1680–1755). The other half had made their
careers in other Government departments and had not been appointed
executive members or officials in the Board of Mines until they were in
their mid-thirties: Gustaf Bonde (1682–1764), Urban Hiärne, Robert
Kinninmundt (1647–1720), Jonas Cederstedt (1659–1730), Anders
Strömner (1646–1730) and Magnus von Bromell. It should be pointed
out that there was no difference in education between the two groups.
The difference between them was that the new type was an official who
had come up slowly through the hierarchy of the Board, whereas the
older type was a political appointee from outside. Here Emanuel Swedenborg occupies an intermediate position to which we shall have cause
to return. It was also a time of change as far as the birth of the executive
members was concerned. During the seventeenth century the Assessors
had always been of noble birth, but during the eighteenth century it
became increasingly common for them to come from the ranks of the
untitled.[32] The Assessors who were not of noble birth during the period

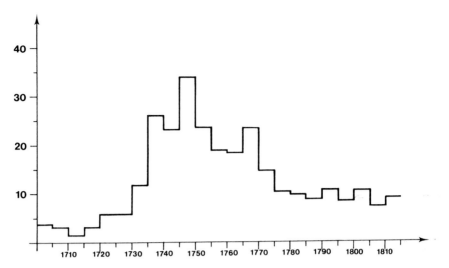

Fig. 5.5. Average number of mining statutes (*Bergsordningar*) issued per year in 1701–1815 (five-year average). The decade 1715–1725 marked the beginning of the period when state control of mining was at its strongest.

1715–1725 were all raised to the nobility a year or two before or after their appointment.

But the most characteristic feature of the period 1715–1725 is that it saw the mining industry starting to be brought under much closer supervision. During the period 1637–1715, as has been mentioned, about four mining statutes were issued per year, but during the period 1735–1775, the corresponding figure was twenty-three (Fig. 5.5).[33] This is more than at any other period in the history of the collegium, and the diagram also reflects the general industrial policy (*"Manufaktur-politiken"*) during the Era of Liberty.[34] Like the diagrams in the previous two chapters, it manages to mirror the spirit of the age (see Figs. 3.10, 4.12–13).

The attitude to the state's control of the mining industry is allegorically expressed in the frontispiece to the collection of mining statutes issued between 1347–1735 which the Board of Mines published in 1736 (Fig. 5.6).[35] In the centre of the picture, higher than all the other figures, sits *Svecia* on her throne, symbolizing the authority of the Swedish state or, more specifically, the authority of the Board of Mines. Above her head shines the North Star, which was an ancient symbol of illustrious Scandinavia and its perpetually shining prestige (*"Nescit occasum"*) but which by the 1730s had also become a symbol of intellect and genius. Around *Svecia* are grouped various figures, whose actions and gestures invariably lead the eye back to the personification of the state and who thus appear as her loyal subjects. In the upper right corner can be seen a mine, with miners and hoisting machinery. Around *Vulcan*, the god of metalworking, are depicted hammer, anvil

Fig. 5.6. Allegorical representation of state supervision of all aspects of the mining industry. Frontispiece to the collection of mining statutes issued 1347–1735 which was published by the Board of Mines in 1736. Engraving by Jacob Folkema in 1736. (Photo Bengt Vängstam)

and bellows, ore skip, "*Schlegel und Eisen*", ore, charcoal and two *putti* engaged in assaying a mineral on a balance. To the right of the throne stand three women. One is holding out a crucible containing the molten, refined metal. Behind her is a second, holding a quill and ready to take down the statutes which *Svecia* sees fit to promulgate. A third woman is carrying a waterwheel as a symbol of a stage in the refining process. To the left, *Neptune*, the protector of seafarers, and *Mercury*, the god of commerce, together symbolize trade in mineral products across the seas. Beneath *Neptune's* outstretched arm can be seen a ship departing with all sails set (loaded with bar iron for England?). The picture is intended to imply that state control extends to every stage in the mining industry, from prospecting for new deposits to export of the finished products.

Of the statutes issued during the period 1715–1725, there is one in particular which we have reason to note. Ever since the Middle Ages, the state had claimed proprietary rights to all mineral deposits in Sweden.[36] In those cases where a mine was worked by private individuals, they paid taxes, often tithes, to the state for the right to utilize what was regarded as the property of the state. The import of the concept "crown prerogative" (*regale*) varied, however, and it was often somewhat obscure, particularly with regard to the right to exploit newly discovered mineral deposits. A statute of 1649 had laid down that all mineral deposits were the property of the state, but this had led to a reluctance on the part of landowners to report new deposits, for fear of losing forest and property if the state chose to begin mining.[37] A new statute, issued on August 27, 1723, attempted to clarify and modify the "crown prerogative" concept in order to encourage the search for and mining of new deposits.[38] In the new sense, "crown prerogative" meant only that the state claimed a percentage of the output of a new mine, and it was stated explicitly that no one need fear the loss of their property if a mineral deposit were discovered:

[We are] graciously pleased to declare that by the term "Crown Prerogative" is henceforth to be understood only that if any person should either conceal and suppress or be unwilling to bring up and work such metals, minerals and fossils of use to the Realm, that We then, so that the riches and bounty, which God hath thus laid down in the earth, shall not remain lying useless there but be brought to the surface to the lasting advantage of the Land and its People, reserve the right through our Board of Mines to privilege and grant the use of such metals, mineral lodes and mines to those who have the disposition, inclination and wherewithal to maintain and continue them in their proper operation, observing good mining practice and under the superintendence of Our Board of Mines.[39]

A landowner who did not himself wish to work a deposit on his land was promised full compensation for his property and for forest lost in the sinking of the mine, and was also guaranteed a return of one percent of the profits. The natural resources of the country were to be sought out and exploited to the greatest extent possible, and this also meant that no riches and bounty which God had laid down in the earth were to lie unutilized instead of being brought out "to the lasting advantage of the Land and its People".

But the section of the statute of 1723 quoted here also had another consequence. If a mine had been closed, anybody who had the "disposition, inclination and wherewithal" could obtain the permission of the Board of Mines to resume operations. The problem in many Swedish mines was of course, as has been pointed out in Chapter 4, that it proved impossible to keep them free from water. But if anyone succeeded in draining an abandoned mine and keeping it free from water, then he himself had the right to work it and to enjoy the profits (always provided, of course, that he paid the prescribed taxes and observed the regulations of the Board of Mines). The effect of the statute of 1723 on Swedish mining has not been generally studied,[40] but it played a crucial part in the introduction of steam power technology into Sweden. The Newcomen engine offered an opportunity to bring up the riches and bounty which God had laid down in the earth, and the statute of 1723 ensured that most of the profits earned would go to the person who undertook the venture.

In 1730, the economist Eric Salander (1699–1764) published his ideas on how wealth could be increased in an "impoverished nation, where capital has drained away, industry lies idle and profit has vanished".[41] Salander presented his politicoeconomic programme in concise maxims: the poverty of a nation was caused either by natural deficiency or by divine punishment, or because work had stagnated.[42] The most important of these obstacles was the last-named, but it could be overcome by "industry and genius".[43] Thus the first precept for the economic salvation of a nation was "work and diligence". God himself had given man the rule in the words of the Bible: "In the sweat of thy face shalt thou eat bread".[44] According to Salander, work needed to be done in four areas: (1) Farming in its widest sense (including fishing), (2) mining, (3) manufacture and handicraft, and (4) commerce and shipping.[45] The most important area was the third, manufacture and handicraft, by which all the others could be improved.[46] The main purpose of politics should therefore be to promote and supervise industry. Promotion of industry entailed granting privileges, executing applications promptly, providing protection of inventions etc.[47] No innovation or proposal which might benefit industry was to be rejected until it had been carefully examined and found to be either useful or useless. How else, asked Salander, have so many once poor nations acquired their present wealth?[48] In order to prepare the way for industrial development, politicians should:

let talented persons travel abroad, and support them, in order that such and other knowledge might be imported. As well as to attract and accommodate foreign artisans, and help them with suggestions and employment to expand the industries more and more.[49]

Salander's book reflects the politicoeconomic ideology underlying the official activity which brought knowledge of steam power technology to Sweden during the period 1715–1725. In Chapter 7 it will be shown how the Board of Mines sent talented individuals abroad with a brief to report on new technology which might be useful to the mining industry. Chapter 8 will describe how foreign artisans were attracted to Sweden and granted privileges on an invention, i.e. the Newcomen engine. Chapter 9 will show how carefully these proposals were investigated by the Board of Mines with respect to their potential usefulness or otherwise, and also how promptly these applications were dealt with by the Swedish civil service.

6. The Technology: Early Newcomen Engines

The Newcomen engine was the first established heat engine, i.e. the first practical working engine capable of converting thermal energy (heat) into mechanical energy (work). In his book *From Watt to Clausius: The Rise of Thermodynamics in the Early Industrial Age* Donald S. L. Cardwell wrote in 1971 of the Newcomen engine:

It was the first really successful prime mover to be put to work apart from the immemorial ones of wind, water and muscle [...] The machine itself was magnificently adapted to its end. Notwithstanding the genius that went into its design, especially into the automatic valve mechanism that was operated by the motion of the great beam, it made no demands beyond the technical resources—the very limited resources—of the early eighteenth century. It was safe, reliable, and basically simple; it was powerful, and economically competitive with other means of pumping water. It belongs to that small but select group of inventions that have decisively changed the course of history.[1]

The early history of the Newcomen engine has attracted great interest among historians, especially during the last twenty years. Since the founding in 1920 of the Newcomen Society for the Study of the History of Engineering and Technology, historians of technology have attempted to show the importance of the Newcomen engine, which was for a long time overshadowed by its successor, the steam engine of James Watt. In the marketing of new technologies, as sometimes in politics, it is common practice to discredit older competitors as inefficient and obsolete. The marketing of the Boulton & Watt engine was highly successful,[2] and it may well have been that adverse comments on the Newcomen engine by earlier historians were merely uncritical repetitions of late eighteenth and early nineteenth century marketing claims.

Our knowledge of the diffusion of the Newcomen engine has increased over the last decades, the turning point being an article in 1967 by John R. Harris in *History*, in which he questioned the traditional assertion that the Newcomen engines were far less numerous than those of Watt.[3] Recent findings were published by John S. Allen in a revised edition in 1977 of L. T. C. Rolt's book on Thomas Newcomen, and a quantitative assessment of steam engines in eighteenth-century Britain was presented by John Kanefsky and John Robey in *Technology and Culture* in 1980.[4]

Historians of science have long been intrigued by the fact that this engine was invented by a "common ironmonger". There is a chain of consecutive scientific discoveries, almost beautiful in their intrinsic logic, running from Galileo, through Torricelli, Guericke, Huygens and Papin to Thomas Newcomen (1663/4–1729).[5] A myth was initiated in 1797 by a friend of James Watt, in an article on steam engines in the third edition of *Encyclopedia Britannica*, that Newcomen had actually received information from Robert Hooke on Papin's work and that documents proving this could be found in the archives of the Royal Society.[6] No such documents have ever been discovered during the 200 years that have passed since then, but historians of science still discuss this question energetically as if it were vital to our general understanding of the relationship between science and technology. Eugene S. Ferguson has remarked that Desaguliers' description in 1744 of Newcomen as an ironmonger has been misinterpreted by modern writers as meaning a pedlar, or even a junkman: "Thus Newcomen is likely to be thought of as a ragged, gaunt pusher of a handcart, waiting for Dickens to be born so that he could get into one of his novels."[7] It would be more accurate to picture Newcomen as a man well versed in trade, and with access to brass and iron foundries, smithies and the assistance of skilled metalworkers.

The growing need for thermal energy in England since the sixteenth century had led to increased coal mining.[8] As the mines were sunk ever deeper the problem of keeping them free from water became more pressing, and by the end of the seventeenth century it was no longer possible to cope with the expanding need for pumping machinery using traditional sources of power. Flooding threatened the non-ferrous mines of Cornwall and Derbyshire and was in fact a problem in all branches of mining. The potential mineral resources could not be made available with existing technologies.[9]

Thomas Newcomen of Dartmouth sold and perhaps manufactured equipment for the Cornish tin mines, so that he, like everyone else in mining, was acutely aware of the problem.[10] We are not concerned here with the story of the invention of the steam engine, its antecedents, or the possible influence of the scientific community.[11] It is sufficient to state that during the first decade of the eighteenth century, Thomas Newcomen and his assistant John Calley (?–1717) developed a practical

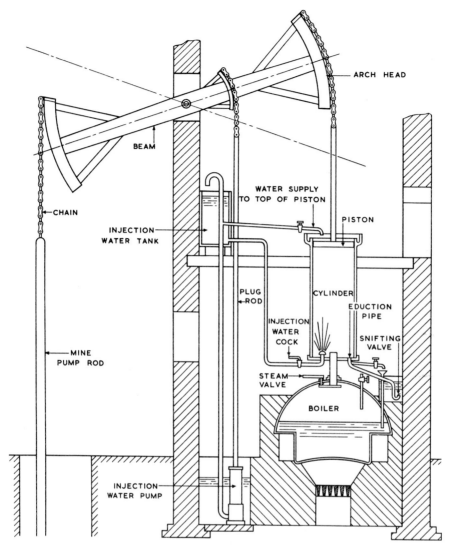

Fig. 6.1. Diagram of the principal features of the early Newcomen engine. Reproduced from Henry W. Dickinson, *A Short History of the Steam Engine* (Cambridge University Press: Cambridge, 1939). (Photo Science Museum, London)

working steam engine which proved to be of great benefit to the British mining industry. The first engine that is known to have been used successfully in industry was built in 1712 at a colliery near Dudley (see Fig. 7.3).[12] A remarkable feature of the Newcomen engine was that the basic design of 1712 changed little, and that by about 1717 it had acquired the form that it was to retain for the next fifty years.[13] It is possible therefore to define with fair precision the state of the art during the period with which we are concerned, and the reader is referred to the bibliography for more detailed descriptions of the early Newcomen engines.[14]

The *working principle* of the Newcomen engine is, as Cardwell has written, "easy to understand even if the theoretical principles underlying it are rather more subtle than they seem to be at first sight".[15] The engine consisted basically of a boiler surmounted by a large cylinder containing a piston (Fig. 6.1). When the cylinder had been heated up and filled with steam from the boiler, cold water was injected into the cylinder. The condensed steam left a partial vacuum which caused atmospheric pressure to force the piston down the cylinder. Thus power was produced by the pressure of the atmosphere and not by the pressure of the steam. The piston was connected by a chain to one end of a pivoted beam, and the pumps in the mine were similarly connected to the other end. Once the cylinder was heated up and filled with steam again, the weight of the mine pump shafts pulled the beam down on one side and lifted the piston up on the other. The spent injection water was discharged through a valve, and the cycle was repeated. The valves that alternately admitted steam and cold water were opened and closed by the oscillations of the beam, and this made the engine self-acting.

The *performance* of a Newcomen engine was impressive by eighteenth-century standards. An engine with a 24 inch piston could lift a weight of about 1.5 tons.[16] Operating at approximately 14 strokes per minute and with a working stroke of about 7 feet it would develop a power of near 10 hp.[17] This was about the same as that of a large windmill or waterwheel, but the advantage over a windmill was that it was not subject to unpredictable changes in weather and over a waterwheel that it could be located wherever there was a need for power. A power of 10 hp may seem modest, but it must be appreciated that at least 50 horses or 300 men would have been needed to achieve the same output with muscle power if continuous operation was required as indeed it was in mines threatened by flooding.[18] The main disadvantage was low efficiency. Since the steam was condensed in the cylinder during each working cycle, this mass of metal had to be heated and cooled at every stroke. The large heat capacity of the cylinder resulted in a thermal efficiency of only 1%, but this mattered little in British coal mines where there was an abundance of cheap, low-grade coal that could be used to fire the boiler.[19] The engine was used almost exclusively to pump water out of the mines, but it is important to realize that it was not "a pumping engine only".[20] It was a prime mover developed to fill a specific need in the British mining industry. The principle could be employed to provide mechanical power for other purposes; the only problem was a technical one of adaptation.

The *construction* of the Newcomen engine was the most remarkable feature of the new technology. As an invention it was an ingenious solution to a specific problem by "an assemblage of elements, some adopted but most adapted".[21] It was a synthetic invention where the whole was much greater than the sum of the parts, but as a construction it demanded only a combination of already existing manufacturing techniques. Since the Newcomen engine operated at a steam pressure just slightly above atmospheric there was no need for a pressure boiler. This was important, since it was not possible to make vessels that could withstand high internal pressure.[22] The boiler was basically nothing more than an ordinary brewer's boiler, which was something any coppersmith could build. Nor was there any difficulty in casting a brass cylinder six feet long and weighing one ton, as guns of that size had been made for a long time, and the manufacturing technique was well established in a belligerent Europe.[23] (It was in fact, as we shall see in Chapter 13, the Royal Gun Foundry in Stockholm that was entrusted with the casting of the cylinder for the Dannemora engine.) Since the pressure was low the piston needed only to be a rough fit in the cylinder—like a cannon-ball in a gun.[24] Only with the Watt engine, operating at a higher pressure, did it become necessary for the bore to be perfectly cylindrical. The automatic valve motion was mechanically the most advanced component of a Newcomen engine, yet it was no more complicated to manufacture than the clock of a church (which was another mechanism controlling its own movement).[25] So the manufacturing know-how for building a Newcomen engine already existed; it was ready and available. All that was needed was the availability of facilities to build a brewer's copper, a fieldpiece, and a church clock, plus those parts that any millwright, plumber or mason could build (Fig. 6.2).

The *economics* of the early Newcomen engine are difficult to establish since figures are scanty.[26] The total cost of building an engine might range between £700 and £1 200 depending on the size.[27] But it was not the mineowners who made the investment. Thomas Savery, the inventor of an earlier and less successful steam engine, held a patent for the period 1698–1733. He possessed, as Harris has described it, "the interesting combination of a long monopoly period and an unsaleable invention".[28] Newcomen entered into partnership with Savery, and thereby gained a longer period than the fourteen years that was normally granted for a patent. After Savery's death in 1715 the patent rights were acquired by a joint stock company of businessmen in the City of London, known as "The Proprietors of the Invention for the Raising of Water by Fire", which gave the innovation the backing of the mercantile capital and the financial expertise of London.[29] Thus during the period 1715–1733, the Proprietors had a monopoly on the building and use of Newcomen engines. A mineowner who wanted to use a Newcomen engine had to come to an agreement with the Proprietors, and the usual arrangement was that the Proprietors built the engine and the mineowners paid a rental for the use and maintenance of it. The rental was negotiable and depended on the profits of the mine. The mineowners were thus saved the expense of the large initial investment as

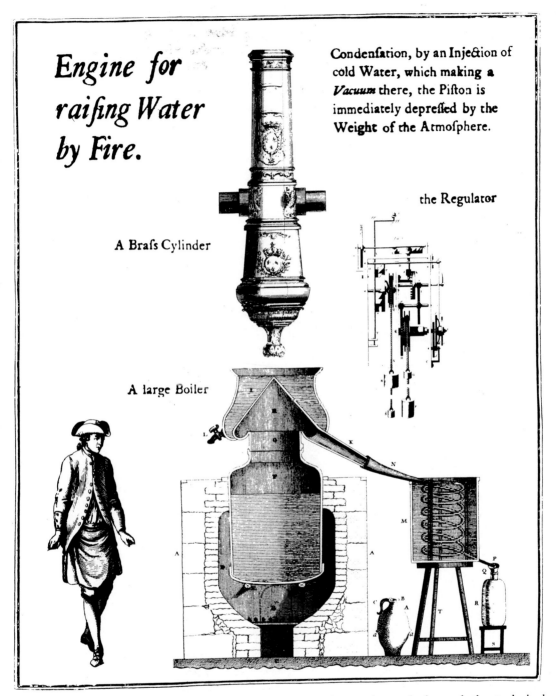

Engine for raifing Water by Fire.

Condenfation, by an Injection of cold Water, which making a *Vacuum* there, the Pifton is immediately depreffed by the **Weight** of the Atmofphere.

the Regulator

A Brafs Cylinder

A large Boiler

Fig. 6.2. The Newcomen engine made no demands beyond the technical resources of the early eighteenth century. The manufacturing techniques needed were basically the same as for a brewer's copper, a fieldpiece, and a church clock. (Collage by the author)

well as the responsibility for the engine's operation and maintenance. This was, as we shall see in Chapter 13, a situation very different from the agreement between Mårten Triewald and the owners of the Dannemora Mines.

The engine won immediate acceptance in the mining industry and spread fairly rapidly. In the period 1712–1725 some 70 Newcomen engines were built in England, mainly in the collieries of Northumberland and Durham and of the Midlands.[30] Altogether, about 1 700 Newcomen engines were built in England during the eighteenth century, and they continued to be built and used long after the Watt engine and other more efficient types had been introduced. (One Newcomen engine, now at the Science Museum in London, was used from 1791 until 1918.[31])

The first Newcomen engines on the Continent were built in the 1720s.[32] It should be observed that most of our knowledge of these engines is of recent date, and that there may have been others not yet discovered. The first was built near Liège in 1721 (see Chapter 8),[33] the second (maybe) at Kassel in the electorate of Hessen-Kassel in 1722,[34] and the third in Königsberg near Schemnitz in what was then a part of Hungary, also in 1722.[35] The fourth was built in Vienna in Austria in 1723,[36] and the fifth at Passy near Paris in 1725.[37] The Dannemora engine built in 1727–1728 was, as far as we know today, the sixth Newcomen engine outside England (and certainly not the first as has often been claimed in Swedish literature). The four characteristics of the new technology which determined its diffusion were the following.

(1) The *low efficiency* of the Newcomen engine made it uneconomical to use except where coal was cheap. To fire a Newcomen engine with wood was of course possible, but the engine had not been developed for wood fuel and the economics of this were unknown. Graham J. Hollister-Short has written on the diffusion of the Newcomen engine into Europe that it "was so symbiotically linked to the mining of coal that [...] it could not in any significant or lasting way break clear of the technological matrix in which it had first come to maturity".[38]

(2) The *reciprocal motion* of the Newcomen engine was well suited to the powering of mine pumps—a purpose for which it had originally been designed. But the reciprocal motion limited its usefulness as a prime mover for other purposes, and it was not easily adapted for the turning of wheels. For a mechanical technology based on wood to convert the reciprocal motion to rotary remained a problem.[39]

(3) The comparatively high power of the Newcomen engine made it a *centralized prime mover* since it was capable of replacing a large number of traditional, less powerful prime movers such as horse whims and windlasses. This was—as for all centralized, large-scale technologies—a disadvantage as well as an advantage. It promised economic and administrative benefits, as well as the possibility of central control

of the power system. But centralization also creates more vulnerable systems. If a mine were to be made entirely dependent on an engine for the draining of water it was essential for continuous operation to be guaranteed. This was a situation very different from a traditional mine that could rely on a number of separate power units. Thus the Newcomen engine offered the mining industry an efficient but vulnerable alternative. But in what respect were the early Newcomen engines less than dependable? This may appear to contradict the praise of their mechanical simplicity and reliability discussed earlier in this chapter.

(4) The Newcomen engine was based upon, to use the terminology of Lewis Mumford, the new paleotechnic *coal-and-iron complex* as opposed to the traditional eotechnic water-and-wood complex.[40] The manufacturing techniques needed to build a Newcomen engine may have been simple enough, even by eighteenth-century standards, but the task was still beyond the ordinary millwrights employed in the mines. They were used to building their prime movers out of wood, and had not been trained to design, build, operate, and repair machines of iron. But more important, they were not used to thinking in terms of steam, vacuum and atmospheric pressure. Although they must have had, as discussed in Chapter 4, an intuitive feeling for concepts such as force, power, work and efficiency, they were used to more palpable sources of power —working men and horses, wind and running water. It should be pointed out that this was not a difference in the level of theoretical knowledge between the millwrights and the pioneers of the Newcomen engine. The latter often had a very hazy and confused grasp of the underlying principles and the forces at work. But they were experienced in keeping their engines in operation, and they could identify and repair their most common malfunctions. It was simply a difference in training and experience—not unlike the difference between a coachman and a chauffeur in the early twentieth century.

Practical experience of Newcomen engines was rare during the early eighteenth century, when the number of engines was still relatively small. There was also little information in print on the new technology, and this illustrates a difference between science and technology in the way information is disseminated. Scientific discoveries are quickly published in detail, for the satisfaction and prestige of priority. In technology, on the other hand, innovations are kept secret for as long as possible for reasons of commercial or military competition. There is, of course competition in both science and technology but the rules and rewards of the games are different. To be sure, anyone could buy one of the published prints of the early Newcomen engines, like the one by Henry Beighton in 1717 which is reproduced here (Fig. 6.3).[41] But these were advertisements rather than complete working instructions published for the benefit of all. Most people would have been able to understand the working principle of a Newcomen engine if they saw one of these engravings and either read or heard a description of it. But this

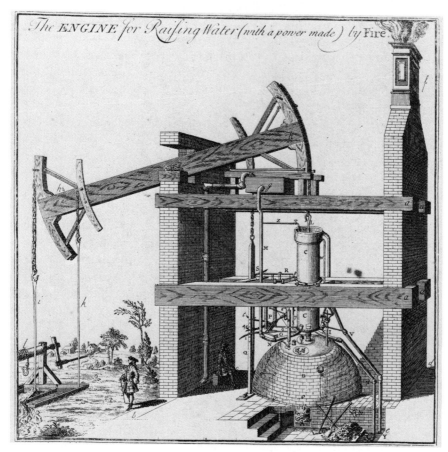

Fig. 6.3. The first published image of a Newcomen engine. Engraving by Henry Beighton, 1717. (Photo Science Museum, London)

feeling of understanding was quite different from the knowledge and experience needed in order to design and operate a Newcomen engine. It was very similar to seeing a sectional drawing of an internal combustion engine and having its principle explained. Most of us have experienced this feeling of understanding, yet few of us would be able to design and build a working internal combustion engine on the basis of it.

The knowledge needed in order to design a Newcomen engine will be discussed in the following chapters, when we examine the extent and level of the knowledge that was brought to Sweden. But the practical experience required in order to keep a Newcomen engine in operation must be emphasized.[42] The setting of the valves was most critical, and the adjustment of the self-acting mechanism could be learned only by experience. If wrongly set, the stroke of the piston might knock the bottom out of the cylinder and ruin the engine. If the engine happened to stop there could be several possible explanations. The faults were easy to repair, but it took practical experience to identify them. A mineowner had to be able to rely on the assistance of people with such experience if he was to employ the new technology successfully. The weakness of the early Newcomen engines thus lay not in the hardware, but in the scarcity of competent personnel. Accordingly, the diffusion of the new technology depended on the particular individuals concerned, and it is to them that we will turn our attention in the following chapters.

7. The Travellers, 1715–1720

EMANUEL SWEDENBORG

Emanuel Swedenborg (1688–1772) travelled abroad in 1710–1715 on his own initiative.[1] The tour was paid for—not without objections—by his wealthy father, Bishop Jesper Svedberg. Swedenborg spent the first three years of his time abroad in London and Oxford. His main subjects of study were mathematics and astronomy, but he also wanted to learn and master different crafts. In London he took lodgings with various craftsmen—a watchmaker, a cabinet-maker and an instrument-maker —to find out the secrets of their skills.[2] While he was away, Swedenborg sent books and scientific instruments to his friends in Uppsala. Among the scientists he met was Francis Hauksbee, the experimental physicist and instrument-maker, and Swedenborg sent a book by Hauksbee on experimental physics to his brother-in-law Erik Benzelius, the Uppsala University Librarian.[3] In several letters he also mentioned an air pump ("*Antlia Pneumatica*") constructed and sold by Hauksbee, and suggested that Benzelius should buy one for the library.[4]

After going on to spend two years in Holland and Paris, Swedenborg began making his way back to Sweden in 1714. He made a lengthy halt in Rostock, from where he wrote a letter to Benzelius, dated September 8, 1714.[5] Swedenborg told how he had promised his father that he would publish a dissertation ("*Specimen Academicum*"), in which he intended to describe some of his mechanical inventions. In the letter, he listed in Latin fourteen of them. It was indeed a remarkable catalogue, including ideas for a flying machine and a submarine. Item No. 5 was the following:

An engine that operates by fire to raise water; and to be constructed at those blast furnaces where there is no waterfall but the water is still. The fire in the hearth should be capable of supplying the waterwheels.[6]

This was the principle of the steam engine: mechanical energy from thermal energy. These very lines in the letter from the young Emanuel Swedenborg seem to have been how the idea itself first reached Sweden. The engine was to be used to raise water to power a waterwheel, and the height the water was to be raised could not therefore have been more than about four metres. Nor could the quantity of water have been very large, because the excess heat from the hearth was to power the engine. This description fits a steam engine of the Savery type. Although Newcomen's first engine was erected near Wolverhampton in 1712, it is unlikely that Swedenborg saw it, as he spent all his time in London and Oxford. A description of Savery's engine had been published in the *Philosophical Transactions* of the Royal Society in 1699,[7] a publication Swedenborg is known to have studied.[8] He may also have seen the Savery engine erected in or about 1712 at Campden House, Kensington, to raise water for domestic purposes.[9]

It is noteworthy that the engine mentioned by Swedenborg was not primarily a pumping engine. It was to raise water only as a means of obtaining rotary motion from waterwheels to power the bellows of a blast furnace. This was what Savery had intended, although it is seldom noted that the full title of his patent in 1698 was for a "new Invention for Raiseing of Water *and occasioning Motion to all Sorts of Mill Work* by the Impellent Force of Fire, which will be of great use and Advantage for Drayning Mines, Serveing Towns with Water, *and for the Working of All Sorts of Mills where they have not the Benefit of Water nor constant Windes*" [my italics].[10] Swedenborg, too, saw in this engine a solution to the problem of providing mechanical energy for works lacking waterpower. In his letter, Swedenborg wrote that he had the descriptions of his inventions, together with the calculations, ready for publication, and that he was now fully occupied in completing the drawings.[11] When they were all ready he sent them off to his father, but on his return to Sweden in 1715 he learned to his dismay that the work at which he had laboured so diligently had been carelessly lost by the bishop.[12]

Emanuel Swedenborg soon found other things to occupy his mind. In 1711 the first scientific society in Sweden, *Collegium curiosorum*, had been founded in Uppsala by Erik Benzelius on the initiative of Christopher Polhem.[13] The fellows met for discussions which centred around the inventions of Polhem. The activity soon declined, but one result was the publication in 1716–1718 of the first technical and scientific journal in Sweden, *Daedalus hyperboreus*, which Swedenborg undertook after his return, on his own initiative and at his own expense.[14] Swedenborg was a devoted admirer of Polhem, and the aim of the journal was primarily to make the inventions of the "Archimedes of the North" known and respected.[15] A few months after his return, Swedenborg wrote to Christopher Polhem on the subject of waterpower.[16] Polhem told him about the experiments performed in *Laboratorium mechanicum* (see Chap-

Fig. 7.1. Emanuel Swedenborg in the first of the two roles he played in the introduction of steam power technology into Sweden: a young man travelling as the inclination took him, observing and recording whatever happened to attract his interest. Cf. Fig. 9.1. Portrait of a young man believed to be Swedenborg by an unknown artist. (Photo Swedish Portrait Archives)

ter 4). But all this, Polhem wrote, was far too complicated to be discussed in a letter. If Swedenborg was at all interested in mechanical studies ("*Mechanske studium*"), he was welcome to come and stay with Polhem at Stjernsund to discuss the subject.

Swedenborg was not only the editor of *Daedalus hyperboreus*, but also the contributor of several articles of his own. There is, however, no mention in any article of his engine to raise water by fire, although he published articles on some of his other "inventions" during his stay in England.[17] In the first volume, in 1716, Swedenborg published two designs for hoisting machinery, but in both cases the power was to be provided by waterwheels.[18] It is clear that Swedenborg had come under the influence of Christopher Polhem, and that he had quickly adapted himself to the traditional power technology: waterpower. Nevertheless, we know that in Swedenborg there was in 1715 at least one person in Sweden who had realized the possibility of converting thermal energy into mechanical energy by means of an engine, and who was aware that this mechanical energy could be transformed into rotary motion to solve the energy problems of industry.

THE BOARD OF MINES AND HENRIK KALMETER

The Board of Mines was in the habit of sending promising young men abroad to study mining and metallurgy. Anders Svab, for example, had travelled in England in 1712.[19] In a letter to King Karl XII in June 1718, the Board pointed out the usefulness of this practice: it provided the Board of Mines with capable and experienced persons who could effect necessary changes in the mining industry.[20]

> But since those who have acquired experience of mining operations in this manner have one after another been promoted, a shortage of such persons would arise in the Board, had one not the foresight to encourage further able and willing individuals similarly to improve their capabilities.[21]

The Board therefore asked for one of its Auscultators, Henrik Kalmeter, to be given a travel grant of 1 000 *copperdaler* for a three-year journey abroad. Kalmeter had expressed a wish to undertake such a journey, and the Board considered him highly suitable.[22] Henrik Kalmeter (1693–1750) was born in Falun, where his father was a Head Clerk at the Great Coppermine.[23] He studied at the University of Uppsala, and was appointed Auscultator in the Board of Mines in 1714. He spent the next four years with the Board, and Anders Berch, writing Kalmeter's biography for the Royal Swedish Academy of Sciences in 1752, observed of this period of his life:

> He diligently attended the meetings of the Board, to learn from the members' deliberations how matters should be dealt with. He spent his free time educating himself: in the Board's mineral collection, in the room of mechanical models, in the assaying laboratory, and in the archives [...] He had during his four years as Auscultator in the Board the opportunity to learn its business, and by travelling in the mining districts to familiarize himself with mines, blast furnaces, smelting works, and all that pertained to this subject.[24]

This is not just another example of the panegyric style of the Academy's obituaries. The Board of Mines passed a similar judgement in their letter to the King in 1718, adding that Kalmeter also had the advantage of being born and bred at the Great Coppermine.[25] The Board was of the opinion that Kalmeter had shown "an extraordinary inclination, aptitude and diligence in learning those things that concerned the operation and economy of mines".[26] Kalmeter is known in Sweden for the exhaustive diaries he kept during his travels abroad, and in particular for his account of mining and metallurgy in England.[27] Of all the Swedish mining officials who visited England during the Era of Liberty, perhaps no one became more familiar with the country than he. Sven Rydberg numbers Kalmeter among the elite of Swedish experts on England in general during the period.[28] He spent no less than four years in England and Scotland, visiting many different parts.

Kalmeter is to be regarded not as a man travelling as the inclination took him, observing and recording what happened to attract his interest, but as an Auscultator with the Board of Mines, supported by the Board as a part of its traditional practice of sending promising young men abroad in order to secure a continuous supply of competent and experienced staff. Furthermore, he received clear and detailed orders concerning where he was to go and what he was to observe. Kalmeter was in some financial difficulty before he started his journey, which makes clear his dependence on the support and instructions of his employer.

In 1718, the public finances of Sweden were in a poor state. Twenty years of war had strained the economy of the nation far beyond its resources. Yet the King tried to organize another campaign, this time against Norway, for which his favourite, Baron Görtz, raised the funds by issuing Government bonds and emergency coinage. Görtz' Finance Department Extraordinary (*Upphandlingsdeputationen*) had only been able to pay out to Kalmeter a part of the travel grant he had been allotted.[29] In the spring of 1719 Kalmeter had arrived in Edinburgh from Rotterdam, and he was running out of money. He complained to the Board of Mines,[30] but the political situation had changed since his departure the previous autumn. The King had been killed in Norway, and Görtz, hated by the civil service, had been executed as a scapegoat for the Absolute Monarchy.

The Board of Mines wrote to the Government that it considered it important for Kalmeter to be given the means to proceed to the mines in England and other countries.[31] It was essential to obtain information on mining practices and conditions abroad. However, it was realized that there was no likelihood of any cash being forthcoming from the Paymaster-General's Office during the present period of public distress, so the Board asked for permission to redistribute its own budgets, which was granted.[32] The money was sent to Kalmeter,[33] and he was told that the orders of the Board of Mines were that he should proceed to England and learn as much as possible about operations, conditions and production in the country's mines.[34] He was to send back regular progress reports, as well as a full report once his journey was completed. The Board of Mines was particularly anxious to have information on copper smelting in England, but Kalmeter was also to ascertain for himself what things most deserved his attention. No one in the Board had studied England's mining industry since 1712 (*N.B.!*),[35] and the Board was probably anxious to know what major technical innovations had been introduced since then.

By the time the Board had settled the financial difficulties and sent Kalmeter his orders, he had already had the opportunity to discover what he obviously thought was one of the most remarkable innovations in the British mining industry. He had first spent six months studying the mines of Scotland. Then in the autumn of 1719 he travelled south down the East Coast of England to London, from where he sent his report on Scotland to the Board in January 1720.[36] Kalmeter finished his report with the promise of another on the mining industry in England, in which he also would describe some new or improved engines and industries. He gave only *one* example:

among which is the engine set up at some places in Newcastle, which by steam from hot water, or rather, by the creation of a vacuum and the pressure of the air, raises weights out of mines.[37]

Kalmeter had arrived in Newcastle on August 20, 1719.[38] The next day he visited some of the collieries on Tyneside. At Norwood he visited Sir Henry Liddell's colliery. He wrote in his diary that:

since the mine was troubled by water, they were now in the process of building an engine which by means of fire or rather vapours from the steam would draw the water out of the pit [...] I will describe this invention, which is the most remarkable ever discovered, later, when I have had an opportunity to see it in operation.[39]

At the time, there were several steam engines of the Newcomen type operating and under construction in the Newcastle area. There was a great demand for the new invention in the district, and Thomas Newcomen and his partners were fully occupied in exploiting this market.[40] During the next few days, Kalmeter saw another engine under construction at Elswick colliery, and a third at Byker colliery.[41] The construction of the third "fire-engine" had proceeded so far that "its so-called Boyler and Cylinder were put in".[42] The first notes by a Swedish traveller on the function, use, and terminology of the new technology had thus been recorded.

Kalmeter evidently managed to obtain more detailed information about the new invention. His report on Scotland, in which he had promised to describe the new engine, was dated January 22, 1720. It was probably at about that time that he received the Board's orders to report continuously on anything remarkable that he saw, and he must have thought that the new engine merited a special report without delay. His report from Scotland had reached the Board on February 16, and two weeks later Emanuel Swedenborg wrote from Stockholm to Erik Benzelius in Uppsala. Swedenborg finished his letter with an extract from a letter "which recently arrived from Kalmeter who is in *Neucastel*, regarding a curious new pumping engine".[43] Since this was, in all probability, the first detailed report on the new technology to reach Sweden, it will be quoted in full:

Close to the city is a newly invented pumping engine built for their collieries, which are badly troubled by water, which is their greatest incommodity. This engine was completed only 6 weeks ago, an exceedingly beautiful invention, powered by fire and water, with a huge iron boiler completely covered above except for a small hole. In this boiler the water is boiled and the whole engine is powered by the vapour, coming from the little hole on top, which is extremely powerful and pushes up the balance of the pump; and as soon as the atmosphere is lost, the opening sucks the balance or the pump, which is on the other end of the balance, down again, and causes its movement. It goes like a churn or a drum, by which one makes butter, made in metal, and fits so tightly that no atmosphere can penetrate at the sides of the piston which runs in the drum. This construction is scarcely to be described, such an engine would be highly useful in Sweden at mines where there is no waterfall. It pumps 400 *oxhufwud* of water per hour, and can be made to run even more strongly. It consumes about 9 *tunnor* of coal in twenty-four hours, and can work to any depth desired. Secretary Triwaldz's brother, who serves here at Messrs. Redley, has promised to send the drawing with a detailed description to his brother in Stockholm.

My thoughts on the matter contained in this letter and on the drawings of such engines that were published a couple of years ago including how this is to be put in practice in Sweden, I intend to present at a later occasion.[44]

Kalmeter's original letter has not been found. It is known, however, that he sent private letters to members of the Board of Mines, as well as his official reports.[45] Swedenborg, who was an Assessor Extraordinary in the Board and a friend of its Vice President, Urban Hiärne, is likely to have had access to any letter from Kalmeter.[46] The working principle of the engine was described almost correctly by Kalmeter. When there is a vacuum in the cylinder ("the atmosphere is lost") atmospheric pressure pushes the piston down into the cylinder ("the opening sucks the balance or the pump down"). But Kalmeter made one mistake in thinking that it was the pressure of the steam that pushed the piston to its top position ("the vapour [...] which is extremely powerful, and pushes up the balance"), when in fact it was raised by the weight of the pumps acting on the other end of the beam. He did not yet fully comprehend the principle of the Newcomen engine, that is, that it was atmospheric pressure alone that produced the force, and not the pressure of the steam.

Kalmeter used Swedish units for cubic measure when he stated the pumping capacity and coal consumption of the engine. In the eighteenth century there was an abundance of different units of measure, which varied not only between nations but also between different cities and regions in the same country. Measures of volume and of weight were particularly complex because the value represented by a unit depended on the kind of substance measured. This situation bedevilled commerce, and necessitated the publication of detailed tables for calculating the equivalents of various units.[47] But it is also relevant to the process of technology transfer. There is a quantitative, as well as a qualitative, side to technological information. If technical knowledge is to be transferred correctly across cultural boundaries, the characteristic quantities defining the technology in question must be transformed into the units of the recipient culture.

The round numbers—four hundred and nine—cited by Kalmeter suggest that he had restated the British units as their nearest Swedish equivalents without converting them. Did this cause a serious error in his report? For the pumping capacity he used the Swedish liquid measure, *"oxhufwud"*, which probably was a translation of its approximate British equivalent "hogshead". In the eighteenth century, one hogshead equalled 238.5 litres,[48] and this unit is known to have been used to state the pumping capacity of Newcomen engines.[49] Since one Swedish *"oxhufwud"* was equal to 236 litres,[50] the difference between the two units was only about 1 %. For the coal consumption, Kalmeter used the Swedish unit of capacity, *"tunna"*. As a measure for charcoal, this was equal to 165 litres (full measure).[51] The common unit of measure for coal in the Newcastle area was the boll, and one boll equalled 159 litres.[52] The difference was only about 4 %. In other words, the error caused by this uncorrected translation of the units was negligible.[53]

Since Kalmeter did not mention the depth, it is not possible, in spite of this information, to calculate the power and efficiency of the engine. He seems, in fact, to have thought that the engine could pump this amount of water from *any* depth. Kalmeter did not mention that the steam was condensed by the injection of cold water into the cylinder. Nor did he describe the different stages in the work cycle, or mention that it was regulated automatically by the motion of the beam. Except for the information that the boiler was made of iron and the cylinder of brass (*"metall"*),[54] Kalmeter said nothing about materials, manufacturing processes, assembly or dimensions. His report was merely a description of the engine's principle, use and advantages. He gave no information about the cost of building it.

Kalmeter was the first Swede to be faced with the problem of describing the new technology to his fellow-countrymen, who had never seen or heard of anything like it. The difficulty presented by the absence of an established terminology is obvious. The word *"cylinder"* was not yet in common use in Sweden,[55] and Kalmeter must have taxed his mind to find a simile that made the concept comprehensible. The Swedish mining engineer's simple but effective solution was to describe

Fig. 7.2. Churn, made in 1685, from the province of Dalarna, where Henrik Kalmeter grew up in the 1690s. (Photo Dalarnas museum, Falun)

the engine's huge, bronze cylinder, weighing almost 2 tons, as a churn ("like a churn or a drum, in which one makes butter"). It was an excellent comparison! It made clear that this was a vertical cylinder in which a piston moved up and down (Fig. 7.2). By this analogy, Kalmeter succeeded in describing two machine elements, their mutual relationship and their movement. When he had added that the whole object was made of bronze, and that the fit was airtight ("fits so tightly that no atmosphere can penetrate at the sides of the piston") the description was complete. In the absence of an established Swedish terminology for the new technology he managed to convey vital technical information by a comparison readily understood by his compatriots in rural Sweden.

The drawings referred to by Kalmeter "that were published a couple of years ago" were probably the engravings by Henry Beighton in 1717 and Thomas Barney in 1719.[56] The information at the end is noteworthy. Kalmeter had evidently met Mårten Triewald in Newcastle during his visit there in August–September 1719, although he did not record this in the diary of his journey. He was obviously well aware of the clandestine nature of his mission. He was not simply a young traveller eager to learn what there was to see, but an industrial spy on an official mission. The secrets of new technology were carefully guarded, and he had to proceed accordingly. In 1719, Mårten Triewald had been in Newcastle for a year or two (see Chapter 10). He had been hired by the Ridley family to build a Newcomen engine at Byker colliery together with Samuel Calley, the son of Thomas Newcomen's assistant, John Calley. It is likely that this was the engine described by Kalmeter in his letter, and that he got his information from Triewald. Kalmeter reported that Mårten Triewald had promised to send a drawing with a full description to his elder brother Samuel Triewald, at the time a well-known author and politician in Stockholm. If Mårten Triewald's intention in 1719 had been to reveal the secret of the Newcomen engine to his fellow-countrymen in Sweden without any personal compensation, he was later to change his mind.[57]

The Board of Mines had sent Kalmeter to England with instructions to report on anything he thought important, and he was chosen because he was experienced in mining and metallurgy. As such he was also well aware of the major energy problem in Swedish industry, namely that of providing power at mines lacking access to waterpower. In the steam engine Kalmeter saw a possible solution to this problem ("... such an engine would be highly useful in Sweden at mines where there is no waterfall"), and he therefore dutifully reported back on the new technology. Thus, the first knowledge of steam power technology was brought to Sweden, not as a result of a random diffusion or a bold private initiative, but as a matter of routine by a Swedish civil servant on official duty.

JONAS ALSTRÖMER

Six months later, another Swedish traveller had the opportunity to learn about the Newcomen engine and its advantages. This was Jonas Alströmer (1685–1761), later to become one of the founders of the Royal Swedish Academy of Sciences. Alströmer is representative of the Era of Liberty and its industrial policy, which was founded on the principles of the Mercantile System.[58] In 1707 Alströmer had moved to London, where he eventually established a commercial firm and made a fortune as an agent of the Swedish Government and of various Swedish companies and individuals. He became a British citizen, and held dual citizenship. As a merchant, trading in goods between Sweden and England, Alströmer became acutely aware of how Swedish raw materials and semi-manufactured goods were exported to England, only to be bought back at a higher price after having been processed in English factories. Would it not be possible, he wondered, to establish industries in Sweden that could produce the finished articles? Alströmer's ambition became to create a Swedish textile industry, and for that purpose he travelled extensively in England, France, Germany and Holland in 1714–1723, filling his notebooks with copious information.[59] In order to enlarge Sweden's textile resources he wanted to improve not only the technology but also the raw material, and he studied both textile machinery and sheep farming.

England was the chief object of Alströmer's studies, and he was later to recommend it as the goal for any young Swede who wanted to travel abroad to study: No nation had more to offer when it came to industry and commerce.[60] He travelled in England in 1714–15 and 1719–20.[61] On March 3, 1720, Alströmer visited Wolverhampton, where he heard something new from a man called Thomas Barney, whom he describes as knowledgeable and clever.[62] His diary continues:

A little way from this place is a fire pump, the first of its kind built in England, which draws up water from some coal mines and ... [Alströmer then describes a horse-powered pumping machine supplying Wolverhampton with water] ... These works are worth seeing. Thomas Barney, a file manufacturer, has engraved the fire engine on a copper plate with a description of the same printed on two sheets of paper which together cost 2 sh.[63]

This was the first Newcomen engine ever built for industrial use: Thomas Newcomen's famous Dudley Castle engine of 1712.[64] The print by Thomas Barney is equally well-known, and is one of the first published descriptions of the new invention.[65] The print is dated 1719 (Fig. 7.3), and it must therefore have been recently published. Alströmer had, in fact, happened to come across the most exhaustive source of information on the new technology that was available at the time. He not only saw Thomas Newcomen's first steam engine, but also met Thomas Barney, the perfect guide, and he bought the engraving.

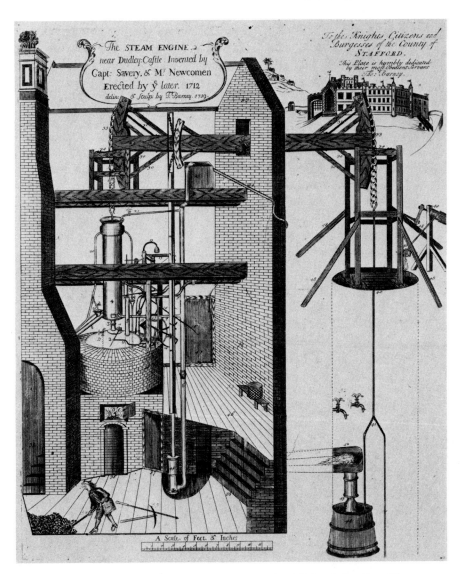

Fig. 7.3. The engraving by Thomas Barney in 1719 of the first Newcomen engine for industrial use. Jonas Alströmer bought a copy of the engraving when he visited Wolverhampton in March 1720 and met Barney. (Photo Science Museum, London)

But the note in his diary is extremely brief, and his only comment is that the engine was "worth seeing". He seems to have left it at that, for he took no further interest in steam engines and there is no evidence that he passed on what he had learned.[66] How could Alströmer, determined as he was to import new technologies to Sweden, stumble on such detailed information about a new kind of prime mover, and yet fail to follow it up?

The answer is that Alströmer was in no need of alternative sources of power. At this period there were virtually no power-driven machines for spinning or weaving wool, only for fulling and raising the surface of the woven cloth. The textile mills he had in mind for Sweden were to be operated by manual labour and waterpower. The problem was not power, but how to import prototype textile machines, skilled workers, and suitable raw materials. Alströmer had also already decided on the location of his future works—the industry that was to be kept alive for forty years by large subsidies from parliament in the futile hope that it would develop into "the manufacturing matrix" of Swedish industry. In 1716, Alströmer had paid a visit to Sweden and his native town of Alingsås, where he had met Christopher Polhem. They had agreed that this was a suitable location for the works.[67] Alingsås, like many other Swedish communities, had grown up around a mill on a small river. Alströmer had long since changed his original family name to take the name of this river ("*Al*" from Alingsås, and "*ström*" from the Swedish word for river).[68] Thus, his very name reflected the traditional source of power in Swedish industry: waterpower fixed in the geographical setting.

Jonas Alströmer's reaction to the first Newcomen engine may be of some general interest. There was an abundance of technological information to be gathered in England, but Alströmer seems to have recorded only what he had need for. He was not an enthusiastic observer, taking notes on and reporting every new innovation that had proved its usefulness. These developments were obviously of no interest to him except as novelties "worth seeing". His reaction is an example of what could be called a *social filter* in the process of technology transfer. The technological information that was transferred before the engineering journals of the nineteenth century was often selected by individual observers. Their choice was determined by the particular nature of their mission, and was thus a reflection of the needs in the recipient culture rather than of all the technologies that were available in the supplying culture.

THE SOCIETY OF SCIENCE IN UPPSALA

The first scientific society in Sweden, *Collegium curiosorum*, was revived by Erik Benzelius in 1719 under the name of *Bokwettsgillet*.[69] The fellows met every week to discuss scientific innovations and newly published books. They published articles and book reviews in their journal, *Acta literaria Sveciae*.[70] Emanuel Swedenborg was elected a member in 1720, and accepted the nomination in a letter dated February 5, in which he promised "to communicate on occasion what might come to my notice".[71] Only three weeks later something did come to his notice—something that he described as "remarkable and of value" in a letter to Erik Benzelius.[72] This included the extract, quoted and dis-

Fig. 7.4. The petrified corpse of the miner and the spot where his remains were found in the Great Coppermine in Falun in 1719. Engraving in *Acta literaria Sveciae* in 1722. (Photo Royal Library, Stockholm)

cussed above, from Kalmeter's letter. Benzelius read this to the society at its next weekly meeting.[73] The note in the report of the proceedings probably reflects the fellows' understanding of Kalmeter's report after their discussion:

From Assessor Swedenborg's letter dated February 29, it was reported that Auscultator Kalmeter of the Board of Mines had provided information from New Castel in England about a pumping engine that had recently been set up in their coal mines, viz. how the strong vapour or atmosphere, which is given off by a big boiler with water inside, and fire beneath, pushes or drives up and down by a kind of heaving 400 *tunnor* of water per hour in brass pumps with pistons also of brass. This machine can be used to any depth, and consumes only 9 *tunnor* of coal in twenty-four hours.[74]

Kalmeter had given the fundamental principle of the Newcomen steam engine: the pressure of the atmosphere forces the piston down in the cylinder when the steam has been condensed to create a vacuum. This was lost in the proceedings of the society, and replaced by the incomprehensible explanation that "the strong vapour or atmosphere [...] pushes or drives up and down by a kind of heaving". The fellows of the Society of Science either did not understand its principle, or did not bother to record it correctly. Furthermore, Kalmeter had given the pumping capacity as 400 "*oxhufwud*", but this was recorded as 400 "*tunnor*"—understating the power by 40 %.[75] The fact that the fellows did not take the trouble to get their units right implies that they took no interest in the practical potential of the new technology.

There were so many other curiosities to catch their fancy! Their discussions ranged over a wide field, and they let their minds roam hither and thither: the finding of a tooth from a giant, a tapeworm 20 feet in length, a werewolf sighted in Bohuslän, the possibility of a subterranean connection between Lake Constance in Switzerland and Lake Vättern in Sweden, the killing of eight cows at Österby Bruk by a thunderbolt, the fact that the American Indian lacked a beard, or that the Lapps held the bear sacred and therefore forbade their womenfolk to eat bear meat while they were menstruating and were unclean.[76]

One novelty in particular seems to have captured the imagination of the fellows at this time. A miner named Matts Israelsson had been killed and buried in a cave-in in the Great Coppermine in Falun in 1676 or 1677. The story of "*Fet-Matts*", whose almost totally preserved body was found more than 40 years later in 1719, is well known.[77] The fellows discussed this petrified corpse ("*Petrefacto cadavere*") at several meetings in 1720 and 1721. In November 1720 they decided to send a letter enquiring about (1) the length of the corpse, (2) its weight, (3) whether it had decomposed or been preserved, (4) the hardness of the flesh, (5) whether the hair had come off, and (6) whether they could have a sketch.[78] In February 1721, they listened to a letter answering some of their questions.[79] But their thirst for information was still not satisfied, and in March Per Martin promised to carry out a chemical analysis of parts of the body.[80] At the next meeting he performed various chemical experiments to show how the vitriolic water in the mine had preserved the flesh, and he promised a full report on his experiments.[81] In June, they listened to a letter from Lars Benzel-stierna, in which he promised to communicate some interesting obser-vations on the corpse.[82] In September, Adam Leijel sent a detailed report to the Society on the condition of Matts Israelsson's remains.[83] Leijel's report was published in the next volume of the transactions of the Society, and illustrated with a whole-page engraving (Fig. 7.4).[84] At a meeting in November, Per Martin was able to display a piece cut from one of the thighs of the indurated *"Cadavere humano"*.[85] At last, the fellows seem to have been satisfied.

Their interest in this bizarre case was not just idle curiosity. The matter was in fact of some scientific importance. If the human remains really had been petrified, it followed that they must be classified as a mineral. The corpse had undergone a transformation that required it to be transferred from one system of classification to another. It presented a taxonomic anomaly. Well-known metallurgists such as Johan Gott-schalk Wallerius, Torbern Bergman and Axel Fredrik Cronstedt were also to include human bodies, salted in vitriol, in their mineralogical systems during the eighteenth century.[86] The point is, however, that the interest of the fellows in this novelty also sheds light on their lack of interest in the simultaneous news of the new technology. Had they been as interested in the Newcomen engine as in the *"Petrefacto cadavere"*, they would have sent a letter to Kalmeter asking for equally detailed information on the new technology and requesting a drawing. They would have performed physical experiments to demonstrate its princi-ple, i.e. that the pressure of the atmosphere could be made to produce a force by condensing steam. They would have welcomed reports giving additional information and printed them in the transactions, illustrated by engravings. But judging from the report of the proceedings and the journal, the pumping engine in Newcastle was only one of many innovations to be passed over during their discussions. They were more interested in knowledge for the sake of knowledge, such as the classifica-tion of matter, than practical knowledge, such as the Newcomen en-gine, that might help to solve urgent problems in industry. This may have been consistent with their definition of science, but it is neverthe-less worthy of note since the Newcomen engine ought to have posed many intriguing questions to the scientists of the eighteenth century.

SUMMARY

Emanuel Swedenborg seems to have been the first Swedish traveller to come into contact with the new technology. In a letter written in 1715 he reported the principle of the steam engine: the conversion of thermal energy into mechanical energy by means of an engine. But on his return to Sweden he came under the influence of Christopher Polhem, and turned his attention to traditional sources of power.

The Board of Mines made a practice of sending promising young men abroad in order to gather intelligence of foreign developments, and to secure a steady supply of competent and experienced staff who could ensure that the industry kept abreast of the times. Since no one from the Board had visited England since 1712 Henrik Kalmeter was sent there in 1718 to find out what major innovations had been introduced in the mining industry since then. In 1720 he dutifully reported back on the principle, use and advantages of the Newcomen engine.

Jonas Alströmer happened to see Newcomen's first engine, the famous Dudley Castle engine, met Thomas Barney and bought his print of the engine in 1720. But the engine was of no interest to Alströmer save as a novelty "worth seeing". The work in his textile mills was to be done manually and by waterpower, and he had no need of alternative power sources. His reaction is an example of a *social filter* in the process of technology transfer.

The Society of Science in Uppsala received a copy of Kalmeter's report in 1720, but they took little interest in the new technology. They were more interested in knowledge for its own sake than in practical knowledge, even when the latter was based upon a new scientific principle.

Thus, basic knowledge of steam power technology was brought to Sweden only a few years after it had proved its usefulness in the British mining industry. It was brought here as a matter of routine by the Swedish civil service, and it could be said that its successful transfer was due to the institutional structure rather than to individual initiatives. This knowledge was collected and assimilated by the Board of Mines.

8. The Adventurers, 1723–1725

DE VALAIR

On March 14, 1723, a Frenchman calling himself Colonel de Valair handed in an application to the Swedish Government.[1] He said that he had been called to Sweden after it had become known that he "possessed knowledge of several arts and crafts" which might be useful to the Swedish nation, and that he had therefore immediately made his way to Sweden to offer his services.[2] Perhaps de Valair had received an offer from a Swedish ambassador somewhere in Europe, or perhaps he had heard rumours of the opportunities that Sweden offered to foreigners at this time.

De Valair's application was twofold. First, he wanted "to build and construct an engine, by which an incredible amount of water can be drawn from the deepest shafts and mines".[3] He said nothing about the working principle or construction of his engine, but he emphasized its value to mining: with the aid of this engine, water could be pumped from waterlogged mines "by a much easier and more direct way" than by the traditional methods, which involved great inconvenience, high costs and a waste of time.[4] The engine would also make it possible to resume work in mines that had been abandoned due to flooding. "The Nation's riches and natural wealth, which consist principally of metals noble and base, could thus be enlarged and improved to a goodly extent".[5] Secondly, de Valair wanted to introduce a method of "converting bar iron into the choicest steel without reduction of the material, and in any quantity desired".[6]

In a detailed memorandum de Valair applied for patents for his two innovations, and stated the terms he sought. For the engine he applied for a "*Privilegium exclusivum*" for himself and his associates (who were not specified). No one, except de Valair and his associates, should be permitted to import or copy such an engine; the penalty for doing so should be a fine of 4 000 *copperdaler* and confiscation of the engine. Anyone desiring to use such an engine had first to come to an agreement with the partners on a fee, and he was not to use other labour than that assigned by them. The application dealt specifically with the concessions that ought to be granted to the partners should they, thanks to this engine, succeed in resuming work in an abandoned mine. In that event they wanted a guarantee of tax relief, and access to labour and material. De Valair's application was referred to the Boards of Mines and Commerce for a joint opinion,[7] and it was discussed at several meetings at the beginning of May, 1723.[8] De Valair was summoned to some of these meetings for detailed questioning. Right from the first meeting the discussion focussed on his second proposal, the process for manufacturing steel, and this was to throw into the shade his project for an engine to raise water. Regarding the engine, the Boards only had one question: if such an engine happened to be invented in Sweden, or brought here from abroad by someone else, would de Valair insist on its being forbidden? Yes, if it was of the same kind as the one he intended to build. However, he did not wish at all to prevent or prohibit "the arts and methods that are presently used and known here in the Realm".[9]

On May 10 the Board of Mines decided to buy "a bar of good fine iron" to let de Valair demonstrate his method of making steel under controlled conditions.[10] They were probably eager to see if he really could "convert bar iron into the choicest steel without reduction of the material" as he had promised. The idea of eliminating the weight loss occurring when steel is made from bar iron—which was between 5 and 12.5 per cent—was an attractive one.[11] De Valair's second proposal was only mentioned in passing:

... and regarding the requested patent for a water-engine, that he intends to construct, the Board has nothing to remark, and may anyone who intends to use such an engine come to an agreement with the partners as he cares.[12]

In their joint report to the Government, the Boards of Mines and Commerce endorsed de Valair's application for his two projects.[13] Of the two bodies, the Board of Mines had played the more active part. Its President in 1723 was Count Gustaf Bonde (1682–1764), and this prompts a comment. At the beginning of the Era of Liberty, the alchemical tradition had forever lost its impact on chemistry in Sweden. Only a few advocates of hermetic philosophy and believers in transmutation of metals remained, but the most prominent among them was Gustaf Bonde.[14] He was to become a successful politician and he rose to be a privy councillor and also the Chancellor of Uppsala University. Sten Lindroth has described him as something of a psychological enigma: in public a politician of sound knowledge, but privately devoted to the most esoteric alchemical speculations, which he published anonymously.[15] Bonde was a true alchemist, detesting gold-making as the vulgar misuse of an ancient knowledge that should be used for the spiritual rebirth of man.

De Valair's suggestion that bar iron could be converted "into the choicest steel without reduction of the material" was a true alchemist's idea: transmutation of metals. This may explain the great interest shown by the Board of Mines in de Valair's application. It appealed to both sides of its President's nature, to the civil servant and to the alchemist.[16] De Valair's steelmaking project was to occupy the Board for several years, and even reached the stage of plans for industrial production near Sundsvall in the province of Medelpad.[17] His other proposal, the "water-engine", obviously benefitted from this great interest. It was approved at the same time, almost as an afterthought. The ease with which de Valair succeeded in acquiring a monopoly on the Newcomen engine in Sweden may thus have been influenced by the alchemical beliefs of the President of the Board of Mines.

De Valair's application for a patent for a "water-engine" was observed from abroad by someone who had a personal interest in developments in Sweden. This was Mårten Triewald, employed at Ridley's collieries in Newcastle (see Chapter 10). In a letter dated September 24, 1723, Henrik Kalmeter wrote from London to Jonas Alströmer in Amsterdam (he wrote in English, as was their habit since Alströmer had lived in England for so long):

One Colonell Valaire has proposed in Sweden to convert the worst of Iron, nay Pigg Iron, into excellent Steel without waste, which is a Secret indeed, as Mr. Triewald sayeth. He has likewise proposed an Engine to draw water out of mines, and all this to the State.[18]

It is not clear from the letter whether Kalmeter's information came from Triewald, or whether Triewald had merely commented upon it. But it is interesting to note that Triewald was able in Newcastle to follow developments in Stockholm. De Valair's application had been discussed by the Boards of Mines and Commerce in May, and Triewald knew about it a few months later. This is something that this study illustrates on numerous other occasions: *the mobility of technical information in terms of speed and distance.* Letters and books reached destinations all over Europe within a couple of weeks, and a traveller could cover a great distance within a month. There were other and more severe constraints on technology transfer than the problem of communication.

On October 23, 1723, de Valair was granted a patent for his "water-engine" ("*wattumachine*").[19] The patent gave no details of the working principle or construction of the engine, but stated that it applied to "a new and hitherto unknown water-engine".[20] In all essentials the concessions followed those in his application, and even echoed his reasons: watery mines could be drained in a simpler, cheaper and quicker way, and work could be resumed in abandoned mines. But there also seems to have been genuine appreciation in Sweden of the need for a new technology:

> As We are very well aware of the relief and advantage to the mines in this country if such an engine could be constructed and used for draining water with the effect that Colonel Valair has promised ...[21]

Six months after the patent had been granted, de Valair submitted a new application to the Board of Mines.[22] He told the Board that he had looked carefully into the question of where his engine could be of most use to the public as well as to himself, and that he had finally decided to demonstrate its effect in the mines of Öster Silvberg, in the parish of Tuna. He had formed a company for this purpose, together with "several men of rank and other experienced persons".[23] There they intended to build an engine and, once they had removed the water, to investigate whether the mines contained any ore that made them worth reopening. He pointed out that the mines in Öster Silvberg were known to have yielded large amounts of silver in earlier times, and that he therefore assumed that water in the workings, not lack of ore, had been the reason for abandoning them. De Valair referred to the recently published Royal Statute (see Chapter 5) dealing with the concessions that were to be granted to anyone who took up work in a new or abandoned mine. He was aware that the proximity of the Great Copper-mine in Falun might make it difficult to supply the planned works at Öster Silvberg with firewood, charcoal and building materials, and he asked the Board of Mines for its help. The Board of Mines instructed the Mine Inspector for the district, Göran Wallerius, to investigate the present state of the mines in Öster Silvberg and to see how de Valair could be furnished with what he needed.[24] If de Valair were to appear at Öster Silvberg, Wallerius was to give him all the assistance that his office enabled him to provide.

Öster Silvberg, in Dalarna, is considered the oldest silver mine in Sweden.[25] The deposit was probably worked in the late fifteenth century. It was rich in galena at the surface, but the ore decreased rapidly lower down. The mines had been abandoned in 1641, when the cost of extraction began to exceed the value of the yield. The ore contained a high percentage of silver. It was also auriferous, and the silver obtained from the ore held as much as 10–15 percent gold. Awareness of these high percentages of gold and silver had prompted several attempts to resume work in the mines after 1641, although by then the deposit had probably been fairly well exploited. In the summer of 1695 Baron Nils Gripenhielm (1653–1706), Governor of the County of Kopparberg, had amused himself by assaying samples of minerals from the mining district of the county.[26] He was a passionate amateur metallurgist, a collector of minerals with his own private assaying laboratory at the Governor's residence in Falun. He collected a few samples of minerals from gangue mounds at Öster Silvberg, where the

Fig. 8.1. One of the commemorative coins minted by Nils Gripenhielm from gold from the gangue mounds in Öster Silvberg in 1695. Natural scale. (Photo Antikvarisk-Topografiska Arkivet, Stockholm)

mines had lain abandoned for more than fifty years. To his delight, the samples turned out to be rich in gold and silver. More ore was brought back to Falun, and he obtained enough gold to mint a hundred commemorative coins (Fig. 8.1).[27] Encouraged by this, Gripenhielm seems to have continued, for in 1701 he made a solid silver cup, gilded inside. He put an inscription on the cup, of which a few lines read in translation:

Although my meagre rock by all had been rejected,
Yet Gripenhielm my ore with diligence extracted
and made the world to see that it should not disdain
what Öster Silfverberg in Dala-Tuna doth contain.[28]

De Valair and his associates, at least, did not disdain what Öster Silvberg contained. Only two of the associates are known, and one of them was Baron Otto Reinhold von Strömfelt (1679–1746),[29] He had been appointed Governor of the County of Kopparberg in 1719,[30] and must have known of the work of his predecessor. It was probably reports of Gripenhielm's success some twenty years earlier that tempted de Valair and his associates to try to resume work at Öster Silvberg. Perhaps they had even seen one of Gripenhielm's ducats, felt its weight in their hands and admired its lustre. Hence it was possible to extract gold from Öster Silvberg, and the "water-engine" was to be their means.

De Valair had claimed that he had not considered where his "water-engine" was to be built before the privilege was granted on October 23, 1723.[31] Not until later had he decided on Öster Silvberg, after a company had been formed for this purpose. But it is more likely than not that Strömfelt and the other associates had joined forces with de Valair in the first months of 1723, or earlier, with the definite intention of extracting gold and silver from Öster Silvberg with the aid of his engine. De Valair's original application to the Government in March, 1723, had been written in Swedish, a language of which he had no command, and it was evidently composed by someone well acquainted with Sweden and its legislation.[32] The application had dealt explicitly with the concessions that de Valair and his associates considered they ought to be granted if they succeeded in resuming work in some abandoned mine. The assumption that the company had been formed

well before the patent had been granted, and that they had had a
specific project in mind from the outset, is confirmed by another letter
in English from Henrik Kalmeter in London to Jonas Alströmer in
Amsterdam, dated October 12, 1723:

> Mr. Törner has writ some time ago to Mr. Rönling to get a Man over to
> Sveden, that could erect a fire Engine there of the same kind as at New Castle
> and otherwhere. He has with the last Post iterated the same and recommended
> a great deal of secrecy, and at least desired to have a modell of the Engine. I
> think both will be impossible, few or none of the workmen being let into the
> secret of building it. But as Mr. Rönling has spoke to me about it at the desire
> of Mr. Törner, I'll see to send a draught of the Machine to him, if that will do,
> and as it may in a great deal, if any one in Sweden is of humour or capacity to
> undertake it. Mr. Triewald, when he goes there, can best bring it about.[33]

Mr. Törner was none other than Olof Törne (von Törne after being
ennobled in 1726), the second of de Valair's known associates.[34] He
had been in England in 1715–1716, and was now Assessor Extraordinary in the Board of Commerce.[35] Rönling was Jonas Alströmer's
agent in London.[36] From the letter it is clear that the "water-engine"
de Valair had in mind was a steam engine of the Newcomen type.[37]
But de Valair and his associates seem to have had little knowledge of its
construction, although they knew that such engines were used successfully in Newcastle and other places in England. This information
seems, in fact, to have been their only asset. But even this slender
knowledge could be a profitable commodity if handled with care.

Their aims seem to have been twofold. Firstly, they wanted to secure
a monopoly of the new technology, but in broad and general terms since
they could not describe it in any detail. It seems that de Valair and his
associates were certain that they were the only ones in Sweden who
knew about the new invention. They therefore tried to get a monopoly
as quickly as possible. De Valair had said that he wanted to make it
illegal for anyone else to import an engine of the same kind as the one
he intended to build, but that this did not apply to "the arts and
methods that are presently used and known here in the Realm".[38]
They realized this objective, being granted a patent without ever
having to describe the engine in greater detail than as "an engine, by
which an incredible amount of water can be drawn from the deepest
shafts and mines".[39]

Secondly, they wanted to get the knowledge they needed in order to
build such an engine. This would enable them to exploit all the mineral
deposits in Sweden that could no longer be worked with traditional
technologies. Since Törne repeated his request in a second letter, they
seem to have been anxious to commence work as soon as possible.[40]
Their lack of knowledge of the engine's construction is evident from
Törne's request: if they could not get a worker or a model from
England, at least they wanted a drawing. Until the patent had been

granted, however, they had to proceed with some caution; hence the appeal for "a great deal of secrecy".

Mine Inspector Wallerius did not report back to the Board of Mines until November 20, 1724.[41] He had been ordered to examine the feasibility of building de Valair's engine at Öster Silvberg, and now reported that he had made inquiries concerning the whereabouts of de Valair, and whether he and his associates intended to carry out their project. In Falun he had heard rumours that they had changed their minds. They had brought an Englishman to Sweden, he had been told, to build the engine ("to reveal the invention of this new engine").[42] But according to Wallerius' information the Englishman had left Sweden. Wallerius had not heard a word from de Valair and his associates. De Valair's plan to build a Newcomen engine at Öster Silvberg in 1723 was the first project for a Newcomen engine in Sweden, but it has hitherto escaped notice.[43] The fate of the project, and the identity of the mysterious Englishman, will be told in the following section. The Board of Mines laid Wallerius' report *ad acta*.[44] Although technical development met with benevolent support from the authorities, it was left to individuals to take the initiative. It was not long, however, before a similar initiative was taken, but this time from another quarter.

O'KELLY'S PROPOSAL

On March 2, 1725, the Swedish Ambassador in The Hague, Joakim Fredrik Preiss (1667–1759), wrote a letter to the Government in Stockholm.[45] He had a strange story to tell, and it began:

> Last night I was visited by a Catholic Irish Nobleman and Colonel by name Johan OKelly, Esquire, Seigneur d'Aghrim, who said that he possessed the knowledge to build the fire-engine to pump water out of flooded mines. That he had constructed such an engine in Liège, and that Baron Wansul and other distinguished gentlemen were his partners. That this engine was capable of drawing about 14 000 barrels of water a day out of the mines.[46]

O'Kelly had come to Preiss to offer Sweden his services. Carl Björkbom gave an account of O'Kelly's proposal in 1936,[47] and it has more recently been studied by Graham J. Hollister-Short.[48] Björkbom does not seem to have studied the archives of the Board of Mines, which contain additional material, and his account can therefore be supplemented and corrected.[49]

Some comment on John O'Kelly, and on his experience of building Newcomen engines, is called for before we return to his proposal to the Swedish Ambassador in 1725. Björkbom was not able to identify "the mysterious O'Kelly", nor to verify his claim to have built a Newcomen engine near Liège. Björkbom suggested that the John O'Kelly who visited Preiss might have been identical with a captain of the same name in the English army, who, according to Björkbom's informant, had settled in Liège in the 1720s.[50] His assumption was proved correct in an article by Georges Hansotte in 1950, in which Hansotte gave an account of the introduction of the steam engine to the Liège area in 1720.[51]

John O'Kelly d'Aghrim was born in 1672 in Galway, Ireland, and was a descendant of the renowned sept of Ui Máine with a pedigree going back to the eleventh century.[52] He had served as a captain in the English army until 1711.[53] His whereabouts between then and 1720 are not known, but during this time he obviously learned how to build Newcomen engines. He knew Henry Beighton, one of the pioneers in the development of the Newcomen engine, and from O'Kelly's account of the early history of the steam engine, it is clear that he had an intimate knowledge of the new technology.[54] Hollister-Short has discussed whether it was difficult for a foreigner in the days of the first Newcomen engines to contrive to see an engine and learn how it worked, and how O'Kelly had acquired his knowledge of steam engine construction:

> It is clear, however, that such knowledge was fairly widely disseminated in England, at least among those of the *cognoscenti* with a taste for experimental philosophy: that O'Kelly is to be numbered among such men appears from an elegant and lucid description (in French) of the machine and its mode of operation drafted by him, in early 1725, for the members of Bergskollegium [the Board of Mines] in Stockholm.[55]

The description for the Board of Mines referred to here was the one published by Björkbom in 1936, and will be discussed later.[56] But in that description O'Kelly spoke with scorn of "*Mess. les Philosophes*", and he must have possessed practical experience of constructing Newcomen engines, for he knew where to order parts and hire workmen from England.[57] It seems likely that O'Kelly had participated, in some capacity, in the construction of one or more of the twenty-five or so Newcomen engines that were built in England before 1720.[58] One of the points illustrated in this chapter is precisely this difference between *theoretical* and *practical* knowledge of the new technology. It was one thing to see a small model operating on a table in a drawing room and to have its principle explained. Or even to see an actual engine at work and to buy a print of it, as Alströmer did in 1720 when he saw Newcomen's first engine and met Thomas Barney. But it was quite a different matter to know how to dimension, choose the material for, manufacture, and assemble its many reciprocating parts; how to adjust it to make it run smoothly; how to identify and repair its most common malfunctions—in short, to have the practical knowledge to realize, operate and maintain the new technology.

In 1720 John O'Kelly had been engaged by a private company in Liège to build Newcomen engines for the mining industry.[59] Their first commission was at a coal mine in Jemeppe-sur-Meuse near Liège.

Fig. 8.2. A gentleman, late for a ball, having his gaiters removed by his servant. The watch in his hand would have represented several years' income to his servant. In the absence of any portrait of John O'Kelly and his workman Saunders, this picture may serve to illustrate not only the two individuals but also the difference in their social positions. Detail from the engraving "Analysis of Beauty" by William Hogarth in 1753. (Photo Bengt Vängstam)

Hansotte has described how this engine—the first Newcomen engine on the Continent—was completed by February 1721, having taken only six months to build, and how the partners ran into severe financial difficulties and quarrelled with O'Kelly over the terms of his contract. O'Kelly was not reimbursed for his expenses, and he complained bitterly in several letters to Baron Berthold de Wanzoulle, one of the partners.[60] In a letter in February 1721 he wrote that he had invested so much money in the engine that he had been forced to borrow 100 guineas from his friends in England:

and I am not only ashamed to ask them for more but afraid that I would be refused. I have in my youth got into the bad habit of eating which I can't rid myself of, and I can't see any opportunity to continue in it. I have just sent my watch to be sold and when I have got rid of my goods, my partners here have so much regard for me and enough generosity to let me die.[61]

Apart from the temporary personal discomfort of being broke, there could be a more serious lasting consequence: O'Kelly complained that he would be disgraced if his workman ("*Ouvrier*") noticed his lack of money, and that the workman would then hire himself out to others, who would be only too ready to engage him. Without his watch—a status symbol that distinguished gentlemen from workmen in the eighteenth century—he was socially degraded, and unable to keep up appearances as an affluent entrepreneur in front of his hired workman.[62]

In 1723 O'Kelly gave up, and sold his shares in the company.[63] He remained in Liège, negotiating for another contract to build a Newcomen engine, but he was deeply in debt and harassed by his creditors, who threatened to put him in jail.[64] It is known that towards the end of 1723 he married Marie Albine Angelique van der Moere, widow of Count Jean de Rochefort, in Brussels.[65] This is practically all that is known about O'Kelly before he came to pay his visit to the Swedish Ambassador in The Hague on the evening of March 1, 1725.

O'Kelly told Preiss that Prince Wilhelm of Hessen, brother of the King of Sweden and Governor of Maastricht, had heard about the engine he had built near Liège.[66] The Prince had made enquiries about O'Kelly, in order to advise him to go to Sweden and build such an engine. But his partners had kept his name secret from the Prince, and had not told the Prince that he was still in Liège. However, O'Kelly's workman had heard about these enquiries, and had introduced himself to the Prince. He had told the Prince that he had mastered the art of building such an engine, and the Prince had then given him 100 *pistoles* to go to Sweden. O'Kelly had heard, he told Preiss, that the workman had built an engine at the silver mine ("*Silfwergrufwan*") with royal support, but that the engine had not had the promised effect. This had distressed O'Kelly, when he considered how much money the construction of such an engine must have cost. He had therefore come to Preiss to offer his services to Sweden.

O'Kelly's information about an engine at the silver mine was obviously a distorted rumour of de Valair's project for an engine at Öster Silvberg. No engine was ever built there with royal support, but it was true that de Valair and company had received a royal patent for their "water-engine", and that the Board of Mines had given them permission to start building one at Öster Silvberg. From Göran Wallerius' report, mentioned above, we also know that de Valair and his associates had brought an Englishman to Sweden, presumably in 1724, but that he had soon left the country.

Björkbom, who knew about de Valair's patent but had not studied the archives of the Board of Mines, guessed that O'Kelly's workman and de Valair were identical.[67] This is quoted by Hollister-Short who adds, on basis of the date of ratification of de Valair's patent, that Prince Wilhelm's search for O'Kelly and his workman's departure to Sweden, had taken place in early 1723.[68] The date is not known, but it was probably later since we know from Kalmeter's letter to Alströmer that Törne was still looking for someone who could construct an engine in October 1723.[69] Björkbom misinterpreted Preiss' original letter at one essential point. Preiss wrote that O'Kelly's workman, who was reported to have built an engine at the silver mine in Sweden, was a man called Saunders. Björkbom thought, however, that Saunders was the name of a workman that O'Kelly offered to send to Sweden during his conversation with Preiss, and thus that Saunders was still in O'Kelly's service in March 1725.[70] But it is quite clear from Preiss' letter that Saunders was the name of O'Kelly's former workman, who had left him earlier to go to the silver mine in Sweden.[71] This may appear a trivial point, hardly worth the space devoted here to its correction, but it is of some relevance if we are to examine the structure and sociology of technological knowledge.

When referring to Saunders, Preiss used the Swedish word *"Werkdräng"*, which has been translated here as "workman".[72] The Swedish word *"werk"* meant mechanism, apparatus or factory.[73] *"Dräng"* meant manservant, but it also had a pejorative meaning in the sense of "inexperienced and unskilled labourer".[74] *"Werkdräng"* thus implied an inexperienced man in a subordinate position doing unskilled work on the steam engine. This must at least have been the spirit of O'Kelly's description (presumably in French), since Preiss chose this Swedish word. The implication was that O'Kelly himself was Saunders' experienced superior who did the theoretical work. O'Kelly made this point very clear, since Preiss reported:

And regarding this workman, he is an honest, diligent and sensible fellow, who is well trained in executing what he is set to do since he [O'Kelly] has trained him for a couple of years, and he has earned 100 pounds sterling a year. But except for that, he has no knowledge of the fundamentals of science and does not in the slightest understand the calculation of the proportions to which this engine must be designed in all its parts.[75]

O'Kelly must still have doubted whether his point had been taken, since he added in a short letter in his own hand that *"Mon Valet Saunders est tres Capable de ce qui regarde la Manœuvre de la Machinne a feu, mais il ignore la Theorie"*.[76]

"La manœuvre" and *"la théorie"*: here was the essence of his point. The *practical* work, the labour that was required in order to apply the new technology, could very well be left to honest, diligent and sensible fellows like Saunders if they had been well trained. But to make the engine function, there was also some *theoretical* work to do. The calculation of its proportions demanded the skills of an experienced superior

who had mastered the fundamentals of science, someone like O'Kelly himself, for example. O'Kelly took great care to make the theoretical part of the new technology seem difficult, even mysterious, while stressing at the same time that he, John O'Kelly, was one of the few who fully understood it. He offered to Preiss to reveal the complete science of building steam engines.[77] The secret lay in dimensioning the engines to the correct proportions, and in counterbalancing the beam to equal the weight of the atmosphere with regard to the friction (*"frottement"*). Preiss reported that O'Kelly had told him:

that it all depended on a rule of proportion, which in England they usually try to establish through elaborate calculations, but which he had brought to perfection through experience so it could be worked out by means of a few lines. Its basis was the numbers seven and nine, with the aid of which a number of square roots could be calculated and from which the proportions could then be established.[78]

This piece of blarney served the purpose of making the art of building steam engines seem highly theoretical, while at the same time projecting O'Kelly as someone who had a sound scientific grasp of it. He, alone, had been able to deduce a simple, mathematical law from the empirical data. This was the hallmark of science, but that was not all—his law could be applied in the designing of steam engines. That is, the steam engines that he offered to build would not be crude machines based simply on practical experience, but would be properly designed on what would today be called scientific technological principles.

When Preiss had asked where O'Kelly could be found, should it turn out that there was an interest in Sweden in his proposal, O'Kelly had replied that he was about to travel to Spain.[79] The Spanish ambassador in London, Marquis de Posso Boueno, had persuaded him to go to Biscay to build an engine for the waterlogged mines there (Preiss mentioned that he had verified this information with the Spanish envoy in The Hague). O'Kelly thought, however, that he could delay his departure for a month because he had business to attend to in Amsterdam. Preiss finished off his letter by saying that "This is, Most Gracious Sovereign, what this OKelly has presented, most modestly and without boastful embellishment of his knowledge".[80] Preiss asked what he should reply to O'Kelly if it turned out to be true that the engine built by Saunders did not function, and if such an engine was considered to be of use to Sweden.

In a postscript Preiss reported that, after he had finished writing the letter above, he had received another visit from O'Kelly, who had come back to present his rule of proportion "out of reverence for Your Majesty and devotion to the Swedish people".[81] On second thoughts, he had probably realized that he stood to gain more by showing his knowledge more explicitly. O'Kelly gave the rule in a short letter, which Preiss attached to his own letter of March 2.[82] The rule gave the diameter of the cylinder of the steam engine as a function of the depth from which the water was to be pumped. If the depth was 50 fathoms, the weight of the pump rods and the water in the pumps should be divided by 9, but if the depth was 100 fathoms the weight should be divided by 7. The square root of the quotient gave the diameter of the cylinder.

In general, the power of a Newcomen engine is directly proportional to the area of its cylinder, i.e. to the square of the radius of the cylinder. Anyone who had realized this simple basic principle of the Newcomen engine would also be able to calculate the diameter of the cylinder *theoretically* needed in order to pump water out of a mine of a certain depth with pumps of a certain diameter. O'Kelly's rule of proportion stated this fundamental relationship, but it also included a proportional factor that was based on *practical* experience. In theory, the area of the cylinder should be twice as large for a depth of 100 fathoms as for 50. But according to O'Kelly's rule, the area of the cylinder should be made not twice but 2.6 times as large. This was probably mainly due to the friction in the pumps, whose increase with depth was not linear and had to be compensated for.

O'Kelly only gave the proportional factors for two depths, 50 and 100 fathoms, but it would, however, be possible to make a rough interpolation and extrapolation for depths of up to some 100 fathoms in order to estimate the diameter of the cylinder.[83] This would be highly useful, but it applied only to engines and pumps of the kinds O'Kelly was used to building. It did not apply to engines of a different effective power and connected by a different arrangement to pumps made of a different material with different friction characteristics. Moreover, O'Kelly's rule of proportions was only useful if one knew what units of weight and measure he was working in, because the proportional factors were determined by the units he used for the weight of the pump rods and the water, and for the diameter of the cylinder. Units of weight, in particular, varied considerably in those days between different countries and cities. (The importance of units as a quantitative factor in the process of technology transfer across cultural boundaries was discussed in Chapter 7.) O'Kelly's rule of proportion was thus nothing more than a rule of thumb. It was a summary of his own practical experience as an engineer, valuable as such, but presented to Preiss and the Swedish authorities as a universal physical law, a scientific principle that could be applied to the new technology. Only a week later, Preiss reported that O'Kelly had come back for a third time.[84] Since his last visit O'Kelly had been told, in a letter from London, that Saunders was back in London. O'Kelly did not know whether Saunders intended to remain in England, as he had told some people there, or if he planned to return to Sweden. In either event, O'Kelly thought that there was now nothing preventing him from being of service to the Swedish people, and he offered to send his son, "who had also studied physics, and understands the theory as well as the practice of this physical engine", to Sweden.[85] The son must have been Lucas O'Kelly, the only son of John O'Kelly's first marriage, of whom nothing else is known except that he ended his days as a barefoot friar in the Carmelite order in Dunkirk in 1778.[86] In 1725, he was probably in his twenties, and since John O'Kelly relied on his competence and co-operation, we may assume that he had helped his father to build a steam engine, either near Liège in 1720–1721 or perhaps even earlier in England.

Here, again, O'Kelly stressed the importance of theoretical knowledge; to be specific, knowledge of physics. That this was essential was implied by his reference to the steam engine as "this physical engine". All engines, obviously, obey physical laws in their working, but the design of the common waterwheel, for example, was based on the practical experience gained by this time. What O'Kelly implied was that the steam engine was different from any ordinary engine in that its design represented a conscious application of the principles of physics. Hence, theoretical knowledge was essential if one wished to build a successful engine. On this third visit, O'Kelly also presented Preiss with a detailed description of the working of the engine and promised to deliver a drawing a fortnight later.[87] He was at present working on the drawing:

on the pattern of the one which is printed and sold in England, many details of which are drawn less to be of use, for they are not required, than to confuse strangers in their desire to discover its construction.[88]

The description and the drawing will be discussed in the next chapter, but it is interesting to note how O'Kelly tried to belittle the value of the only information on the new technology that had been published, and that he thus thought might have been available to the Swedish authorities. We have noticed earlier that these prints were indeed available to foreigners, since Alströmer bought one in 1720 and Kalmeter knew where to get one in 1723.[89] The accuracy of the early Newcomen prints, and whether they were designed to deceive, is an open question.[90] But regardless of whether O'Kelly's remark was true, it may be worth discussing its intention. If O'Kelly had hoped to receive an offer from Sweden, he had first to prove that he was more authoritative than his two supposed competitors. The first was Saunders, whom O'Kelly had already discredited on previous visits to Preiss by presenting his own knowledge as more theoretical. The second was the published prints of the Newcomen engine, which were spreading far too rapidly for O'Kelly's liking. He tried to undermine confidence in these prints by saying that some details were incorrect, but without saying to which of the prints or what details he was referring. This would serve the purpose of casting doubt on *all* published accounts of the Newcomen engine, and the conclusion would be that reliable knowledge of the new technology could only be brought to Sweden by an individual who had proved himself knowledgeable. If this was his gambit, it worked nicely. From then on, the Board of Mines only discussed transfer of the new technology to Sweden by individuals who "possessed the correct knowledge of the construction of this engine".[91]

Regarding the engine he had built near Liège, O'Kelly said that the fuel could be half wood, and half coal.[92] It consumed as much fuel as "six horses could pull".[93] This is the first time fuel consumption is mentioned in the Swedish sources, albeit vaguely and in undefined units. More important, perhaps, was that this point was essential if the new technology was to be adapted to conditions in Sweden, a country lacking fossil fuels. Any steam engine would have to rely upon the native resources of wood for fuel. But O'Kelly pointed out that coal was preferable to wood, since "the fire can thereby be kept in a more steady state".[94]

O'Kelly called on Preiss again a month later, on April 7, to deliver his drawing of the engine.[95] He told Preiss that the King of Spain had now approved and ratified the agreement with him to construct an engine in Biscay, and that the Spanish Ambassador in London had summoned him for further discussions. O'Kelly would return to The Hague on his way by land to Spain to inquire whether Preiss had any news from Sweden. O'Kelly had earlier promised Preiss to postpone departing for Spain for as long as possible in the hope of a reply from Sweden. O'Kelly's frequent visits to Preiss, the increasingly detailed information on the construction of the engine, and the references to his impending departure were probably all part of a plan to elicit a quick, affirmative answer from Sweden. But the Swedish bureaucracy was not to be hurried. When O'Kelly's proposal arrived, it was processed through the normal channels.

THE VERDICT OF THE BOARD OF MINES

Since the engine was to be used in mines, the Swedish Government automatically referred Preiss' letters as they arrived for an opinion to the Board of Mines, its advisory body on anything that concerned mining.[96] On April 11, 1725, the Board summoned de Valair to its meeting.[97] He was told that there had appeared—to the Board's astonishment, no doubt—another foreign colonel, who also claimed to be able to build a new type of engine to pump water out of mines (according to biographical sources, John O'Kelly was never promoted above the rank of captain, and we may assume that "Colonel" was an honorary title that both he and de Valair had bestowed upon themselves).[98] Since de Valair had already been granted a patent for an engine "with such a function", the Board wondered what he had to say.[99] De Valair requested permission to reply in writing, which was granted. He received extracts from Preiss' letters—with the exception of everything that concerned the design of O'Kelly's engine. After Wallerius' report the previous autumn the Board had reason to suspect that de Valair did not know how to build his engine. If he was told of O'Kelly's design in detail he might pretend that it was just such an engine that he had intended in his application for a patent. O'Kelly had also requested that his rule of proportion should be kept secret, so this, too, was omitted from the extracts de Valair received.

Almost three weeks elapsed before de Valair handed in his reply, a short, indignant letter in German.[100] He wrote that as soon as the patent had been granted, he had striven to bring to Sweden *"einen Experimentierten Machinisten"*. His efforts had been successful, for he had been able to recruit the very man who had built and supervised (*"erbauet und dirigiert"*) the engine near Liège. Several reliable persons had testified that this man knew his business well, and that he had been taught by none other than John O'Kelly himself. De Valair had recently sent him to Öster Silvberg, where this engine would be very useful (*"sehr praticabel"*). So—if anyone else wanted to build such an engine in Sweden, they had to negotiate with de Valair and his associates, as stipulated by the patent. This applied to John O'Kelly as well, and de Valair intended to defend his patent vigorously (*"kräfttigist zu schützen"*).

De Valair's reply was a good try, but he was attempting to cover up an untenable position. He did not mention that Saunders had left Sweden, but the Board already knew that from Wallerius' report of the previous November. They also knew from Preiss that Saunders was back in London in March 1725, and unlikely to return to Sweden. De Valair had written that he had brought him here *"mit schweren kosten"*.[101] This contradicted O'Kelly's version that Saunders had received 100 *pistoles* to proceed to Sweden from Prince Wilhelm, brother of the Swedish King and Governor of Maastricht. But since we know that de Valair's associates were trying hard to get someone to Sweden who could build an engine, it is possible that Prince Wilhelm had been engaged in this.

Saunders is most likely to have been in Sweden at some time between May and October 1724, since de Valair was allowed to commence work at Öster Silvberg in April 1724 and Wallerius reported in November the same year that "the Englishman" had left the country. It is not known why Saunders left. Did he receive a better offer from England, or did the project at Öster Silvberg run into difficulties? We do not know, but we may speculate on Saunders' thoughts when he arrived at Öster Silvberg, where de Valair had sent him to inspect the site for the intended engine.

Although Öster Silvberg was situated in the heart of Bergslagen, the ironworking district, it was a desolate spot.[102] The mines are situated

Fig. 8.3. The largest of the mines, "*Storgruvan*", in Öster Silvberg. (Photo Lars Andersson, 1963)

on top of a hill, from which Saunders would have seen dark-green pine forests for miles in every direction. In the distance, he might have seen thin streaks of smoke from one of the few blast furnaces in the valleys. Around him lay the ruins of what had been left when the mines were abandoned some eighty years earlier. In front of him was "*Storgruvan*", the largest of the mines, filled with water (Fig. 8.3). It was believed to be at least 108 fathoms deep, but no one knew for sure.[103] The Englishman was a long way from home, far from the mining areas of

England and the emergent Industrial Revolution. Was this the place where he was supposed to build his engine, an example of the most advanced technology of the day? Was this the mine he was to drain of water—and keep free of water, *nota bene*—so that it could be worked again in the hope of extracting enough gold and silver to make it all worth while? The place was, in modern terminology, devoid of infra-structure, supporting technologies, skilled labour, etc. To build a New-comen engine here would not have been to introduce the new techno-logy in a new social context, but rather to put it in the middle of nowhere. Saunders probably just shook his head, turned his horse downhill again, and headed for England.

De Valair's protest against O'Kelly's proposal was of no avail. It only proved to the Board of Mines that the engine for which de Valair had been granted a patent was identical with the "fire engine", and this was not at all "a new and hitherto unknown water-engine" to them as the patent had stated. They themselves had known about it since 1720, when Kalmeter had sent his report from England. De Valair's protest also convinced them that he did not himself know how to build an engine.

The reply of the Board of Mines to the Government on the subject of O'Kelly's proposal was dated May 14, 1725.[104] It has been discussed by Björkbom, and it was indeed a most thorough and competent assessment of the new technology.[105] But the case had been well prepared when O'Kelly's proposal was on the agenda at the meeting of the Board on May 5.[106] On the table lay not only Preiss' three letters, O'Kelly's drawing, his detailed description of the engine, and de Valair's reply, but also a reasoned analysis of the advantages and disadvantages of the new technology. This was "Assessor Swedenborg's thoughts on the conveniences and inconveniences of this engine".[107] That Emanuel Swedenborg, Assessor at the Board of Mines at the time, was the first in Sweden to make an official assessment of steam power technology has not attracted the attention of historians, but it was his analysis that formed the basis of the Board's reply of May 14. The documents in the case were presented to the Board on May 5, "where-upon there was a good deal of discussion".[108] The Board agreed that its primary concern was to decide whether de Valair's patent really could bar the way to someone else who had knowledge of the function and construction of the engine:

if Valaire himself does not possess this knowledge, but intends to use O'Kelly's workman, who does not have complete knowledge of this engine's construction either.[109]

In their reply of May 14 to the Government, the Board wrote that although de Valair had been granted a *"Privilegium exclusivum"* for his water engine, he had never made clear, either in writing or verbally, that this engine "was the same as that which works by fire, and was long before invented and used in England, and already known at many places".[110] Furthermore, he had made no preparations for building such an engine, let alone drained the water out of any mine. The Board doubted whether he himself knew how to build an engine, especially since he had told them that he had brought to Sweden an "experienced mechanic", whom they happened to know to be O'Kelly's apprentice. Therefore, they suggested that O'Kelly, or any one else who had real knowledge of the engine's construction, should be free to come to Sweden to build one "if he finds a good opportunity and earns his living".[111] This last passage makes clear once again that although technical development met with benevolent support from the authorities, it was left to individuals to take the initiatives. In this case, the Board of Mines was nothing but an advisory body. Its task was to examine whether O'Kelly's proposal infringed de Valair's patent, not to implement technological change. This was left to individual mine owners, or groups, such as de Valair and his associates, who wanted to resume work in abandoned mines.

The Board of Mines had suggested that de Valair should be given one last chance to prove himself, before his patent was declared invalid.[112] On June 21, the Board summoned Törne to their meeting as they knew that he was one of de Valair's associates.[113] They informed him about O'Kelly's proposal and requested him, Baron Strömfelt, and the other associates, to ask de Valair if he:

within a certain time could give proof of his engine, since he has been granted a patent, so that if he cannot achieve anything, others, who may be interested in negotiating with Colonel OKelly, would not be prevented by his patent.[114]

Törne promised to do so, but he must at the same time have realized that the game was up. Without Saunders they had no hope of building an engine "within a certain time". And in fact this was the last time de Valair's project for a steam engine at Öster Silvberg was mentioned. His patent was thereby rendered invalid. But de Valair had many other projects up his sleeve. At the same time he was also occupied with his plans for a steelworks south of Sundsvall, with production based upon his method of "converting bar iron into the choicest of steel without any reduction of the material".[115] Two years later, he reappears in Germany, whence he sent two proposals to Gustaf Bonde, President of the Board of Mines: one for an improved copper smelting process, and another regarding a method of extracting gold from copper.[116] He wrote that he had met an *"artist"* in Holland, who, by means of a secret powder, had managed to extract gold from Swedish copper. Once again, de Valair succeeded in appealing to the alchemical interests of Gustaf Bonde. In 1730, a man named Hechtenschantz was allowed to demonstrate this method at Avesta before Bonde and other prominent persons, but it came to a sorry end.[117] De Valair seems to have been an ingenious deviser of projects, and something of an international adventurer. He must have possessed poise and powers of persuasion, for his proposals were listened to with respect by prominent persons. But he was less successful in realizing his schemes, which were always dependent on the practical knowledge of others. As an agent of technological change, de Valair was an impresario rather than an entrepreneur.

As mentioned before, the Board of Mines had made an assessment of O'Kelly's engine in their reply of May 14.[118] This will be discussed in the following chapter, but it was essentially an account of possible advantages and disadvantages. The Board came to no definite conclusion about the desirability of such an engine in Sweden. This depended on what it would cost to build, and they suggested that Preiss should be requested to find out from O'Kelly. The Government ordered Preiss to do this in a letter dated May 31, and requested him also to ask O'Kelly "whether he thinks that this engine can be used and kept running here in winter".[119] This question was not mentioned in the discussions in the Board of Mines, and it may well have been King Fredrik's own reaction to the new technology when he envisaged waterwheels covered with ice and troughs frozen to the bottom in winter in his Nordic country.[120] The Board obviously had a better understanding of the principle of the Newcomen engine, and had focussed on the more important geographical factor bearing on the transfer of new technology to Sweden: the lack of coal, and the necessity of using wood as fuel.

O'Kelly was still in London when Preiss wrote to him and received an answer on July 13.[121] O'Kelly quoted two figures, 9 000 *gulden* for a large engine and 6 000 *gulden* for a smaller one. To the last question he gave a reassuring answer: unless the water freezes in the mine, the cold cannot prevent the use of this engine in winter. The Board was obviously not hurried by O'Kelly's talk of an impending departure for Spain, for it did not comment upon his figures until September 9.[122] It was merely decided to discuss the matter further, but O'Kelly's proposal was not mentioned again in the minutes of the Board. The Board was, as has been mentioned, primarily an advisory body. O'Kelly's proposal had been most useful, because it had provided additional information on the new technology, including a drawing and a description of the engine. It had also enabled the Board to solve the puzzle of de Valair's "water-engine", and to straighten out the question of the patent. A legal intricacy had been untangled, and the paper work tidied up. What more could a government office desire?

It is not known whether John O'Kelly ever built an engine in "Biscay" or not. Björkbom suggested that he might have been responsible for a Newcomen engine reported by Conrad Matschoss to have been built in England in 1726 for Toledo.[123] As mentioned earlier, O'Kelly had married for the second time in Brussels in 1723.[124] He settled there, had two sons in his second marriage, and died in 1753. He still remains something of a mystery. Where did he learn to build

Newcomen engines before 1720, the year in which he went to Liège to build the first steam engine outside England? What did he do between July 1725, when Preiss received his last letter from London, and his death in Brussels in 1753? But as far as we know, he played an important part in the early dissemination of steam power technology into Europe.

O'Kelly's importance as an agent of technological change in Sweden was threefold. *First*, through his detailed description, rule of proportion, drawing, and conversations with Preiss, a comprehensive knowledge of the new technology was brought to Sweden. Due to the structure of the Swedish civil service, this knowledge accumulated in the Board of Mines. *Second*, his proposal compelled the Board to make an attentive assessment of the new technology. This systematized their knowledge, and made them aware of the potential advantages and disadvantages of the Newcomen engine. *Third*, de Valair's patent of 1723 was rendered invalid. It had been a legal hindrance to technology transfer, and may have prevented or discouraged others from attempting to build the first steam engine in Sweden. The way was now open again to anyone who wanted to take the initiative. The level of knowledge within the Board of Mines had, however, been raised considerably since 1723. It was no longer sufficient to apply for a patent for "a new and hitherto unknown water-engine".

CONCLUSIONS

In the previous chapter it was shown how basic knowledge of steam power technology had been brought to Sweden by 1720 in the course of the routine work of the Board of Mines. This initial stage in the process of technology transfer had taken place as a result of the institutional structure of the Swedish civil service. By 1725, Swedish knowledge of the new technology had increased considerably. Foreign entrepreneurs, or adventurers if one prefers, played a major role in this second stage of technology transfer although they failed to introduce the Newcomen engine into Sweden.

De Valair and O'Kelly both saw Sweden as a possible market in which to exploit their knowledge of the new technology. De Valair had entered into partnership with a group of prominent Swedes who had a personal interest in introducing the new technology, as well as the social position and financial means to do so. He had been able to place the new technology in the right social context, but the weakness of his position was that he lacked the practical knowledge to apply the technology and that he failed to secure the assistance of Saunders.

O'Kelly, on the other hand, had the necessary practical know-how, but failed to place the technology in its right social context due to an insufficient understanding of the Swedish society of the 1720s. The story of how Prince Wilhelm had given Saunders 100 *pistoles* to go to Sweden had perhaps misled O'Kelly into believing that this was how technological change was initiated in Sweden. Did he imagine that King Fredrik would give him a purse of gold and a slap on the back, and command him to hurry on to the mining district? If so, he had been strengthened in this misconception by his experience of negotiating with Spain. But his proposal to Sweden was only referred in the customary manner to the Board of Mines, whose primary concern was to determine the legality of his proposition in the light of de Valair's patent, so that "others, who may be interested in negotiating with Colonel OKelly, would not be prevented by his patent". O'Kelly made a mistake in not first securing the support of a private group with the means to realize his project.

Saunders also possessed the practical knowledge, but not the social position necessary to establish himself as an entrepreneur in his own right. To O'Kelly he was "*Mon Valet*", and to de Valair "*einen Experimentierten Machinisten*". He had earlier been in the service of O'Kelly, but had left him in about 1723. O'Kelly had already complained in 1721, when forced to sell his watch and unable to keep up his appearance as an affluent entrepreneur, that he was afraid his workman would hire his services to others. Saunders had been hired by de Valair and his associates, presumably in 1724, and brought to Sweden. But he had soon left and returned to England. Among the pioneers of the Newcomen engine in Sweden, he was the wage-earner—dependent on offers from others, but free to accept or reject the offers that came his way.

Unlike Saunders, the two "colonels" had been able to establish contacts at the right social level. Preiss had introduced O'Kelly in his first letter to the Swedish Government as a "Catholic Irish Nobleman and Colonel by name Johan OKelly, Esquire, Seigneur d'Aghrim", and de Valair, whatever his background may have been, was held in esteem by Count Gustaf Bonde, President of the Board of Mines. They were both gentlemen in the eyes of the Swedish civil service, and their proposals were treated with respect. Any foreign adventurer who wanted to exploit his knowledge of the new technology in Sweden and establish himself as an entrepreneur of Newcomen engines, would thus have needed at least the following qualifications:

1. Practical knowledge in order to apply the new technology
2. The support of a private group, which had the social position and financial means to make it possible
3. The social position of a man of rank

Each of the adventurers failed to satisfy one or another of these qualifications; hence their failure to introduce the new technology in Sweden. De Valair possessed the second and the third, but failed to acquire the first (practical knowledge). O'Kelly possessed the first and the third, but did not think that he needed the second (the support of a private group). Saunders possessed the first and the second, but according to Swedish social values, not the third (the social position of a man of rank). The failure of the adventurers to introduce the Newcomen engine into Sweden was a personal one. From the Swedish point of view, their importance as agents of technological change lies in the knowledge that their activities brought to Sweden. The information they gave in their applications and communications added to the Board's knowledge and led the Board to make assessments and take decisions that systematized this knowledge. This process, an interaction between individual agents and the institutional structure, generated new knowledge, the extent of which will be discussed in the following chapter.

9. Technology on Trial, 1725

EMANUEL SWEDENBORG'S ACCEPTANCE BY THE BOARD OF MINES

Emanuel Swedenborg's career as an Assessor in the Board of Mines has never been fully examined. Here, only a brief outline of his acceptance by the Board will be given. As has already been described (see Chapter 7), Swedenborg began corresponding with Christopher Polhem after his return to Sweden from abroad. The young man and his ideas quickly made a strong impression on Polhem, even more so when in 1716 Swedenborg began to publish one volume after another of his journal, *Daedalus hyperboreus*. Swedenborg found an influential patron in Polhem and soon he also came under royal patronage. In the autumn of 1716, he met Karl XII in Lund. The King so approved of the journal that he appointed Swedenborg Assessor Extraordinary in the Board of Mines with instructions to assist Polhem. Swedenborg reported what had happened in a letter to his brother-in-law, Erik Benzelius, in December 1716:

But since my enemies intrigued too much against my letter of appointment, and drew it up ambiguously, it was sent back to the King with some objections, although they knew very well what support I had. Then a new letter of appointment was immediately granted me [by the King], and a Royal Command to the Board of Mines, which my opponent had to sit and write at His Majesty's own desk in two different versions, of which He [the King] then chose the most favourable. So that those who sought my ruin were happy to escape with honour and reputation, and they came very near to burning their fingers.[1]

The Board of Mines read the King's letter at its meeting on January 7, 1717, and laid it *ad acta* without any comment.[2] These were the closing years of the Absolute Monarchy, and the King could still impose his will on a reluctant bureaucracy. The Board of Mines was an institution with a strict hierarchy based upon competence and experience (see Chapter 5). It was probably very much opposed to this swift promotion of an outsider to its ranks; and the Swedish civil service, which, to Swedenborg's malicious delight, had been humiliated at the King's desk in Lund, was to have its revenge.

Swedenborg was sworn into office on April 6, 1717,[3] and attended meetings of the Board during the following two weeks.[4] On April 17 he applied for leave to assist Polhem in southern Sweden, which was granted with alacrity.[5] For the next two years he was fully occupied in assisting Polhem with the construction of docks in Karlskrona, salt production in Bohuslän, and sluices for a project foreshadowing what a hundred years later was to become the Göta Canal.[6] During these years he intended to improve his standing in the Board of Mines, and in January 1718 he wrote to Erik Benzelius:

As regards the Board, I will diligently study mechanics, physics and chemistry, at least learn the fundamentals of it all, with the hope that in due course there will be no reason to consider me completely unworthily admitted, for I have no wish to be called "well versed in law".[7]

"Well versed in law", or *legis-consultissimus* as Swedenborg wrote, was probably the scornful epithet applied by the Board to one who had demanded admittance to their group on the strength of a piece of paper signed by the King. He began to publish his scientific ideas in pamphlets which, according to Lindroth, "were of varying quality, deficient matters mingled with brilliant ideas and convincing conclusions".[8] Dogmatism was a characteristic feature of his early scientific work, and that is perhaps why he was least successful in the exact sciences of mathematics and astronomy. After his collaboration with Polhem had come to an end in 1719, he carried out some studies intended to demonstrate his credentials to the Board of Mines.[9] The reason that he did better in technology than in science is probably that his conclusions were based on systematic accumulation of practical experience or, as he wrote in a letter to Erik Benzelius on November 3, 1719:

What I have been working on is first a careful description of our Swedish blast furnaces and their operation. The second is a theory or investigation of fire and ovens, where I have first collected everything I have been able to find out from blacksmiths, charcoal burners, roasting-furnace workers, blast-furnace masters etc., whereupon the theory is based, and hope to have made several *decouverts* there that will prove useful. For example how to make fire in new tile stoves so that the wood and coal that are consumed in one day can be used for six days.[10]

He also wrote a handbook on ore prospecting,[11] and sent all this to the Board of Mines, where he had gained a friend in the Vice President, the ageing Urban Hiärne. In November 1719 he attended the meetings

Fig. 9.1. Emanuel Swedenborg in the second of the two roles he played in the introduction of steam power technology into Sweden: a civil servant in the Board of Mines, and its most competent scientist. Cf. Fig. 7.1. Engraving by M. Bernigeroth in the first volume of *Opera philosophica et mineralia*, published in 1734. (Photo Royal Library, Stockholm)

of the Board for the first time in over two and a half years.[12] His standing had obviously improved, because on December 21, 1719, "Assessor Extraordinary Mr Emanuel Swedenborg entered and was sworn into office".[13] We can only speculate on why Swedenborg was sworn into office for a second time. It might have been because the Board now began to accept him in his own right, thanks to the experience he had gained during his year as Polhem's assistant, his scientific publications and his investigations into blast furnaces, ore prospecting and a tile stove to consume less fuel. (The last-mentioned of these was, as was shown in Chapter 3, a matter of great interest to the Board of Mines during the eighteenth century.) But he was not yet thought of as a fully fledged member of the Board, and in 1720 he applied for a vacancy as a full Assessor, but was passed over.[14]

In 1721–1722 Swedenborg travelled to the Netherlands and Germany to study developments in mining and other industries.[15] This is not the place to describe his journeys, which were of great importance to his future development as a speculative philosopher. Suffice it to say that while abroad he published several books in Latin on geology and chemical mineralogy, and that he made valuable foreign connections, so that when he returned to Sweden in 1722 it was as a renowned scientist of international standing. For our purposes we must also recall that Swedenborg had passed on a copy of Kalmeter's first report on the Newcomen engines in Newcastle to the Society of Science in Uppsala in February 1720 (see Chapter 7). He therefore knew of the new technology when he went abroad the following year. We also note that he was in Liège in December 1721.[16] Even if he did not see O'Kelly's engine, which had been completed earlier that year, it is likely that a foreigner visiting Liège at this time to learn about new innovations in the mining industry would hear of its installation. Finally, we must also note that when Swedenborg was in The Hague he renewed his acquaintance with the Swedish Ambassador, Joakim Fredrik Preiss, and that the two had several discussions on the economic situation of Sweden and possible means of improving it.[17]

Between 1722, when Swedenborg returned home, and the early 1730s, when he began to renounce the practical affairs of the world, followed what Lindroth has described as "the unknown decade in Swedenborg's life. It was a time of silence; he published nothing but quietly attended to his work in the Board of Mines."[18] The minutes of the Board are indeed filled with references to his work and to his many visits to the mines and ironworks of Sweden during these years, but only one aspect of his service will be discussed here. On April 11, 1723, Swedenborg finally received formal permission to take his seat as Assessor Extraordinary in the Board—six years after his appointment by the King.[19] When de Valair was summoned to the meeting of the Board on May 3, 1723, to be questioned about his "water-engine" and his other project,[20] Swedenborg had only been participating in the meetings of the Board as a fully accepted member for a couple of weeks. He was appointed full Assessor on July 15, 1724—eight months before Preiss' first letter arrived.[21]

This brief outline of Swedenborg's acceptance by the Board of Mines is given in order to show how he came to take part in the Board's discussions on the new technology right from the beginning. He had been appointed by Karl XII at the time of the transition from the Absolute Monarchy to the Era of Liberty, and to the Board of Mines of the early 1720s this Assessor Extraordinary foisted on them at the whim of the late King was probably an undesired relic of *l'ancien régime*. Swedenborg had to work hard to overcome this resistance, but it also tells us something about the level of competence that the Board demanded of its officials. It was not enough to be young and brilliant and to have an influential patron, even a royal one. No, to take a seat on the Board of Mines demanded proven knowledge and experience of mining matters. Rather than saying that Swedenborg finally succeeded in defeating the intrigues of his enemies, we might say that by 1723 he had managed to give sufficient proof of such knowledge and experience. Due to his unconventional enrolment, his credentials had to be immaculate, and what he lacked in practical experience of the Swedish mining industry he made up for with his knowledge of the latest developments in science.

Swedenborg's interest in science was demonstrated in February 1725, when he proposed that the Board should buy an air pump (*"antlia pneumatica"*) from the renowned instrument-maker Francis Hauksbee in London.[22] He had met Hauksbee during his first journey abroad in 1712, and had suggested at that time that Erik Benzelius should buy one for the University library in Uppsala.[23] He now told his fellow-members of the Board:

with such an *antlia pneumatica* many kinds of experiment can be performed, which concern the atmosphere, fire and water.[24]

It would cost 1 000 *copperdaler* he reported, and the Board discussed his proposal at length and immediately.[25] It is clear that Swedenborg was by now held in esteem by the Board. They agreed that it would cost three times as much to have it built in Sweden, and since the means were available in the funds set aside for the Board's *Laboratorium mechanicum* it was resolved that Swedenborg should order such a "useful machine" (but that he should take care to have it properly insured before it was dispatched, so the Government would not risk any financial loss).[26] The air pump arrived by ship later that spring and the Board objected strongly when the Swedish Customs imposed duty on an instrument that was to be of public use.[27] It had been brought from England "for the utility and good of the Realm in general", and public funds would in any case be used to pay the duty.[28] In this argument about the duty on Hauksbee's air pump in 1725 we hear an example of a justification for science that was to echo throughout the eighteenth century: the public utility of science.

EMANUEL SWEDENBORG'S TECHNOLOGY ASSESSMENT, 1725

Two months after Swedenborg had suggested the purchase of the air pump, the Board received the first two letters from Joakim Fredrik Preiss regarding O'Kelly's proposal. The Government had asked the Board of Mines to pronounce on the status of O'Kelly's proposal in relation to de Valair's privilege of 1723, and to express its general opinion of the new engine. In the previous chapter it was shown how the Board succeeded in carrying out the first of these two tasks, i.e. in untangling the legal intricacies. As for the second, it is now clear, after this outline of Swedenborg's gradual acceptance at the Board of Mines, why he was the one to be asked to evaluate the new technology: O'Kelly had stressed that the design of the engine was based on scientific principles, and Emanuel Swedenborg was the Board's most competent scientist. Who, after all, could be more suitable than Swedenborg, who had only recently requested an instrument to perform "many kinds of experiments ... which concern the atmosphere, fire and water", to assess an engine that was operated by fire and the atmosphere to raise water?

The basis for Swedenborg's assessment of the new technology was Preiss' letters, together with O'Kelly's drawing (Fig. 9.2) and description of the engine. The contents of Preiss' letters were discussed in the previous chapter, and Swedenborg could rely on them as an accurate account of what O'Kelly had said, because he knew Preiss to be an intelligent and knowledgeable man. As for the drawing, it arrived at the Board's offices on April 28,[29] in good time to be studied by Swedenborg, who did not present his technology assessment until the meeting of May 5.[30] As mentioned earlier, Preiss took the trouble to verify O'Kelly's claim that he had been commissioned to build an engine in "Biscay". The scrupulous Preiss also checked O'Kelly's drawing of the engine. A Swedish engineer (*"Ingenieur"*), Anders Fischer, who was passing through The Hague at this time, was asked by Preiss to look at the engine when he reached Liège, and to make a drawing of it so that it could be ascertained whether it was the same machine as that drawn by O'Kelly.[31] Preiss reported that Fischer had done so:

although he had great difficulty, for the people there were so nervous and vigilant that they would not allow him to inspect the engine as closely as otherwise would have been possible.[32]

Fischer's drawing is not appended to Preiss' letter in the archives of the Board of Mines, and a note on the letter records that it was

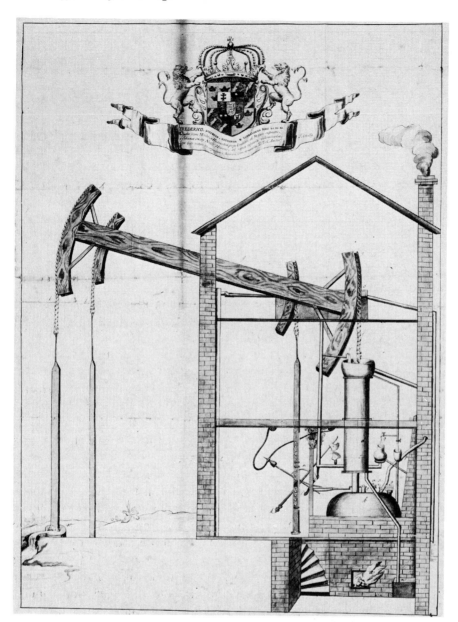

Fig. 9.2. The drawing by John O'Kelly of the Newcomen engine he had built near Liège in 1720–21—the first Newcomen engine on the Continent. The drawing was sent by the Swedish Ambassador in The Hague to the Government in Stockholm, and referred to the Board of Mines for an opinion. It was available to Emanuel Swedenborg when he made his assessment of the new technology in April–May 1725. (Photo The National Record Office, Stockholm)

forwarded to Lars Benzelstierna, Assessor in the Board of Mines, on June 14, 1725.[33] In the archives of the Board of Commerce there is a copy of O'Kelly's drawing made by Daniel Bergenstierna, Auscultator in the Board of Mines at the time.[34] Since this is an exact copy of O'Kelly's original drawing, we may assume that Fischer's drawing did not lead Bergenstierna to make any changes. O'Kelly's drawing is thus probably in all essentials a correct representation of the Liège engine, the first Newcomen engine outside England.

O'Kelly's description of the engine in French gave an account of its principle and function. It was not intended as a construction manual, and O'Kelly gave neither the dimensions nor the materials for its various parts, but it is clear that he had more than a passing knowledge of its construction. Of its principle, he wrote:

the principles underlying its movement are the rarefaction and condensation of the air contained in the cylinder, for when a vacuum (or a kind of vacuum to avoid vexation to *Mess. les Philosophes* who maintain *quod non datur vacuum*) has been created, the atmosphere, meeting no resistance, exerts all its weight on the piston; and if the weight of the atmosphere be greater than that of the column of water, which one intends to raise by this means, it is certain that one will succeed.[35]

The problem was thus to dimension the engine so that the pressure exerted on the piston by the atmosphere equalled the weight of the water column to be raised by the engine. It had been established by experiments, O'Kelly wrote, that the pressure of the atmosphere on one square *pouce* was 18 *livres*.[36] On a cylinder with a diameter of one *pouce* the pressure was some 14 *livres*. He had forgotten the exact figure for the moment, he said, but it could easily be established by an experiment with mercury. He continued:

But one can easily find the atmospheric pressure on any cylindrical body by the simple rule of three. For a cylinder with a diameter of one inch is to a cylinder with a diameter of, say, 30 inches such as you are using, as a weight of about 14 pounds is to the weight you seek.[37]

But this was wrong. The power of a Newcomen engine is directly proportional to the area of its cylinder, i.e. to the square of the diameter of the cylinder. If the unknown pressure was to be calculated by means of a simple rule of three, one had to use the *areas* of the cylinders, and not the *diameters* as O'Kelly wrote. If the diameter was to be used, one had to take the square to express it as a linear function of the pressure. The interesting thing is that this is exactly what O'Kelly's rule of proportion had stated (see Chapter 8).[38] His rule gave the diameter of the cylinder as a function of the square root of the weight of the water in the pumps and the pump rods. Furthermore, it would have been unnecessary to carry out an experiment with mercury to discover the pressure of the atmosphere on a circular area with a diameter of one *pouce* if it was known that the pressure on one *square pouce* was 18 *livres*. It could easily be calculated with π, assuming that the figure of 18 was correct.[39]

In the previous chapter it was shown that O'Kelly's rule of proportion was nothing more than a rule of thumb: the fruit of his practical experience as an engineer. O'Kelly said so himself in the sentence following the passage quoted above:

> But, Sir, the rule that I have already sent you is not only short, but also includes friction, and is calculated from long experience.[40]

His rule of proportion was, as far as can be judged, a sound engineering rule, but we must conclude from his description that he failed to express it in the simplest mathematical form. His description was lucid enough as far as the function of the various parts of the engine was concerned, but it was fundamentally wrong on the most essential point: the calculation of the diameter of the cylinder. This is not to imply that O'Kelly did not know how to design a Newcomen engine properly. Proof that he did is provided by the engine in Liège and by his rule of proportion. John O'Kelly obviously had *practical* knowledge of the new technology, but he failed to express it *theoretically*.

Emanuel Swedenborg's report, which was discussed by the Board of Mines on May 5, consisted of five pages and a figure.[41] It was a detailed account of the disadvantages and advantages of the engine, including his comments on how to calculate the diameter of the cylinder. It was rigorously structured, *pro et contra*. He began by listing five disadvantages of the engine. *First*, since the engine required "an even and penetrative fire to divide the water evenly and forcefully into vapour",[42] it would be used to best advantage where coal or "good charcoal" was available.[43] Swedenborg did not have to point out that Sweden was almost completely lacking in fossil fuel deposits, and that charcoal was precious in the mining districts. It seemed possible to operate it on wood, he wrote, but the power would be less and also more uneven. In other words, it was not an engine that could easily be adapted to the geographical environment of Sweden.

Secondly, the design of the engine was "based not only on the laws of mechanics, but also on those of physics".[44] To build it would be expensive and difficult, and it would not be possible to keep it in working order unless it was constantly supervised by its constructor ("*auctoren*"), or someone else who understood physics. This was essential if the engine had to be repaired or modified as the mines went deeper, which would necessitate changing the proportions of all the parts. This would be not only expensive, but also beyond the capabilities of "any common master builder" used to working with traditional waterwheels.[45] In other words, it demanded a new type of engineer, with scientific training.

Thirdly, Swedenborg thought that large engines would be less efficient than small ones since it would be difficult to produce enough steam in large boilers, and because it would be more difficult to make larger valves and pistons airtight. Although this was a misjudgement, it is indicative of the level of mechanical workshop practice in Sweden at the time. In other words, it was difficult to determine the technical limitations of the new technology because there was no experience of a technology based on steam and iron.

Fourthly, the engine could not be used in mines with numerous shafts and tunnels, since it would be difficult to connect the pumps to the engine through *Stangenkunst* systems. Swedenborg probably knew of the failure of Polhem's attempt to introduce a centralized power system, consisting of only a small number of large waterwheels and a *Stangenkunst* distribution network, in the Great Coppermine in Falun at the turn of the century.[46] The traditional power sources of waterwheels, windmills and horse whims, were relatively small prime movers, so there had never been any real need to develop a more sophisticated technology for large-scale power distribution. In other words, the auxiliary technologies put a limit on the utility of the engine.

Fifthly, the major mines in Sweden were already provided with waterwheels that kept them free of water. Swedenborg did not know whether there was any mine where the traditional methods had failed, and where there might therefore be an interest in such an engine, "but in such a case, this engine would be useful and necessary".[47] In other words, the technologies already available were generally sufficient.

Turning to the three advantages that he could think of, he was more brief. *First*, the power of the engine seemed greater than that produced by any conventional method, as it had been shown that it could raise 500 barrels of water per hour. *Secondly*, it might therefore be used at places where there was more water than could be drained by other methods, "as well as at those places where there is no running water or possibility of using waterwheels, and an abundance of good forests".[48] The *third* reason was not strictly an advantage but rather a reflection of Swedenborg's own fascination. He wrote:

3° Furthermore it is of an extraordinary interest, in that a new principle, viz. fire, *evaporatio aqua* and *vacuum* operates the whole movement, which is otherwise done by *moventia mechanica*.[49]

He did not draw up a balance sheet of the disadvantages and advantages to pass judgment on the desirability of the new technology, but we may summarize his comments as they have been interpreted here:

Disadvantages
1. It could not easily be adapted to the geographical environment of Sweden
2. It demanded a new type of engineer with scientific training
3. It was difficult to determine the technical limitations of the new

technology because there was no experience of building machines driven by steam and made of iron
4. The auxiliary technologies put a limit on the utility of the engine
5. The technologies already available were generally sufficient

Advantages
1. More power than from any traditional technology
2. Useful where there was no running water and an abundance of wood fuel
3. A fascinating new technology

The third part of Swedenborg's memorandum consisted of his comments on how to calculate the diameter of the cylinder. On O'Kelly's erroneous method of calculating the diameter "*par la simple regle de trois*", he commented:

But this rule ought to be changed to the extent that instead of the diameter of the cylinder the area or *qvadratum diametri* should be used, since the pressure of the atmosphere on circular bodies [. . .] is not proportional to the diameter of the cylinder but to its area.[50]

On O'Kelly's rule of proportion Swedenborg commented that it did indeed seem to be based on experience since the theoretical value for the diameter of a cylinder for a depth of 100 fathoms was less than that arrived at by O'Kelly's rule. This was correct, he agreed, "since because of the friction, or *frottement* as it is called, it becomes larger, which one can only determine by experience".[51] But Swedenborg thought that there was a simpler method of finding the diameter of the cylinder, "than using these numbers 7 and 9, if one only draws a few circles in the following way ..." (see Fig. 9.3).[52] His method was, in short, a geometrical method of finding the diameter of a circle with an area twice as large as that of a given circle of known diameter. For example, if one knew the diameter of the cylinder for 50 fathoms, it would be easy by his method to find the diameter of the cylinder that would be needed for a depth of 100 fathoms. Swedenborg wrote: "one can make a figure of such circles, from which the diameters for all depths easily can be found".[53] This was not correct, for if the diameter for 50 fathoms was known, the method only enabled the diameters for successive doublings of 50 fathoms, i.e. 100, 200, 400, 800 ... fathoms, to be established. This was hardly the problem, for few mines were deeper than 100 fathoms. What was more important, however, was that Swedenborg's method could only be used to establish the *theoretical* value. (It was merely a geometrical method for finding $\sqrt{2}$.) This was taken for granted in O'Kelly's rule, of which the important aspect was the statement of the proportional factors 7 and 9, which allowed for the losses due to friction. Swedenborg admitted:

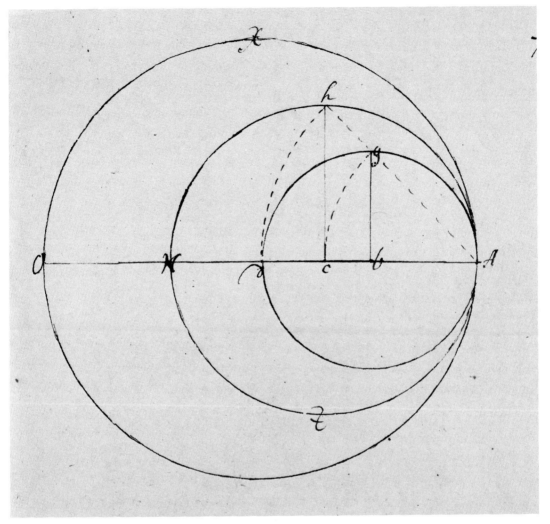

Fig. 9.3. Swedenborg's illustration to his technology assessment of the New-comen engine. It shows a geometrical method of determining the radius of a circle with an area which is twice the area of a given circle. In the figure, the area of the circle with the radius "ch" is twice the area of the circle with the radius "bg". (Photo The National Record Office, Stockholm)

the diameter should always be made a little larger than is shown, because of the friction, or the so-called *frottement*, since it will not matter if the cylinder is larger but it will if it is too small.[54]

"A little larger!" These were not the words of a practical engineer, trying to dimension his engine accurately, but those of a man fascinated by theory, sitting at his desk and playing with a pair of compasses. This little postscript is less illuminating than the rest of his report.

We are now in a position to compare Swedenborg's assessment with the reply of the Board of Mines to the Government, dated May 14, 1725,[55] which was discussed by Björkbom in 1936.[56] The Board considered Swedenborg's memorandum at its meeting on May 5,[57] and spent both the morning and the afternoon session of its meeting on May 14 improving an initial draft of its letter to the Government.[58] The final version was read on May 15 before being signed by the members of the Board (among them Emanuel Swedenborg).[59] The letter followed Swedenborg's memorandum word-for-word for the most part, but those changes that were made illuminate the opinion of the new technology held by the Board of Mines as an institution.

The major change in the letter of the Board of Mines was that it listed the advantages of the engine *before* the disadvantages. The Board obviously took a more positive view of the advantages than Swedenborg did. One minor change was that the letter stated that the engine could pump 600 barrels of water per hour out of a mine, not 500 as Swedenborg had written, and added that this information had been supplied by O'Kelly. The Board knew from Kalmeter's report in 1720 that a Newcomen engine in Newcastle could pump some 600 barrels per hour. The Board repeated the advantages and disadvantages almost verbatim, as well as his criticism of O'Kelly's mistake in using the diameter instead of the area to calculate the power of the engine. But it passed over Swedenborg's own suggested geometrical method, saying "which is all said to be demonstrable with drawn lines".[60] The Board evidently took no interest in his exercises with his compasses, as this was not the point in question. The Board of Mines concluded its letter to the Government by writing:

From all this, Most Gracious Sovereign, may Your Royal Majesty graciously learn on one hand the advantages of this engine, if it can be constructed and have its promised effect, and on the other hand the difficulties to which it seems to give rise.[61]

As an assessment of the new technology, the reply of the Board of Mines was admirably comprehensive, considering the information available. As will be shown later, the case of the Dannemora engine was to prove it right on every point, and we must give credit to Emanuel Swedenborg for having carried out the first technology assessment of the steam engine in Sweden. As related in the previous chapter, the Board did not reach a definite verdict for or against, merely suggesting that the Government should ask Preiss to find out the price of such an engine from O'Kelly. Their verdict was, in effect: "Maybe, but what would it cost?" The advocates of the new technology had made out a *prima facie* case, but more evidence was still required.

ADDITIONAL INFORMATION

Henrik Kalmeter, an Auscultator in the Board of Mines who had been sent to England in 1718, had reported back in 1720 on the Newcomen engines in Newcastle (see Chapter 7). In the spring of 1725, Kalmeter was in London, preparing the final report on his travels in England in 1719–1725. His report was completed on July 5, 1725, and was sent off to the Board of Mines.[62] It arrived too late to influence the Board's assessment of O'Kelly's proposal, but it meant that they had additional knowledge when the members met again in September to discuss the price quoted by O'Kelly for constructing a Newcomen engine. Since writing his first report in 1720, Kalmeter had acquired a more thorough understanding of the principle of the engine. He could also report that, thanks to the invention of this engine, several new collieries in Newcastle were now being worked at a profit.[63] This only served to confirm the opinion of the Board of Mines that the engine would be most useful where coal was available as fuel. He also made a more general comment on the employment of the Newcomen engine:

And regarding this engine, it was first invented some twenty years ago by a man named Savery, and later after many trials perfected and erected at several places in England, where the lack of waterfalls does not allow the use of waterpower. It cannot be denied that it is said that it often has to be repaired, especially at greater depths.[64]

He reported that he had seen an engine in Newcastle which operated at a depth of 50 fathoms and pumped 375 hogsheads of water per hour, and another one in Wales that pumped 575 hogsheads per hour.[65] In Swedish units, these figures corresponded to 560 and 860 barrels per hour, which agreed very well with O'Kelly's figure of 600 barrels per hour for the capacity of a Newcomen engine. Kalmeter noted that "the size of the cylinder is determined by the depth and the weight of the water",[66] but the Board of Mines already knew all about this from O'Kelly and Swedenborg. However, Kalmeter gave one new piece of information that probably did interest the Board of Mines. He wrote that the cost of building in England, though it varied, was reported to amount to £1 500 or £1 600 sterling.[67] The Board had Kalmeter's report available when it discussed O'Kelly's prices of 9 000 *gulden* for a large engine and 6 000 *gulden* for a smaller engine.[68] The equivalent of Kalmeter's figure in Swedish currency was approx. 60 000 *copperdaler*, and of O'Kelly's 30 000 and 20 000 *copperdaler*. It becomes clear why they showed an interest in O'Kelly's quotation for a smaller engine, as this appears to have been a bargain compared to one from England.

During the autumn of 1725 the Board of Mines received further information on the new technology, but this time through informal channels. Having completed his report to the Board of Mines, Kalmeter left England in July 1725.[69] He was now, in secret, helping Jonas Alströmer to recruit skilled workers for Alströmer's many industrial projects. Kalmeter arrived in Liège on September 9, and the day after his arrival he wrote a letter to Jonas Alströmer in Stockholm (he wrote in English, as was their habit). "The Englishman, that was here to build the fire engine, is gone", Kalmeter told him. "I don't know where, & the business came to nothing."[70] A couple of weeks later he was better informed, and told Alströmer in a letter dated October 5 of how he had gone to se an alum works near Liège:

> In my way thither I call'd to see the fire engine, whereof I Spoke in one of my letters being then wrongly informed, as if it was down. 'Tis true, one OKelly, who first undertook the creating of it, miscarried, so that they were oblig'd to make one Saunders come over. At present there is one Massillon, who has the direction of it. We are become great friend, & as he first gave me to understand, that he not only perfectly well knew how to erect such an engine, but even others either to go with wind or water, & I found that he was a litle dissatisfied with his present condition, the wages being low in this country, so I begun to ask him if he should be willing to take a trip into other Countries, & I have found him resolved. If you please to Mention this to the College of Mines or others of your friends, & let me know Your thoughts & theirs, I shall be glad, because I have promis'd the Man an answer. He seems to be a good sober Man, & if so be, that he should come to go, he would save a Carpenter for the Slitting Mill. As for the conditions, I can say nothing, because he would not utter himself, since I could not act & conclude with him. However, I believe he will come to about a Crown a day or rather less when he works & travelling wages. In short I have found him one of the best I have to deal with upon this subject of going abroad, for the rest of them have made me most mad with them [...] As for the fire engine, I forgot to tell You, that it has cost about 1 000 pistoles a building, each pistol of 15 fl. Liège Money, tho' they had done vast attempts at first the charges of keeping it run to a pistole a day, including the coals to burn under the boyler. It works 60 fathoms deep & the builder would fain not go further with it, There is none in Eng[land] as I know, that works upon a greater depth.[71]

This passage from Kalmeter's letter has been quoted in full to illustrate the abundance of information that was available about Newcomen engines at this time, and the ease with which it was disseminated internationally. Kalmeter had asked Alströmer to tell the Board of Mines "or others of your friends" that Massillon, whom Kalmeter referred to as a "Master Builder", was willing to come to Sweden and build a Newcomen engine if he received "about a Crown a day or rather less when he works & travelling wages". No offer from Sweden seems, however, to have reached Massillon, and the reasons for this lack of interest will be discussed below.

Finally we must also note yet another possible source of information. In 1974 Dietrich Hoffmann published some newly discovered documents from the Archiv des Oberbergamtes in Claustahl-Zellerveld that concerned the second Newcomen engine on the Continent, known as the Königsberg engine, which was built in 1721–1724 by Isaac Potter and Joseph von Erlach near Schemnitz in Hungary.[72] In 1725 the Hanoverian Ambassador in Vienna, Dr. Daniel Erasmus von Guldberg, tried to obtain information on the construction of Newcomen engines, which he thought might be useful in the Hanoverian mines in the Harz. In a letter dated June 23, he reported how he had at first failed, but then had a stroke of luck:

> This year a young Swedish nobleman, von Schönström, who is travelling around looking at mines, having already made this his business in Sweden, arrived here from Hungary, and told me, when I was entertaining him and asked about the engine at Königsberg, that not only had he seen it, but he also had a drawing of it with him. I asked this Swede so earnestly about this drawing—since I have not yet been able to obtain one either of the Königsberg engine nor of the one here to study—that he communicated the same to me, and I had it copied.[73]

Guldberg's copies of Schönström's drawings are remarkably detailed, and it seems likely that Schönström himself had copied them. After all, it was said only that he had them with him, not that he had made them himself. In this episode von Schönström plays a minor part in the larger and complex story of the spread of the Newcomen engine on the Continent. But here we are concerned only with the purpose of von Schönström's journey and whether his information ever reached Sweden.

In 1722 Abraham Daniel von Schönström (1692–1758) had been granted permission to travel abroad by the Swedish Government.[74] On June 21 he applied to the Board of Mines to be appointed an Auscultator, because be wished to avail himself of the opportunity, while he was travelling south through Sweden on his way to the Continent, to visit ironworks and mines and to attend sessions in the mining courts.[75] This was granted, but the title was simply a pass to allow him to study the mining industry in Sweden. He had inherited shares in Tolvfors Ironworks in Gästrikland, and he was probably only making the usual *grand tour* of a young nobleman. It was in his interest to learn all he could about the business he had inherited (he became manager of Tolvfors Ironworks in 1739). He therefore travelled on his own initiative, and his connection with the Board of Mines was merely a formal one. But did his drawings of the Königsberg engine ever reach the Board of Mines, or influence developments in Sweden in any other way? It seems that they did not,[76] but it should be observed that von Schönström and Swedenborg were good friends at this time.[77] They had mutual business interests, as they held shares in the same ironworks. The drawings of the Königsberg engine may thus have reached Swedenborg in 1725, but if so they were too late to influence the official assessment of the new technology.

The von Schönström episode illustrates two important things. *First,* it is yet another example of the abundance of technical information that was available at the time. It was not difficult, as Hollister-Short has pointed out, for a foreigner to contrive to see an engine and learn how it worked.[78] *Secondly,* the recent discovery in the Oberbergamt archives implies that other archives might contain other undiscovered documents that will show that the diffusion of technical information on the

Newcomen engine was far more widespread and international than has hitherto been believed. This is of some general interest in the study of technology transfer, for if the limitation on the process of transfer is not the result of unavailability of information, we have to look for other factors.

AGENTS OF TECHNOLOGICAL CHANGE

We can now identify the agents of technological change in this process. There were *Swedish individuals* travelling abroad on their own initiative, such as Swedenborg (1710–1715), Alströmer and von Schönström. They were of little or no importance, and this may have been due to their independence of a larger social context. The *foreign entrepreneurs*, however, were of major importance. But it was shown in the previous chapter that in order to succeed de Valair, Saunders and O'Kelly would have needed practical knowledge, the backing of a private group, and social status. *Private groups* were essential if the entrepreneurs were to realize their projects, because there was no state support, and Strömfelt, von Törne and company were one such group. Of the *institutions*, only the Board of Mines was of importance, but this body was one of the major agents of technological change. We saw in the previous chapter that it was the interaction between foreign entrepreneurs and the Board of Mines that generated the new knowledge. The Swedish Government took no active part, and served only as a "post office" by forwarding letters between Preiss in The Hague and the Board of Mines.

Some observations must be made on the functioning and characteristics of the institutional structure, i.e. the Swedish bureaucracy. It was efficient in that news of a technical innovation, such as O'Kelly's proposal to Preiss in The Hague, was received by the Ambassador and passed through the diplomatic channels to the Government in Stockholm, which automatically remitted it to the Board of Mines. This meant that a technical question, even if intercepted by the extension of the Swedish civil service through its diplomatic service in Europe, reached the correct authority as a matter of routine. One feature of the functioning of the institutional structure was the promptness with which these cases were handled by the civil service.[79] Preiss' letters from The Hague took ten days to reach Stockholm. Three days after a letter had arrived, it had been read by the Government and remitted to the Board of Mines. The Board invited de Valair to come and be questioned, and asked Swedenborg to assess and report on the new technology. Then the Board met and studied the documents in the case, and prepared its reply to the Government. This was then considered by the Government, which wrote to Preiss telling him how he should answer O'Kelly—and the whole process took only three months! Moreover it should be remembered that all the letters were copied by hand at each stage of the process before they were passed on. Preiss, the Government and the Board of Mines—they all kept copies of every document that was exchanged.

An important characteristic of the institutional structure was the competence of the individual officers in the civil service. Preiss, for example, held a doctorate in law from Oxford University and was very knowledgeable in political economy.[80] He was a professional diplomat, but also gave a most accurate report on O'Kelly's description of the new technology. Another example is Göran Wallerius, who was Mine Inspector in the district in which the Öster Silvberg mine was situated and, as such, the local representative of the Board of Mines. When de Valair asked for permission to build an engine at Öster Silvberg, Wallerius was instructed to assist him (although the project came to nothing). Wallerius had studied mathematics at the University of Uppsala, worked as a *Markscheider* in the Great Coppermine, and assisted Polhem with his hydrodynamic experiments (see Chapter 4).[81] In 1708–1710 he travelled widely in Germany, France, Holland and England to study the mining industry and science.[82] After his return he became an active member of Sweden's first scientific society, *Collegium curiosorum* in Uppsala.[83] The Board of Mines could supply de Valair with a very competent local representative indeed.

There were two characteristics of the institutional structure of early eighteenth-century Sweden that seem to have had a decisive influence on the process of technology transfer. One could be described as a strength, the other as a weakness. The officials of the Board of Mines were in most cases also owners of or shareholders in mines and ironworks (Swedenborg, for example, held the majority of shares in Axmar Ironworks in Gästrikland). The Board of Mines was certainly not housed in an ivory tower. Technical information flowed in and out, through personal as much as official channels of communication. It was a very different situation from that of today, when it goes without saying that an official must not have private interests in his field of duty, but it is evident that in the eighteenth century this state of affairs was conducive to technological development.

On the other hand, the Board of Mines was, as has been pointed out, primarily an advisory body. It took an active part in the transfer of new technology as far as the foreign tours of its Auscultator Kalmeter were concerned. But the main purpose of such travels was to provide a continuous supply of competent persons for the offices of the Board, and the Board never gave any financial support to any proposal to build a Newcomen engine in Sweden. This was not its purpose, nor did it have the funds. It was to take another twenty years, until 1747, for the ironworks to get together and create the Swedish Ironmasters' Association (*Jernkontoret*).[84] This was a private credit institution for the mining industry, one aim of which was to assist technological development. It was also very much due to the support of the Swedish Ironmasters' Association that later steam engine projects in Sweden could be realized.[85] It was a major change in the institutional structure of the Swedish mining industry. But in 1725, initiatives for technological change were still left to individuals or private groups.

THE STRUCTURE AND SOCIOLOGY OF TECHNOLOGICAL KNOWLEDGE

This interaction between individuals, foreign entrepreneurs as they were, and the institutional structure had an important bearing on the structure and sociology of the technological knowledge that was generated. As the new technology spread over Europe, individual entrepreneurs like de Valair and O'Kelly tried to use their knowledge of it to obtain monopolies. In order to do this, they had to be constantly at the front line of diffusion of the new technology. Every year they had to present themselves, and their knowledge, farther and farther from England. O'Kelly, for example, had moved from England to Liège in 1720, and in 1725 he was planning to move again. By then he was examining the possibilities on two sections of the front at the same time: Sweden and Spain. De Valair had moved to Sweden in 1723, but he had moved further than his knowledge could carry him, beyond the front line, and he ended up in a sort of no man's land. It was a competition in knowledge, with each trying to demonstrate a greater mastery of the new technology than the next. This involved supplying evidence of their knowledge, as O'Kelly had done when he handed Preiss his drawing and description of the engine. This competition between individual entrepreneurs accelerated the process of diffusion. They were in fact themselves the most important agents of technological change in the very process that they tried to use for their own ends. As they competed for new commissions and monopolies at the front line, they pushed the front line further ahead.

This competition to display superior knowledge of steam power technology was not fought primarily in terms of practical experience in the building of successful Newcomen engines. No, it was a battle to decide who had the most thorough theoretical understanding of the underlying scientific principles. When de Valair applied for his patent in 1723, it had been enough to imply that he and his partners knew how to build such an engine. But when O'Kelly arrived on the scene, he immediately challenged de Valair and his associates on grounds of theoretical knowledge. O'Kelly wrote that Saunders *"ignore la Theorie"*,[86] and he told Preiss that Saunders had no knowledge of the fundamentals of science and did not in the slightest understand the calculation of the proportions to which this engine had to be designed in all its parts.[87] Whereas his own son, whom he offered to send to

Sweden, had also studied physics, and understood the theory as well as the practice of this physical engine.[88] De Valair retorted that several persons had testified that Saunders knew his business well, and had been taught it by none other than O'Kelly himself. The implication was that Saunders' theoretical knowledge was equal to that of O'Kelly's son.

This controversy over theoretical knowledge had not been lost on Emanuel Swedenborg. He refers to the need for awareness of the theory in his technological assessment, and it was subsequently repeated by the Board of Mines in its reply to the Government. Swedenborg wrote that the design of the engine was "based not only on the laws of mechanics, but also on those of physics".[89] To maintain such an engine in operation would therefore require the supervision of a man "who understands the mechanical and also the physical laws".[90] Knowledge of the new technology was thus believed to contain both a *theoretical* and a *practical* component, and this was not as a result of any theoretical development, or because practical knowledge alone had proved insufficient. The stress on the necessity for theoretical knowledge was a product of the struggle to gain monopoly of the new technology. It was an extra dimension deliberately added by its practitioners in a competitive situation.

"Science" was the label of quality that O'Kelly attached to his theoretical knowledge in an effort to outbid de Valair. It was indeed a label of superior quality, since science was a cultural commodity held in growing esteem in eighteenth-century Europe. But science also had a *social* value in the early eighteenth century. It was not yet recognized as a profession, linked to a distinct occupational group, but rather considered as an intellectual interest, a desire for knowledge or an attitude to nature. As such, this interest cut across the divisions between specific professions, but was as yet on the whole confined to the upper strata of society. The way in which the social roles of science and technology were seen in Sweden at the beginning of the eighteenth century is illustrated by a contemporary sketch by Christopher Polhem, his dialogue between "Lady *Theoria*" and "Master Builder *Practicus*".[91] The dialogue begins when the Master Builder comes to pay court to the Lady:

Theoria: Who goes there in his black leather apron? I am not used to receiving social calls from such persons.[92]

The Master Builder tries to mollify her, and begs her to excuse his boldness. He has heard that she approves of his work, which is in mechanics and architecture, and he has come to make a proposal of marriage. She thinks that he is joking, and chides him: "A beautiful Lady, as you call me, and your black leather apron go well together."[93] But patiently he explains that it would be of public utility if her knowledge of science was to be combined with his practical aptitude.

Theoria: Though I have heard that he has several children that he prides himself upon.

Practicus: That I cannot deny, but I regret that they are none too well respected as long as it is known that they are illegitimate. But if my beautiful Lady were to become the mother of my children, they would rise to much higher esteem.[94]

She is soon reassured, and promises to answer any question he has regarding his work. She will do this for the sake of "public utility", but "as for the rest, I must however consider".[95] The beautiful Lady took her time; the marriage of science and technology took another hundred years to materialize. The dialogue illustrates the difference in social status between science and technology in Sweden at the beginning of the eighteenth century. In Swedish, *"Fröken"* (Lady) was the title of a woman of noble family,[96] and "Master Builder" designated a certain craftsman. The first was a high *social position* whereas the second was a relatively lowly *profession*. Science was an interest shared by the higher strata of society; technology was the concern of artisans. Between the two, there was a wide social gap.

This meant that the distinction between the theoretical and practical aspects of steam power technology also led to a *social stratification of knowledge*. If theoretical knowledge was a prerequisite for building Newcomen engines, why then it followed inevitably that only gentlemen were able to undertake such assignments. Gentlemen like, for example, John O'Kelly, nobleman, Colonel, Esquire and Seigneur d'Aghrim. As far as the Board of Mines was concerned, men like Saunders and Master Builder Massillon wore "black leather aprons". They belonged to a lower social group, that of professional men, and *ipso facto* did not possess the scientific knowledge that was thought necessary. Emanuel Swedenborg even said this explicitly in his assessment of the technology. Since the design of the engine involved the utilization of theoretical physical principles, it could not be built or maintained by "any common master builder".

SUMMARY

Emanuel Swedenborg was appointed Assessor Extraordinary in the Board of Mines in 1716, but it was not until April 1723 that he was formally allowed to take his seat on the Board. By then he had been able to give sufficient evidence of his knowledge and experience of mining matters. What he lacked in actual practical experience of the Swedish mining industry he made up for with his awareness of the latest developments in science. His scientific interests are exemplified by his suggestion to the Board in February 1725 that an air pump should be purchased in England. The favourable reception accorded to this suggestion indicates that he was by this time fully accepted by the Board.

The Government had asked the Board of Mines to assess the merits of the Newcomen engine, and the Board referred this task to Sweden-borg as its most competent member on scientific matters. His report listed the disadvantages and advantages of the new technology, and may be summarized as follows:

Disadvantages
1. It could not easily be adapted to the geographical environment of Sweden
2. It demanded a new type of engineer with scientific training
3. It was difficult to determine the technical limitations of the new technology because there was no experience of building machines driven by steam and made of iron
4. The auxiliary technologies put a limit on the utility of the engine
5. The technologies already available were generally sufficient

Advantages
1. More power than from any traditional technology
2. Useful where there was no running water and an abundance of wood fuel
3. A fascinating new technology

Swedenborg also suggested his own method of calculating the diameter of a cylinder for a given depth. However, it could only be used to establish the *theoretical* value. This was a self evident part of O'Kelly's rule of proportion, which gave a value for the diameter of the cylinder based on *practical* experience.

The reply of the Board of Mines to the Government followed Swedenborg's assessment in all essentials, although the Board chose to list the advantages *before* the disadvantages. Thanks to the efforts of their Auscultator Kalmeter and of Preiss, the Swedish Ambassador in The Hague, the Board was able to confirm most of the information O'Kelly had supplied. As an assessment of the new technology, the reply of the Board of Mines was admirably comprehensive considering the information available. The Board did not reach a definite verdict, but wanted to know how much it would cost to build a Newcomen engine in Sweden.

Additional information may also have been provided by Abraham Daniel von Schönström, who obtained drawings of the Newcomen engine in Königsberg in 1725. The von Schönström episode illustrates the abundance of technical information that was available at this time, and suggests that the dissemination of this information was far more widespread and international than has been realized.

On the basis of Chapters 7, 8 and 9, the agents of technological change in the process that brought knowledge of steam power techno-logy to Sweden can be identified. *Foreign entrepreneurs* were of major importance, but the backing of a *private group* was essential to them if they were to be able to implement their projects as there was no state support for technological change of this nature. The Board of Mines was the only *institution* of importance, and it was the interaction of the foreign entrepreneurs and the Board of Mines that generated the new knowledge.

The institutional structure of the Swedish civil service was efficient in gathering news of technical innovations, and these were passed through the bureaucracy with surprising promptness. It took only three months after O'Kelly's visit to Preiss in The Hague for the Government to send back its reply, based on all the information that had been collected over the previous few years and considered by some of the most competent men on these matters in the country.

The strength of the Board of Mines was that its members also had private interests in the mining industry. Technical information flowed in and out through personal channels of communication as well as official ones. The weakness of the Board of Mines as an agent of technological change was that it did not initiate any engine-building projects. It was a purely advisory body in this respect, and initiatives were left to individuals or private groups.

The foreign entrepreneurs competed for monopolies of the new technology at the front line of its diffusion. In so doing, they pushed the front line further ahead, and assisted the spread of the Newcomen engine across Europe. Each tried to get the better of his rivals by claiming greater theoretical understanding of the underlying scientific principles. The Board of Mines was led to believe that scientific knowledge was essential to anyone intending to build a Newcomen engine. The knowledge of the new technology was regarded as having a *theoretical* as well as a *practical* component, and this was a subtlety deliberately introduced by the foreign entrepreneurs in a competitive situation.

But science had a social value in early eighteenth-century Sweden. It was an interest shared by the higher strata of society, and not yet thought of as a profession. If science was a prerequisite for building Newcomen engines, it followed that only gentlemen were able to build them. Craftsmen such as the foreign master builders belonged to a lower social group, and were not thought to possess the necessary scientific knowledge.

In this and the preceding chapters, it has been shown that anyone wanting to build a Newcomen engine in Sweden in the 1720s needed the following:

1. Practical knowledge of how to apply the technology
2. The backing of a private group with the means to realize the project
3. High social status *and* a proven knowledge of science

The following chapter will show how Mårten Triewald came to gain the first and the third of these qualifications during his years in England.

PART III
Transfer of Technology

Fig. 10.1. The parchment cover of one of the volumes containing the records of the Stockholm Guild of Blacksmiths in 1698, showing the tools and symbols of the craft. (Photo Nordiska museet, Stockholm)

10. Enter Mårten Triewald

INTRODUCTION

"Mårten Triewald", writes Sten Lindroth, "was no original or creative genius".[1] His standing is chiefly based on the inspiring part he played in the founding of the Royal Swedish Academy of Sciences and on the lectures in Stockholm with which he contributed to the introduction of modern experimental physics into Sweden. When Mårten Triewald came back to Sweden in 1726 after a long sojourn abroad, he quickly established himself. Within a few years he had lectured at the Palace of Nobility, built the first Newcomen engine in Sweden, become chief engineer at Wedevåg Ironworks, set up a diving and salvage company, published a number of books, been elected to the Royal Society and begun to campaign in his native country for an academy of sciences modelled on the English prototype. How did it come about that he was able to return to Sweden possessing such proficiency and such a wealth of ideas, this modestly educated blacksmith's son from Stockholm, who had left the country as a failed businessman? The answer is to be found in the impressions and knowledge that Mårten Triewald absorbed during his eventful youth and his years abroad.

IN THE GERMAN PARISH

The place in Germany from which the Triewald family migrated to Sweden has hitherto been unknown. In his account of the foundation and early history of the Royal Swedish Academy of Sciences, published in 1939, Bengt Hildebrand conjectured that the family may have had its roots in the East Prussian city of Königsberg,[2] a statement which has been quoted in subsequent literature. Mårten Triewald's father, Martin Triewald (1660–1744), was a blacksmith in Stockholm. In the records of the Guild of Blacksmiths for April 6, 1687, it is recorded that "Marten Triwalt von Saksen and hall" appeared before the company and applied to be set a masterpiece.[3] The implication that the town of Halle, in Saxony, was the previous home of the Triewald family is confirmed by the following quotation from Mårten Triewald's essay on coal mining in the *Proceedings of the Royal Swedish Academy of Sciences* in 1740.

In Germany, some fifty and more years ago, coal was not known. In Hall, in Saxony, salt was boiled with wood for so long that the woodlands were consumed, and they had to seek coal, which they found near Wettin.[4]

The indication "some fifty and more years ago" would point here to the late 1680s. It seems likely, therefore, that it was his father, Martin Triewald, who had told Mårten about conditions in the well-known salt springs, and that at some time in the late 1680s, not later than 1687, he had left Halle for Stockholm. On June 20, 1687, Martin Triewald presented his masterpiece, but another eighteen months were to elapse before he could show his indenture of apprenticeship and his letter of birth to be "in proper order".[5] These had presumably to be obtained from Saxony, hence the delay. On New Year's Eve, 1688, Martin Triewald became a master in the Guild of Blacksmiths,[6] and by January 3 he had already recruited his first apprentice.[7] Martin Triewald wasted no time in pursuing the career which was to lead him to the highest posts in the Guild. As Alderman he was head of the Guild of Blacksmiths from 1727–1734,[8] and in 1731 he represented the burghers of Stockholm in parliament.[9]

The young blacksmith appears to have been a forceful individual, who was determined to carve out a career for himself in his adopted country. In the records of the Guild we read that Martin Triewald enticed commissions away from his colleagues by undercutting them, and that he was sharply "admonished" by the Guild.[10] On another occasion he had first to ask to be pardoned, and after he had handed over a "consideration" to the Guild and 24 *öre* to the poor, "his transgression was remitted".[11] On a third occasion he was heavily fined by the Guild "because it is not the first time".[12] On July 28, 1687, not much more than three months after his first appearance before the Guild of Blacksmiths, and perhaps not much longer after his arrival in Sweden, he married Brigitta Noth, the widow of a blacksmith named Christoff Trogen.[13] On May 1 the following year, their son, Samuel, was born, followed three years later, on November 18, 1691, by Mårten Triewald.[14]

Mårten was baptized in the German Church on November 21.[15] The baptismal register lists eight godparents, the first of whom is Daniel Meisel. He, too, was a master blacksmith, and his fortunes are intertwined with Martin's. It was Meisel who promised the Guild that he

Fig 10.2. The interior of the German Church in Stockholm in 1681. Engraving by J. Cr. Sartorius in Chr. Bezelius, *Die Herrlichkeit des Christenthums* (Stockholm, n.d.). (Photo Royal Library, Stockholm)

would obtain and produce Martin Triewald's letter of birth and indenture of apprenticeship within a year, and it was from Meisel that Triewald received the commission for his masterpiece.[16] In 1706, Triewald and Meisel became officials of the Guild, and when Meisel was Alderman, Triewald was his Assistant.[17] In 1727, Daniel Meisel's son, Johan Diedrich Meisel, became Martin Triewald's indentured apprentice.[18] The friendship between Triewald and Meisel throws some light on relationships in the German parish. Ties of kinship and friendship linked the members of the parish, and it is in this group, and not merely in the Triewald household, that Mårten Triewald's formative environment is to be found.

In the German Church, Martin Triewald and his family sat on pew "*Kurze Seite*" No. 10 from 1691 until 1730.[19] The lists of pew rents collected show that other families also had permanent places on this and neighbouring pews for decade after decade. A few rows from the Triewald family, for example, on pew No. 14, sat the Director of the Royal Gun Foundry, Gerhard Meijer Sr. (1667–1710), and his family.[20] His son, Gerhard Meijer Jr. (1704–1784), was thirteen years younger than Mårten Triewald, and the two boys must have met

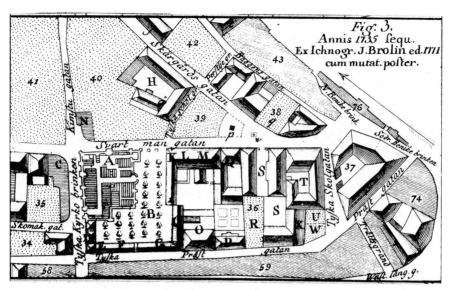

Fig. 10.3. Map of a part of the Old City in Stockholm, showing the outline of the German Church and the arrangement of the pews. Engraving by E. Åkerland in J. A. A. Lüdeke, *Dissertatio historica de ecclesia teutonica et templo S:tae Gertrudis Stockholmiensi* (Uppsala, 1791). (Photo Royal Library, Stockholm)

regularly on Sundays during the first decade of the eighteenth century when the families called on each other after church. In 1727 they met again, when Mårten Triewald gave Gerhard Meijer Jr. the job of casting the huge bronze cylinder for the Dannemora engine. When in about 1713–1715 Mårten Triewald was living in Königsberg and paying occasional visits to Stockholm, and again in 1726, when he returned to his native country after ten years in England, the family still had their place on pew No. 10, and around the Triewalds sat the same families as in Mårten's childhood. Here, in *St. Gertruds Gemeinde*, was a fixed point in his world.

Another fixed point was provided by the home in Åkaretorget. In 1694 Martin Triewald acquired a stone house at no. 47 in the Södra block (today no. 3 in the Medusa block on Triewaldsgränd). Triewald bought the house for 14 000 *copperdaler*, which suggests that he was prospering. In 1694 a third son, Johan, was born, and in 1699 a fourth, Daniel.[21] In his investigation into the history of the Triewald family, Hildebrand mentions Anders Triewald (?–1715), who was Martin's brother and also, eventually, a master blacksmith. However, the archives of the Guild in the Nordiska Museet indicate that there was also a third brother, Zacharias Triewald. Both Zacharias and Anders were apprenticed to their (presumably) elder brother, Martin.

Anno 1693, 29 March Triwalt registered his brother Zahris Triwaldt, he shall be apprenticed for 3 years reckoned from Michaelmas last.[22]

Unlike his brothers, Zacharias never became a master in the Stockholm Guild of Blacksmiths, and perhaps he remained a journeyman for the rest of his life. An application for a passport, signed by Martin Triewald in 1727, points to this being the case:

Whereas my brother Zacharias Triewald has now been lodging with me for fourteen days and now intends to travel down to the south of the country, and has contented me to the full, I can hereby confirm the same.[23]

Anders Triewald worked for his brother Martin for three years, 1697–1700, after which he departed *"nach danzig"*.[24] He returned to Stockholm in 1705, whereupon he, too, appeared before the Guild of Blacksmiths and applied for his mastership, "having learned honestly and well as was to be seen from his letter of birth and indenture of apprenticeship, and then travelled around working and visited places abroad".[25]

The family in which Mårten grew up should therefore be regarded as including both of his father's brothers, Zacharias and Anders. To the young Samuel and Mårten, it must have been exciting to listen to their uncles' tales of their journeyman's wanderings on the Continent. Perhaps it was from them that they acquired their wanderlust. However, the strongest influence on their youthful spirits was presumably that of their father. He was, as we have seen, an energetic and competitive individual. He had come to Sweden as a foreigner but worked his way up to the top of his profession. It was an achievement which both Samuel and Mårten, with varying success, were to attempt to emulate. Martin Triewald must have presented a patriarchal appearance when he presided at his dinner table as the eighteenth century dawned. There sat his brothers, Zacharias and Anders, both apprentices under his jurisdiction. There sat his sons, Samuel, aged twelve, Mårten, nine, and Johan, six. And there sat his wife Brigitta, with their youngest son, Daniel, not yet a year old, on her lap.

Mårten was sent to the German School "under the faithful care and industrious attention of Rector Stenmeyer and Conrector Spöring".[26] Johan Bernard Steinmejer was born in Westphalia and had become rector of the German School in 1688, a position which he was to hold for half a century.[27] Mårten was probably enrolled at the school in the late 1690s.[28] It was the school at which Christopher Polhem had been educated twenty-five years earlier, but, unlike Polhem, Mårten Triewald was to receive no further formal education after his schooling there. Linnaeus, who knew Triewald well, wrote in a well-known letter of 1761 on the origins of the Royal Swedish Academy of Sciences that Triewald "had not himself been a university student".[29]

The eldest brother, Samuel, also attended the German School. When he was only fifteen he was sent to Uppsala, where his name was entered in the roll of the Stockholm Undergraduates' Association on October 1, 1704. Samuel spent only just over a year at the University, but during this time he delivered "a fine oration", opposed one thesis, and defended another. In 1705 he returned to Stockholm, where he became a private tutor in the house of Count Erik Gustaf Stenbock; through Stenbock's wife, Johanna Eleonora de la Gardie, a cousin of Aurora Königsmark, he came into contact with the literary circles of the period. In 1707 he obtained a temporary appointment in the National Record Office, eighteen months later he was granted consent to serve in the Chancery, and in 1711 he received the permission of the Government to travel abroad.[30]

Samuel Triewald went first to Königsberg. He seems to have studied at the university there for a few months, and "at many Disputations showed the progress of his learning and his accomplishment in Latin", and then continued to Berlin, Hanover, Holland and England. By the time of his return in 1713, he had already embarked upon the career as a poet which was to earn him the name of "Sweden's Boileau". He spoke and wrote nine languages, and was "without doubt one of the most widely and modernly educated of contemporary Swedes".[31] "One can understand", writes Hildebrand, "that the well-read poet Samuel Triewald, rising in addition to high public office, must have been a phenomenon in the smith's family from Åkaretorget".[32] His father and his uncles, the three blacksmiths, must have followed his meteoric career to the heights of society with admiration, tinged perhaps with a certain consternation. To Mårten, who was still at school when Samuel left for Uppsala, his elder brother was something of a hero. Hildebrand observes:

That Samuel's younger brother Mårten looked up to him with an admiring and moving affection is hardly calculated to surprise. Without knowing something of the personality and life of Samuel Triewald, one cannot comprehend the story of his brother Mårten, either as a man or as the initiator and founder of the Academy.[33]

Mårten Triewald did not follow his brother to the University of Uppsala. His biographer in the Royal Swedish Academy of Sciences, Lars Laurel, who did not know Triewald personally and who intersperses the facts he has been able to unearth with lavish eulogy, writes in his memorial address:

He could indeed have gone with his contemporaries to the University. He yielded nothing to them in natural or acquired wit. And he had parents with that wealth and willingness for it not to have been difficult for him.[34]

That the Triewald home was a comfortable one is substantiated by Martin Triewald's deed of inventory of 1744.[35] Perusal of the admission and discharge books of the Stockholm Guild of Blacksmiths also shows that Martin Triewald's business was a sound one. After becoming a master blacksmith he took on a steady stream of apprentices throughout his career and appears always to have had at least three in his service.[36] In 1725, Samuel Triewald petitioned the King to be allowed to leave his Chancery post, one of his reasons being that he did not wish to continue to be a burden to his parents, as they had not only financed his studies and his travels, but had also had to meet most of the expense of supporting him during his twelve years in the service of the Crown.[37] The time for Mårten to go to Uppsala would probably have been 1707 or 1708. But it was just at this time that Samuel was embarking upon his political career, while also starting to make a name for himself in the literary circles of the capital. For the smith's son to be able to take the social stride into the salons, the Master Blacksmith Martin needed to furnish him with elegant apparel in the latest fashion and to put enough money into Samuel's pocket for him to participate without embarrassment in society life while still being able to afford his copy of the latest literary work from the city's booksellers.

There is reason to look more closely into the question of why Mårten Triewald did not go off to Uppsala, if only because for the rest of his life he was to express a scorn for universities which, while admittedly typical of the period, was of unusual vehemence. In a letter to Erik Benzelius in 1728, Anders Celsius writes that Triewald has promised to donate his entire collection of instruments to the Society of Science in Uppsala, on condition that the Society be moved to Stockholm "for he has no great love for Uppsala".[38] In his influential prospectus for an academy, which he presented to the Society of Science in 1729 (see Chapter 13), Triewald gives his reasons for following foreign precedent and advocating Stockholm, in preference to Uppsala, as the location for a Swedish academy of sciences:

so that thereby they may not only close it to the pedantry with which the universities are commonly infested and which overgrows the veritable pure goodness like a weed, so that it neither thrives nor receives light nor the repute which it ought by right to acquire ...[39]

In his book on the Dannemora engine, he writes in 1734 with admiration of the inventors of the steam engine, who were "unschooled folk, who had never acquired a certificate at any university".[40] And in the preface to his lectures on the new natural philosophy in 1735, he writes of these noble, pleasing and useful sciences, which had aroused such delight in England "that their pursuit cannot be confined within the walls of the two universities".[41] These pronouncements point to a heartfelt disdain for universities, clearly directed at Uppsala, and they are hard words from one who had never studied at a university. Perhaps Mårten Triewald's contempt for universities, and for Uppsala in particular, stemmed from nothing more than disappointment at having had, in his youth, to curtail his studies for the sake of the expensive education of his elder brother, Samuel.

COMMERCIAL SERVICE WITH SAMUEL WORSTER, MERCHANT

Mårten Triewald "turns from books to commerce", as Laurel relates,[42] and in 1709 we find him employed in the "service" (presumably in the sense of commercial service, i.e. as bookkeeper) of the English merchant, Samuel Worster (?–1746).[43] It is probable that Triewald entered into his employment directly after leaving school around 1705. Samuel Worster was born in London and had settled in Stockholm during the first years of the eighteenth century. He was one of the many immigrant Englishmen and Scots who were the mainstay of Swedish trade during the early part of the Era of Liberty. Worster concentrated particularly on trade with England, exporting Swedish iron, sawn timber, tar and pitch to ports such as London and Hull. The 1730s and 1740s brought great prosperity to Worster and during that period he was to become one of Stockholm's biggest exporters.[44]

During the first decade of the eighteenth century, however, when Mårten Triewald was engaged by Worster, the latter was still in the process of building up his trading house. Despite his fifty years in Sweden, Worster never "mastered the Swedish language",[45] and he must have obtained valuable assistance from Mårten, who spoke both German and Swedish. Mårten was probably entrusted with the job of negotiating with the Swedish ironworks, for he himself tells us that in the early spring of 1713 he visited Leufsta, one of the large ironworks in Uppland.[46] He may also have dealt with the German skippers along the quayside in Stockholm. In March 1713, Worster describes him as "our servant for this purpose, Mr Mårthen Triewalt".[47] Triewald would thus appear to have been in the service of Samuel Worster during the period 1708–1713. Anne-Marie Brötje, who has written about Worster, has observed:

In view of Mårten Triewald's move to England in 1716, his strong Anglo-Saxon orientation later, his zeal for practical economics, his interest in mathematics and science and his collection of instruments, it is worth noting the influence that must undoubtedly have been exerted on him by his surroundings when he was with Samuel Worster.[48]

The seventeen-year-old Mårten stepped into a little piece of England, when he entered the service of Samuel Worster. This was not only because Worster, in his linguistic isolation, found most of his friends in the English colony in Stockholm. Worster came from a cultivated London family of merchants, and when he moved to Sweden he brought with him, and surrounded himself with, a part of his native country's cultural heritage, in the form of an impressive library. The inventory of his estate shows a collection of some 420 volumes, mostly in English.[49] There was religious and philosophical literature representing two contemporary currents: pietism and the philosophy of the

Fig. 10.4. Samuel Worster. Miniature painting by Christian Richter in 1701. (Photo Swedish Portrait Archives)

enlightenment. For example, Worster possessed Bayle's *Dictionnaire historique et critique* in English translation. There were also books on commerce and economics, which preached mercantilism. In addition, Worster had many historical works, mostly concerned with the history of England. Fiction accounted for a smaller section, but it is interesting to find that French literature was represented in English translation, and that Worster owned one of the earliest copies of Shakespeare in Sweden: five octavo volumes, which were valued in the inventory at 25 *copperdaler.*

In her study of Worster's library, Brötje points out that one is surprised to find the sciences so well represented in the library of a merchant as they are here. There are a remarkable number of works on mathematics, medicine and pharmacy. Samuel Worster also possessed a small collection of instruments. Brötje comments that even if we cannot date Samuel Worster's purchases of various books and instruments, Mårten had, in his younger years, albeit in a subordinate position, the opportunity to learn English in the Worster household and also to study scientific writings.[50] She asserts:

In view of the part that Triewald later played as the actual initiator in the founding of the Royal Academy of Sciences, the figure of the Stockholm English merchant, Samuel Worster, with his library and his home, takes on a wider importance. In this environment we have a hitherto unconsidered factor, contributing to the birth of the Royal Academy of Sciences and to the character which the Academy initially assumed and long retained: its Anglo-Saxon orientation and its work for practical economic reform on a scientific basis.[51]

It should be added that Samuel Worster was not the only Englishman with whom Mårten came into contact at an early age. In 1705, an Englishman named William Joye was lodging with the Triewalds. In a letter to Erik Benzelius, dated October 30, 1705, Joye recommended his host's son, "Mr Samuel Triewaldt", to Benzelius' consideration, with a view to his being allowed to borrow books from the university library.[52] And in his book on the Dannemora engine, Triewald writes, in connection with Nicholas Ridley Jr.: "me, whom he had known from my tenderest years".[53] From this we can only infer that Nicholas Ridley, the Newcastle mineowner, visited Stockholm some time in the 1690s, and that he, too, may have stayed at the home of Master Blacksmith Martin Triewald.

In 1711 Samuel Worster rented accommodation in house no. 15 in the Latona block,[54] in other words, only a few doors away from Martin Triewald. By 1720 Samuel Worster had married Margaretha Luther, said to be a direct descendant of the Reformer.[55] Several of their daughters later made advantageous marriages, and Hildebrand has a table showing how these marriages created kinship ties uniting the circle around the Royal Academy of Sciences during its early years.[56] In 1740, one of the daughters, Elisabeth Worster, married her father's onetime commercial assistant, Mårten Triewald, nearly thirty years her senior.[57]

While Triewald was working for Samuel Worster, Stockholm was ravaged by the plague. In 1710 the carts rattled daily through the streets of the city, carrying the victims out to the plague cemeteries outside Stockholm. Two people whom Mårten knew died of the plague: his aunt Margareta,[58] who was married to Anders Triewald, and Gerhard Meijer Sr. Mårten Triewald appears to have remained in the city during this period, for Laurel mentions that "in the pestilence he stumbled over the body of one dead of the contagion and lying at his feet".[59]

In a note dated July 22, 1713, Martin Triewald indicates that his son "Martin Triewald Jr., Burgher of Königsberg", had visited Stockholm and lodged with his father but that he was now going to travel "homewards to Königsberg".[60] Mårten was still less than twenty-two years old, and must therefore have obtained his freemanship at a very young age. If he was established as a burgher of Königsberg by the summer of 1713, and had visited Leufsta Ironworks during the early spring of that year, it may be presumed that at some time later in the spring of 1713 he left Sweden in order to make a career for himself, as his father had done, in a foreign country.

It was an ill-chosen time for setting up as a trader in the Baltic, or for trading from Königsberg with foreign ports. The Great Northern War was raging, and in 1713 the Swedish army surrendered in Holstein-Gottorp. In Russia, Tsar Peter the Great brought all his forces into action against the Swedes, and with the aid of the newly created Baltic fleet the Russians were able to subjugate increasing areas of Finland. By the time winter came, General Armfelt had been forced to yield the whole of southern Finland. At Königsberg, Mårten Triewald was in the eye of the storm. Why did Triewald choose Königsberg? Almost half of the ships arriving at Stockholm at that time came from Danzig, and most of the remainder came from Stralsund, Rostock, Flensburg, Stettin and Hull. Königsberg was not one of the main centres of German trade with Sweden.[61]

In 1713, Mårten had been employed by Samuel Worster for about five years. With the presumptuousness of youth, he no doubt considered himself well versed in every aspect of trade with foreign countries. While with Worster he had learned English, and through him he had acquired a knowledge of English trading firms and their goods. As a Swede, and as Worster's agent in dealings with Swedish ironworks, he had become familiar with Swedish export products. Being of German blood, he could speak German, and he had learned about conditions in Germany from his father and his uncles, the journeyman blacksmiths. It is possible that Mårten Triewald now tried to put his knowledge of English, Swedish and German conditions to good use in three-cornered trade between Königsberg, England and Stockholm, with Königsberg as his base. This assumption is borne out by the information that a ship with his load aboard sailed for England,[62] and by the fact that in the summer of 1713 he had reason and the opportunity to visit Stockholm. On his overseas tour in 1711–1712, the eldest brother, Samuel Triewald, had first visited Königsberg, where he is said to have spent some months at the university.[63] Perhaps it was Samuel who recommended Königsberg to Mårten in 1713 as the place to seek his fortune, and possibly he also provided him with introductions in the city. So much for speculation. Whatever Mårten Triewald's motives and aims may have been, he did not succeed. Laurel says:

For our young Businessman fared so poorly that his own Father tired of making good all his losses. And he suffered especial misfortune through a Skipper who, in his recklessness, lost his Ship which was bound for England for Coal, and on which he had nearly all he owned.[64]

And Laurel comments: "A most difficult beginning, whereat one might well be daunted and heavy of heart!"[65] Mårten Triewald's efforts as a merchant appear to have been on a modest scale, seeing that all his goods went as a single load. His career as a citizen of Königsberg seems to have occupied the years 1713–1715. These three years are the unknown years of Mårten Triewald's life. There is sporadic information about all the other periods of his eventful youth and his stay in England, making it possible to consider what impressions and knowledge he may be expected to have absorbed. But of the years in Königsberg we know nothing, and Triewald himself remains silent in all his many writings. This fact is significant. The years in Königsberg may perhaps have meant nothing but hard work, sacrifice and dashed hopes. Unlike his brother, he had not managed to carve out a career for himself; unlike his father, he had not managed to establish himself in a foreign country.

PENNILESS IN LONDON

"Instead of giving up trade and everything else, as many a person of feeble spirit might have done", writes Laurel, "our Triewald would now go out to England, in order to gain a better understanding".[66] In other words, he had failed to profit from his knowledge of Germany, what remained was England. But now Master Blacksmith Martin said no. Mårten Triewald came to an agreement with his father: if he could have money to go to London, he would never ask his father for anything again.[67] In 1716 he departed for London.[68] Maybe he sailed on one of the first vessels to leave for England in March or April, after the ice broke up. This is the climax to the story of Mårten Triewald's youth and his years abroad. He now leaves his youth and his homeland behind him and sets off, for better or worse, for England. Laurel has provided a graphic description in his bombastic biography, which deserves to be quoted at length:

How was it possible, honest man, for you to travel with so little, and yet happy and contented? How could you, having just suffered misfortune, and receiving such slight relief from your father, who must needs think of his other children, dream again of happiness and fortune in a country which you did not know at all well? That you dared to venture forth! To begin trading with nothing needs more than art. I see your laudable spirit, your great longing, your splendid mettle, but nothing more. I see with disquiet what will happen, when you reach London. My fears are all too well founded. What happens indeed to our Triewald when he gets there? He asks about, but nowhere is he received. The money is gone. Such distress![69]

When Triewald arrived in London, his reflections were no doubt the same as those of other Swedish travellers during the Era of Liberty. Sven Rydberg has summarized the way they saw London as follows:

It was the size, the teeming life, the wealth and the luxury, rather than any beauty, which struck Swedish travellers. The impressions created by the London of the day must undeniably have been confusing. The magnificent edifices did not show to advantage, because the buildings were clustered too closely together; most of the houses were in poor shape, even if the first suggestions of pavements were beginning to appear, and the crowds thronging the streets and alleys formed a very uncertain factor, particularly to strangers in clothes of a foreign cut.[70]

But Triewald did not have time to study the metropolis in any detail; he was far too preoccupied with purely practical problems. When his money ran out, there seemed nothing for it but to sign on aboard a ship "for the most distant parts of the world".[71] It is likely that in later life Triewald came to dramatize his first days in London, but when need was direst, a merciful angel appeared on the scene in the guise of an old Swedish salt. Captain Robert Dickerson, from Stockholm, in the employ of the Grill Trading Company, "with whom he had no other acquaintance than that both were Swedes", saw that all was not well with Triewald, and offered to assist him for as long as he himself remained in London.[72] Triewald obtained a breathing space, and this was probably the point at which he began to study. Perhaps he lived as Spartan an existence as the economist Anders Bachmansson (1697–1772), who studied in London in 1724 and later depicted his student days as follows:

he ate for 4 pence a day, paid 1 penny a day for books, shut himself in a room, lay Swedish, English and Latin Bibles side by side, the better to learn English and a little Latin, decided never more to think of trade, and read in particular [...] books of economics, and some mechanics, [...] and at last paid out 10 shl. to be present at a lecture on experimental Physicks that was held nearby.[73]

It was now that Triewald's enthusiasm for science and for various technical inventions was aroused. In 1716 he witnessed the testing of a diving bell that Edmund Halley (1656–1742) arranged on the Thames.[74] The British astronomer Halley, renowned for the discovery of the comet that bears his name, devoted himself during the years 1716–1721 to the improvement of the diving bell by supplying it with fresh air from submerged air-filled barrels. The attempt that Triewald watched, possibly only a few months after arriving in London, seems to have been one of Halley's first. As Triewald was jostled among other curious spectators on the quay, perhaps he received the inspiration for the first of the many innovations which he was to take home to Sweden. They were inventions which he would later describe himself as having "improved", which he introduced in his own name, about which he engaged in correspondence, and about which he wrote papers that appeared among the proceedings of the newly founded Royal Academy of Sciences.

The diving bell furnished material for a whole book, *Konsten at lefwa under watn* (The art of living under water), which appeared in two editions, an essay in the *Philosophical Transactions* of the Royal Society,[75] a royal privilege granting salvage rights, and, finally, nothing less than a commercial company—"*Norra Dykeri- och Bergnings-Societeten*"—which operated from 1729 until 1799. It was not a bad haul from a day by the Thames! But the example of the diving bell is no exception, for he showed similar enterprise in introducing all the ideas engendered by his years in England. All to the good of his native country and to the renown of their originator. As Carl Forsstrand has remarked, Mårten Triewald knew the art of getting his knowledge acknowledged.[76]

At the same time, another Swede was living in London, also engrossed in his studies, but enjoying very different status—already established and having access to the most distinguished circles. During the years 1714–1720, Johann Helmich Roman, often known as "the father of Swedish music", studied in London on a scholarship from Ulrika Eleonora. He was engaged at the opera house in the Haymarket, drawn to London by Handel's reputation. There is no record of Triewald and Roman having met in London, but in later life, when both were men of renown, Triewald was to manufacture a kind of metronome for Roman.[77]

By the time Robert Dickerson, his "good-hearted skipper", could stay in London no longer and bade him farewell, Triewald had met the Minister of Holstein, Friedrich Ernst von Fabrice (1683–1750).[78] Fabrice engaged Triewald as a kind of secretary, and again he received full board and expenses and "also the privilege of assisting him in his affairs and important correspondence".[79] Fabrice came to London in August 1717,[80] so Triewald cannot have obtained his post as private secretary before that date. He must therefore have depended on Dickerson's benevolence for a period of almost a year in London. It seems a remarkably long time, but the chronology of Triewald's first years in England is uncertain. Fabrice had been the envoy of Holstein to Karl XII of Sweden in Turkey in 1710–1714, and had been present at the incident known as the Fraças of Bender. In the late summer of 1717 he came to England to be briefed for his commission as mediator between Karl XII and George I of England. At an early date, Mårten's brother, Samuel, formed the connections with Holstein which in 1728 were to brand him a traitor and force him into exile. While he was still in the Chancery, for example, he appears to have supplied the Minister of Holstein in Stockholm with information and accepted payment from him.[81] Perhaps by 1717 Samuel already enjoyed such connections with Holstein as to enable a letter of recommendation from him to open the door of Fabrice's house to his brother. If that was the case, it was yet another occasion when Samuel exerted an influence at an important stage in Mårten Triewald's life.

As Fabrice's secretary, Mårten Triewald found himself caught up in international power politics. Fabrice, who has been described as a great friend of Sweden, was in correspondence with the Swedish minister Görtz, whose diplomacy he admired.[82] In the previous year the Swedish ambassador in London, Carl Gyllenborg (1679–1746), and Görtz had conducted secret negotiations with the Jacobites, who had been trying for several years to prevail upon Karl XII to land in Britain and overthrow George I. The plans were discovered, and in January 1717 the English government had Gyllenborg arrested and held him in custody for a couple of months. For a year or two, diplomatic relations between Sweden and England were broken. In this connection it should be pointed out that Samuel Triewald knew both Görtz and Gyllenborg at this time, and that on one occasion he offered his services to Görtz. Mårten Triewald was presumably not entrusted with Fabrice's political correspondence, but he occupied a subordinate position at the centre of events during a dramatic period in Swedish foreign politics.[83]

For Mårten Triewald, the introduction to Fabrice's household was a turning point. He could now concentrate, without having to worry about keeping himself, on all the remarkable new inventions which had captured his imagination in England. He himself writes: "In the year 1717, I travelled to Oxford, Cambridge and Neumarcket, and then had the opportunity to see the finest Fuller's Earth quarries in Engeland."[84] In addition to this essay from the *Proceedings of the Royal Swedish Academy of Sciences*, the visit gave rise to another Triewald invention, later to be published in the Proceedings: "A new clay puddler especially useful for tileworks, invented by Mårten Triewald".[85] He also put his knowledge to practical use at Wedevåg Ironworks.[86] During his first years in England, Triewald appears "very briefly" to have visited Paris;[87] this was presumably during his time with Fabrice, as his attendant or courier. To Triewald, the introduction to Fabrice meant access to the social life and learned circles of London.

THE INFLUENCE OF J. T. DESAGULIERS

Triewald now met the man who was to have the greatest influence on his education, indeed on almost everything to which he later turned his hand. Laurel writes that during his time in London Triewald "had not neglected the society of learned folk, and had made the acquaintance of Desaguliers, who was giving his lectures there on natural science".[88]

John Theophilus Desaguliers (1682–1744) was the son of a French Protestant minister, who was forced to flee to England when the Edict of Nantes was revoked.[89] His father founded a school in Islington, where the boy was educated and later became his father's assistant. In

1710 Desaguliers obtained his B.A. at Oxford, and in the same year he took holy orders. At Oxford he gave lectures in "Experimental Philosophy", and in 1712 he obtained his M.A., married and moved to Channel Row, Westminster, in London, where he continued to lecture. In 1714, Desaguliers became a member of the Royal Society, where Newton is said to have formed a high opinion of him. He was elected curator without salary, but received gratuities of varying size according to the number of experiments he carried out and the number of papers he published. Desaguliers is said to have been one of the first scientists to give public lectures in experimental physics. His lectures were attended by the learned men of the age, and he held his audience rapt with skilful experiments. In 1734 Desaguliers published his lectures as *A Course of Experimental Philosophy* in two volumes (several editions). He also published many other works, in addition to a large number of essays in the *Philosophical Transactions*. In 1741–42 he was awarded the Royal Society's Copley Medal for his successful experiments.

But Desaguliers also had another side, characterized by a deep religiosity and coloured by speculative mysticism. While he was lecturing in London he was chaplain to the Duke of Chandos and later to many other persons of rank, including the Prince of Wales. He also ministered to George I. A portrait painted in 1725 shows him in clerical attire, with a magnifying glass in one hand and a prism on the table before him.[90] (The portrait gives a more flattering impression than the contemporary description of him as "short and thick-set, his figure illshaped, his features irregular, and extremely nearsighted".[91]) Desaguliers was a member of the Freemasons' lodge that met at the Rummer and Grape Tavern. Freemasons have called him "the Father of Modern Speculative Masonry", and it has been said that it was he who by his energy and enthusiasm initiated the revival of Freemasonry that took place in London in 1717 and resulted in the formation of the Grand Lodge of England. Freemasonry in its modern form derives from this particular lodge. In 1719, Desaguliers was elected Grand Master of the Grand Lodge at a meeting at the Goose and Gridiron Tavern.[92] If Triewald came into contact with Desaguliers in about 1717, then he ought also to have been in touch with Freemasonry circles. It has not been possible to establish that Triewald was a member of the Grand Lodge, but the possibility is an interesting one, because it might provide an explanation of how he was able to establish himself so quickly in England.[93]

When Triewald met Desaguliers, the latter was already a well-known lecturer and a member of the Royal Society. Desaguliers was probably the first scientist that Triewald came to know, and Triewald was to model his own life on his. Mårten Triewald was himself to give public lectures, with experiments, to publish them in book form and also to give eager support to learned societies. After he returned to Sweden, Triewald corresponded with Desaguliers, who included some of his letters in the *Philosophical Transactions*. Desaguliers may well also have been instrumental in Triewald's election to the Royal Society in 1731.

Going through the titles in Desaguliers' bibliography may suggest that Triewald obtained a number of ideas from him.[94] In papers written in 1725, for example, Desaguliers discussed cohesion between bodies of lead. It was a problem that Triewald studied, and his letters to Desaguliers containing new findings on the subject were included in the *Philosophical Transactions* in 1729–1730.[95] In 1727, Desaguliers proposed a method of extracting stale and damp air from mines, a problem that Triewald was to deal with in his five essays on coal mining in the *Proceedings of the Royal Swedish Academy of Sciences*. In 1735 Desaguliers presented a machine for indoor ventilation, and in 1742 Triewald wrote of the value and use of a "ventilating machine" devised by him for naval vessels.[96] It is quite easy to portray Triewald in such a manner as unoriginal, a plagiarist who calmly borrowed ideas, even complete projects, from others, and, indeed, this is how he has come to appear to posterity. But the picture is more complex than this; for example: Desaguliers had received the idea for the cohesion experiment in 1725 from Triewald,[97] he quoted Triewald in *A Course of Experimental Philosophy*, and, as we shall see, Triewald had already built a working "ventilating machine" at a colliery in Newcastle in 1724. The contact between Desaguliers and Triewald should be seen rather as a mutual fertilization of ideas. Mårten Triewald's scientific contributions may therefore be worthy of more detailed examination than in Anna Beckman's somewhat polemical study in *Lychnos* in 1967–68.[98]

ENGINEER IN NEWCASTLE

According to Laurel, Triewald remained in Fabrice's service until just after the death of Karl XII, when Fabrice left London and Triewald found employment with the Ridley Company.[99] Triewald himself indicates that he had already been engaged by the Ridleys by 1717.[100] Fabrice left London for a while in January 1718 to travel to Sweden and negotiate with Görtz.[101] Triewald was probably engaged by the Ridleys some time during the winter of 1717–1718.

Nicholas Ridley Sr. (?–1710) came from an old family in the troubled border country between England and Scotland. He had left the countryside and settled in Newcastle, where he purchased collieries around the town and built up an extensive business in coal. He was mayor of Newcastle in 1688, and again in 1707. Richard Ridley (?–1739), the son of Nicholas Ridley Sr., took over the business on the death of his father. He has been described as an independent and pugnacious man, who obstinately refused to be a party to agreements seeking to regulate the coal trade. He was often "scandalously short of money" as a result of his high overheads in the mines and the uncertain

Fig. 10.5. Early eighteenth century view of Newcastle-upon-Tyne. (Photo British Library)

shipments of coal, but he succeeded in building up a sizable fortune, thanks to the newly invented Newcomen engine. With the aid of steam engines he was able to keep his collieries clear of water, and thereby to mine coal at a greater depth than had hitherto been possible.[102]

This was the company into whose service Mårten Triewald entered sometime in late 1717 or early 1718. Triewald himself related how in 1717 "Messrs Ridley in New-Castle" had discussed with Thomas Newcomen and John Calley, the inventors of the steam engine, the erection of such a machine at one of the waterlogged collieries near Newcastle.[103] Only five years previously, in 1712, Newcomen and Calley had built the first steam engine, at a colliery near Wolverhampton. Newcomen and Calley were now fully occupied with other orders, and the Ridleys had to make do with John Calley's sixteen-year-old son, Samuel Calley. "This Calley", writes Triewald, "although he had been as good as brought up in a Fire-Engine, was still rather young and

for all his Practice knew not the slightest Theory".[104] Nicholas Ridley
Jr., Richard's brother, who was temporarily in London, was unhappy
about his "youthful Engineer", and feared that rivals in Newcastle
would be able to entice Calley away from him. Triewald continues:

> It therefore occurred to him through the wonderful providence of God to try
> to persuade me, whom he had known from my tenderest years, and knowing
> with what diligence and inclination I had studied Science and Mechanics in
> London, to assist his young Engineer, and to observe if he should fail to serve
> him with all Loyalty and Integrity, [he] thus persuaded me to travel to New-
> Castle promising that he wished to assist me to knowledge of the erection of
> Fire-Engines, in consideration whereof I should undertake honestly to serve
> him in return for a handsome remuneration.[105]

And so it came about that Mårten Triewald, accompanied by Nicholas Ridley Jr., travelled north to Newcastle to take up an appointment
as superintendent of one of the first Newcomen engines. Triewald, who
was now twenty-six years old, was to remain in the service of the Ridley
family for eight years. When they reached Newcastle, probably sometime during the winter of 1717–1718, the building of the "fire engine"
was in full swing, and Triewald threw himself into his task with
enthusiasm. He began by familiarizing himself with the working principle of the machine, which is simple enough, and Triewald tells us
boastfully that "as soon as I could see this Machine working, I formulated a more complete Theory of it than its own inventors possessed on
their dying day".[106]

During the period 1718–1723 Triewald was involved, according to
his own account, in the building of four Newcomen engines at the
Ridley collieries around Newcastle.[107] In his book on the Dannemora
engine he relates how he went into partnership with the young Calley
and how they drew up a contract stipulating that they should share the
profits from the erection and maintenance of steam engines.[108] The
agreement, which has been preserved in Newcastle, was concluded on
June 1, 1722 between the following four persons:

> Martin Triewald of Newcastle upon Tine in the County of Northumberland
> Gentleman of the first part Nicholas Ridley of the same place Gentleman of the
> second part Samuel Calley of Newcastle upon Tine aforesaid Gentleman of the
> third part and William Prior of Gateshead in the County of Durham Instru-
> ment maker of the fourth part.[109]

The four partners differed in background and social position. Nicholas Ridley Jr., a well-to-do mineowner and member of a family that was
known and respected in Newcastle. Mårten Triewald was a foreigner
and something of an adventurer. Samuel Calley would have been just
twenty years old, admittedly the son of one of the inventors of the steam
engine, but nevertheless still only a youth. He must have been quite
self-assured to be able to hold his own with the others. It is also possible
that he became an equal partner just because of this family connection.

The matter of patents, and the right to build Newcomen engines, was contested, and it is possible that the older partners saw the inclusion of Samuel Calley as strengthening their hand. The fourth partner, William Prior (ca. 1695–1759), was a musical instrument maker in Gateshead, a suburb of Newcastle south of the Tyne.[110] In May 1721 he became "Assay master" in Newcastle's Goldsmiths' Company. He moved into Newcastle, opened his own business, and in 1724 inserted the following advertisement in the *Newcastle Courant*:

At the Sign of the Musical Instrument in the side, Newcastle upon Tyne, liveth Wm. Prior, lately removed from Gate-side, who makes and sells all Sorts of Musical and Mathematical Instruments, Musick, Books, Tunes and Songs, Bows, Bridges and strings, and any Sort of Turn'd-work, at reasonable rates: He also makes and sells Artificial Teeth so neatly, as not to be discovered from Natural ones.[111]

One can imagine that the four partners intended to distribute their responsibilities as follows: Nicholas Ridley put up the risk-bearing capital; William Prior made the control mechanism, valves and cocks; Mårten Triewald and Samuel Calley were responsible for erecting the machines. Triewald's words about having studied the working of steam engines thoroughly and thus arrived at certain improvements are borne out by the preamble to the contract:

Whereas the said Martin Triewald has by much Study and Labour and by and with the Assistance of the said Nicholas Ridley, Samuel Calley and William Prior found out and invented a certain Machine or Engine which by the power of the Atmosphere will effectually draw the Water out of all Mines and Collieries at a very small charge.[112]

On June 29, 1722, barely a month after the contract had been signed, Triewald was granted a patent for "a certain machine or engine, which by the power of the atmosphere, would effectually draw the water out of all mines and collyeries, at a very small charge and expence".[113] The patent gives no information regarding the nature of Triewald's modifications to the Newcomen engine. John S. Allen wrote in 1977 of Triewald's patent, that:

its only novelty lies in the claim that the engine is to work by the power of the atmosphere and that Triewald had appreciated, as others were also to do, that the original Savery patent was not strictly correct for the Newcomen engine, which operated on vacuum and not "the powers of fire".[114]

During the following year, Triewald offered his services to a mine-owner on at least one occasion. In a letter of recommendation dated February 15, 1723, Richard Ridley attests to the fact that Triewald and his partners knew more about erecting steam engines than anybody else, and that nobody could be more "Careful Diligent or Faittfull", and also that they had been in the service of the Ridley family for several years.[115] The letter was addressed to a "Mr. Edw. Wite" of

Lumley Castle, who had been instructed to have a steam engine built at the collieries of his employer, Lord Scarborough, near Sunderland. It is not known whether Triewald and his companions won this order but, as has been mentioned, Triewald took part in the erection of at least four engines. Throughout this period, Triewald was in the employment of Ridley, for the accounts of the mining company include items relating to him as late as 1725.[116] He appears also to have spent the greater part of his time in Newcastle. In December 1723, he was fully occupied with his work in the collieries; in April 1724, the Scottish traveller Sir John Clerk met him in the coalfields (see below); and in the summer of the same year he was a "lamenting witness" to the deaths of thirty-one workmen and nineteen horses in a conflagration in Byker colliery.[117]

In his first essay on coal mining in the *Proceedings of the Royal Swedish Academy of Sciences*, Triewald says that although he was a foreigner, he was allowed to take part in all functions and discussions at the collieries.[118] This was because of the steam engines that he had built and of which he was in charge, work that "had so much to do with the operation of the Collieries". Triewald thus seems to have had full responsibility for the running of the Newcomen engines. When new shafts were to be sunk or new headings driven, it was he who had to see that they were kept free from water by connecting up the engines to new pumps. This assumption is borne out by Triewald's own statement that he often had to go underground: "for I have myself more than once seen the lamp in my hand go out when I have been down the coal mines".[119]

In 1727, as we have seen, Desaguliers proposed a way of drawing stale, damp air out of mines. This was a problem that Triewald had had cause to study during his eight years in Newcastle, and there is evidence that by 1724 he had designed a practical functioning ventilating machine. In April 1724, Sir John Clerk (1684–1755) visited the colliery district of Newcastle.[120] He had recently inherited several collieries, and was eager to study mining methods. In the diary of his journey he relates that he has visited "the 3 coalries of Alderman Ridley about a mile on the est side of the Town [...] The 3 coaleries of Alderman Ridley are manadged with 3 fire Engines".[121] This was at Byker, where Triewald took part in the building of at least one of the Newcomen engines and probably all of them. Clerk records his reflections on the machines, and continues:

> Alderman Ridley keeps always some Engineers about him who take care of his coalworks; one of them a Swede called Mr Denald I was acquainted with. I had occasion to talk 3 or 4 houres at night with this man as to all parts of coalery.[122]

The Swede, "Mr. Denald", was without doubt none other than Triewald, the English pronunciation of whose German name may have been misheard by the Scottish traveller. During the conversation that evening, Sir John told Triewald of the difficulty of bringing coal up from the Scottish pits, and the two talked over different methods of overcoming the problem. This brought them onto the question of providing ventilation underground.

Mr. Denald explained to me the methode which he used to drain ill aire out of Alderman Ridley's works & likeways to carry a mine a great way without the expence of letting down a shaft.[123]

Triewald told him that he had had a furnace built at the pithead. From the draughthole he ran a wooden pipe down the shaft and along to the end of the heading. The effect of this was that when a fire was made in the furnace, air was drawn in through the pipe at its opening, which caused fresh air to be drawn down the shaft to the end of the heading. Sir John relates that the pipe was put together of planks, but that it could equally well have been made of lead piping, and that it could be extended as the heading progressed. The cost of maintaining the fire in the furnace was "very inconsiderable" when a ready supply of waste coal was available. The editor of Sir John Clerk's travel diary observes that the method employed by Triewald is virtually the same as that used in Belgian collieries near Liège as described in a report in the *Philosophical Transactions* of the Royal Society in 1665.[124] It is therefore possible that as early as in the 1720s Triewald's studies included a perusal of back numbers of the *Philosophical Transactions*, and that it was from them that he obtained the idea of his ventilating machine.

During his years at the collieries, with their oppressive and sometimes noxious atmosphere, he had ample opportunity to study and to speculate upon the properties and the effects of the air. And many of his later "inventions" were to be concerned with that field: the diving bell, the ventilating machine and the bellows. The well-known portrait in the Royal Swedish Academy of Sciences depicts him beside an air pump (Fig. 10.6). In the receiver of the pump a doomed bird is vainly flapping its wings, as it tries to rise in the vacuum.[125]

SCIENTIFIC LECTURER: "HAVE INSTRUMENTS, WILL TRAVEL"

All of Triewald's biographers have asserted that he had already made a name for himself as a man of science while he was in London—because he won the favour of no less a personage than Newton himself during that period. As far as intellectual contact is concerned, vague claims of this kind can be totally misleading. Does it mean that Triewald and Newton had long and detailed discussions about physics alone together? Or does it merely mean that on some occasion he had the chance to sweep off his hat and bow to Newton? The story of Triewald's contact with Newton has its origin in Triewald's own account. In the book on the Dannemora engine he writes:

as soon as I could see this Machine working, I formulated a more complete Theory of it than its own inventors possessed on their dying day, which even the great Newton confirmed before the Secretary of the Commission Mr Skutenhielm. And confessed that he had never been able to gain a clear picture of the Fire-Engine from the Inventors, because they always ascribed the power to Steam, which is only the means by which the Force is obtained.[126]

Anders Skutenhielm (1688–1753) was secretary to the Swedish envoys in London, Carl Gustaf Sparre and Carl Gyllenborg. He took up his post in the summer of 1719, so the claimed conversation must have taken place later than this.[127] We have evidence that Triewald and Skutenhielm knew each other in 1723.[128] Mårten Triewald was in London in the spring of 1719, when he attended Desaguliers' physical experiments in the dome of St. Paul's Cathedral,[129] and in 1723 he sailed along the Thames from London to Gravesend.[130] It is therefore probable that he visited London a number of times during his eight years in Newcastle. When he did, he would certainly not miss the opportunity to attend scientific gatherings, and perhaps he met Newton on one of these occasions. In the preface to his published lectures, Triewald wrote: "I soon made the acquaintance of the most eminent and literary of men, in particular that of Sir Isaac Newton, who so willingly and kindly helped me with word and deed."[131] The expression "in particular" should here be interpreted as referring to the degree of Newton's eminence rather than the intimacy of Triewald's acquaintance. Triewald had a rare capacity for presenting himself —without downright departures from the truth—in the most advantageous light. That at the age of almost eighty Sir Isaac Newton would have been so very impressed by the young Mårten seems unlikely. But Triewald was a faithful admirer of Newton all his life. He was, says Lindroth, an ardent Newtonian.[132] When only eighteen, Triewald had had the opportunity to acquaint himself with Newton's scheme of the universe, in Samuel Worster's library.[133]

Triewald's position with Ridley gave him time to continue the studies he had commenced in London. "One saw him no longer. Now he sat night and day in his reading room."[134] Whether the first generation of builders of steam engines possessed theoretical knowledge or were merely practical individuals with no formal training who proceeded by trial and error is a moot point. Triewald belonged to this first generation, and if he can be regarded as representative, the answer is that they were a combination of both.

Triewald lacked higher education; indeed, he had declared his contempt for the pedantry of the universities. He was active for many years as a practical builder of steam engines and, as has been shown, he often went down the mines himself. But he also had a strong desire to study, and to assimilate the theoretical background to his trade. His prolific writings and the excellent library that he left bear witness to the amount that he read in later life,[135] but even while he was a mining engineer he devoted himself to wide-ranging and conscientious study. An illustration of this is provided by the fact that as early as 1722 Triewald acquired Robert Boyle's *Experimentorum novorum physico-mechanicorum*. (This copy of Boyle's book is now in the library of the Royal Institute of Technology, Stockholm, and bears Triewald's signature on the title page, together with the date "1722/23".[136]) Triewald himself also provides a glimpse into his studies and interests during this period in a letter to Henrik Kalmeter (see Chapter 7) from Newcastle, dated December 6, 1723.[137] Triewald thanks Kalmeter for sending him a book by Réaumur, and says: "I have had but just Leisure to look into Monsr. Reaumur's book, & the little I have read has given me no small Satisfaction." The book must surely have been *L'art de convertir le fer forgé en acier*, which was published in Paris in 1722.[138] It is interesting that Triewald could have Réaumur's book in his hands only the year after it had been published in Paris. But Newcastle was not isolated as far as news of recent publications in science was concerned. The local newspaper, the *Newcastle Courant*, published a weekly list of 10–20 books "lately publish'd in London", including scientific works.[139] Triewald appears to have mastered French and Latin, as well as English and his native languages German and Swedish. (He shared this aptitude for languages with his brother Samuel, who spoke and wrote nine.)

Triewald did not study merely for the sake of studying. When after several years of self-tuition he considered that he had mastered the new natural science, he wished to follow Desaguliers' example by giving public lectures illustrated with experiments. In the preface to his published lectures, Triewald tells us that the sciences have aroused such enthusiasm in England "that their pursuit cannot be confined within the walls of the two universities, but in the Capital itself, in more than one place, and also in most of the largest provincial cities, are even more men of skill, who as it were compete in lecturing and in teaching them in their mother tongue; for which there is commonly a large audience".[140] John L. Heilbron writes in his study of early modern physics, *Electricity in the 17th and 18th Centuries*, of these independent lecturers:

Their chief goal was popularization and entertainment, the reduction of the latest discoveries to the level of "the meanest capacities" able to afford the service. One offered to explain everything "in such a plain, easy and familiar Manner, as may be understood by those who have neither seen or read anything of the like Nature before". The meanest capacities could choose from a wide range of purveyors. At the top were the public lecturers associated with learned societies [...] A second class of lecturer consisted of members of learned societies who set up independently of their institutions. A notch lower, perhaps, came the unaffiliated entrepreneurs, who taught in rented rooms, and

Fig. 10.6. Mårten Triewald. Painting attributed to Georg Engelhardt Schröder. (Photo Swedish Portrait Archives)

the itinerant lecturers, who performed in public houses. At the bottom of the heap were the hawkers of curiosities, the street entertainers, and the jugglers who held forth at the fairs of Saint Laurent and Saint Germain.[141]

It was in the third class, as an unaffiliated entrepreneur who taught in rented rooms, that Mårten Triewald began his career as scientific lecturer. It seems obvious that when Triewald returned to Sweden he must have already possessed extensive lecturing experience, when one reflects that in 1728 he gave lectures at the Palace of Nobility that were followed not only by personages such as the physicist Anders Celsius

but by all the society of Stockholm: by councillors, generals and diplo-
mats. To lecture and perform experiments before an audience such as
that facing Triewald at the Palace of Nobility called for experience and
confidence. Triewald was confident, almost to the point of being cock-
sure, but he also had the experience. It has been known that Triewald
claimed to have lectured in Edinburgh, but no details of these lectures
have been published.[142] It is now established that Triewald did indeed
lecture there in the winter of 1724 (and probably also in the spring of
1725), and that his well-known lectures in experimental physics in
Stockholm in 1728 can almost be regarded as a repeat of this course.
But his career as a lecturer in science had begun earlier in 1724. The
Newcastle Courant of July 18, 1724, contained the following advertisement:

> NOTICE is hereby given to all such Persons as have subscribed to a
> COURSE of NATURAL and EXPERIMENTAL PHILOSOPHY, to be
> performed at Newcastle upon Tyne by JOHN THOROLD and MARTIN
> TRIEWALD, That the said COURSE will begin on THURSDAY the 23d
> Day of this present Month of JULY, at Three of the Clock in the Afternoon, at
> the NEW ACADEMY next Door to the MAYOR's House in the CLOSE, and
> will be continued there Weekly on every THURSDAY until the Ending of the
> same COURSE, and all Persons that are desirous to be present at the
> COURSE aforesaid, may subscribe before that Time, but not afterwards.[143]

What these lectures covered is not known, but it is reasonable to
assume that they were roughly the same as those he gave later in the
same year in Edinburgh and a few years later in Stockholm. Triewald's
lectures at the Palace of Nobility in Stockholm would appear therefore
to have had their origin in the series that he and John Thorold (see
note) gave "at the New Academy next door to the Mayor's house" in
Newcastle upon Tyne, starting on July 23, 1724.[144] In Stockholm, the
merchants of the city made up a large section of his audience (see
Chapter 13), and similarly it was in the commercial city of Newcastle
that he was first able to attract a sufficient number of persons "desirous
to be present at the Course". The lectures seem to have been a success,
for in December the pair held a course in Edinburgh. This is clear from
the advertisement in the *Caledonian Mercury* (Fig. 10.7). The whole of the
back page of the newspaper is taken up by the advertisement in which
they announce their lectures—and not just once, but in every number of
the newspaper for three weeks in March and April.[145]

Triewald and Thorold had given a series of twenty-four lectures, with
three lectures a week, beginning in December 1724. The first series
must have lasted into the month of February. In response to popular
demand, they announced a second series of lectures which would begin
in April. It is not known whether this second course was also held; if the
number signifying their desire to attend is not sufficient, the advertisement concludes "so as the Proposers may meet with a proper Encour-

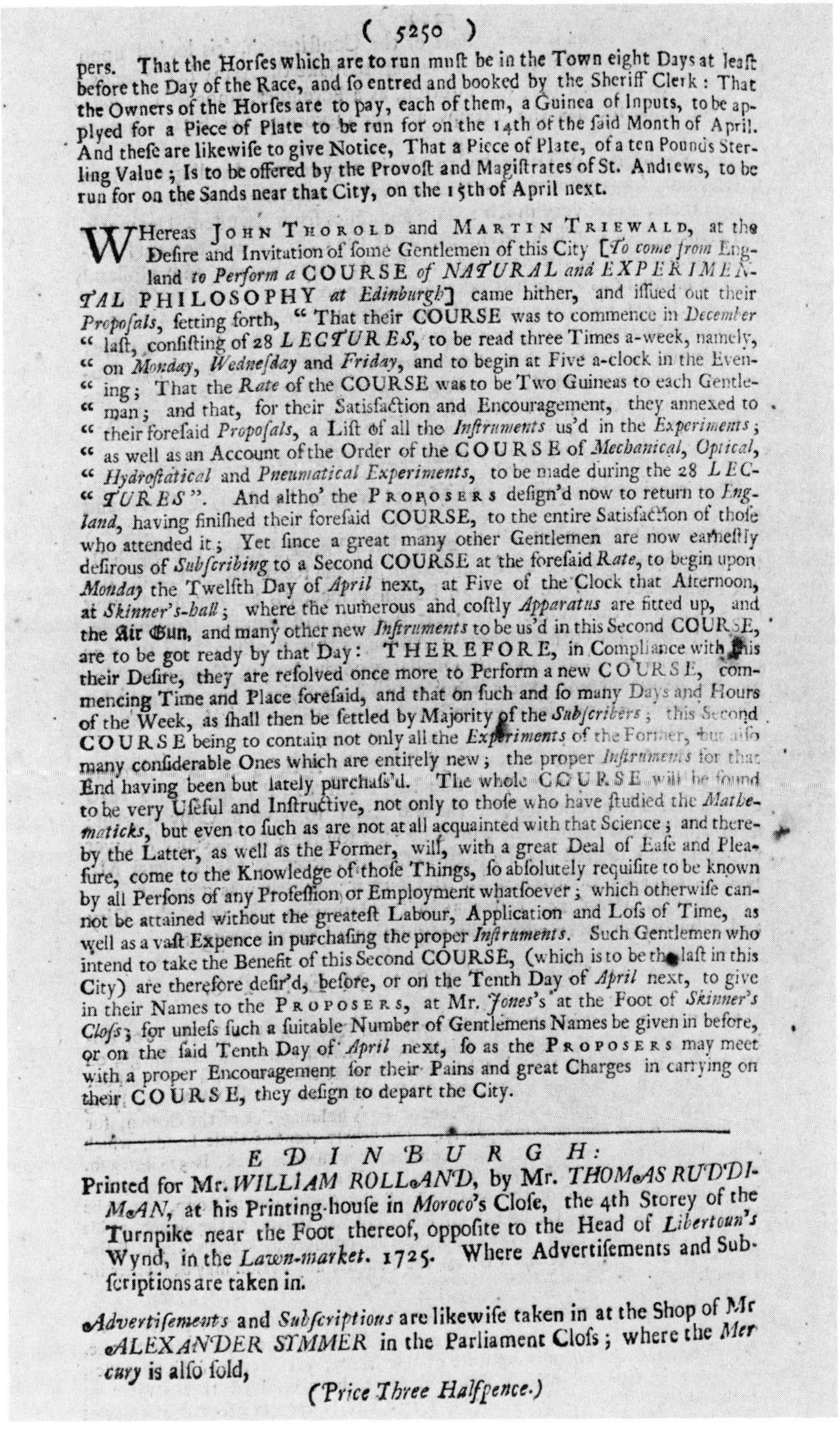

(5250)

pers. That the Horfes which are to run muft be in the Town eight Days at leaft
before the Day of the Race, and fo entred and booked by the Sheriff Clerk : That
the Owners of the Horfes are to pay, each of them, a Guinea of Inputs, to be ap-
plyed for a Piece of Plate to be run for on the 14th of the faid Month of April.
And thefe are likewife to give Notice, That a Piece of Plate, of a ten Pounds Sterling Value ; Is to be offered by the Provoft and Magiftrates of St. Andrews, to be
run for on the Sands near that City, on the 15th of April next.

Hereas JOHN THOROLD and MARTIN TRIEWALD, at the
Defire and Invitation of fome Gentlemen of this City [*To come from England to Perform a* COURSE *of* NATURAL *and* EXPERIMEN-
TAL PHILOSOPHY *at Edinburgh*] came hither, and iffued out their
Propofals, fetting forth, " That their COURSE was to commence in *December*
" laft, confifting of 28 *LECTURES*, to be read three Times a-week, namely,
" on *Monday, Wednefday* and *Friday*, and to begin at Five a-clock in the Even-
" ing; That the *Rate* of the COURSE was to be Two Guineas to each Gentle-
" man; and that, for their Satisfaction and Encouragement, they annexed to
" their forefaid *Propofals*, a Lift of all the *Inftruments* us'd in the *Experiments* ;
" as well as an Account of the Order of the COURSE of *Mechanical, Optical,*
" *Hydroftatical* and *Pneumatical Experiments*, to be made during the 28 *LEC-*
" *TURES* ". And altho' the PROPOSERS defign'd now to return to *England*, having finifhed their forefaid COURSE, to the entire Satisfaction of thofe
who attended it ; Yet fince a great many other Gentlemen are now earneftly
defirous of *Subfcribing* to a Second COURSE at the forefaid *Rate*, to begin upon
Monday the Twelfth Day of *April* next, at Five of the Clock that Afternoon,
at *Skinner's-hall* ; where the numerous and coftly *Apparatus* are fitted up, and
the Air Gun, and many other new *Inftruments* to be us'd in this Second COURSE,
are to be got ready by that Day : THEREFORE, in Compliance with his
their Defire, they are refolved once more to Perform a new COURSE, com-
mencing Time and Place forefaid, and that on fuch and fo many Days and Hours
of the Week, as fhall then be fettled by Majority of the *Subfcribers* ; this Second
COURSE being to contain not only all the *Experiments* of the Former, but alfo
many confiderable Ones which are entirely new ; the proper *Inftruments* for that
End having been but lately purchafd. The whole COURSE will be found
to be very Ufeful and Inftructive, not only to thofe who have ftudied the *Mathe-
maticks*, but even to fuch as are not at all acquainted with that Science ; and thereby the Latter, as well as the Former, will, with a great Deal of Eafe and Plea-
fure, come to the Knowledge of thofe Things, fo abfolutely requifite to be known
by all Perfons of any Profeffion or Employment whatfoever ; which otherwife cannot be attained without the greateft Labour, Application and Lofs of Time, as
well as a vaft Expence in purchafing the proper *Inftruments*. Such Gentlemen who
intend to take the Benefit of this Second COURSE, (which is to be the laft in this
City) are therefore defir'd, before, or on the Tenth Day of *April* next, to give
in their Names to the PROPOSERS, at Mr. *Jones's* at the Foot of *Skinner's
Clofs* ; for unlefs fuch a fuitable Number of Gentlemens Names be given in before,
or on the faid Tenth Day of *April* next, fo as the PROPOSERS may meet
with a proper Encouragement for their Pains and great Charges in carrying on
their COURSE, they defign to depart the City.

E D I N B U R G H :
Printed for Mr. WILLIAM ROLLAND, by Mr. THOMAS RUDDI-
MAN, at his Printing-houfe in *Moroco's* Clofe, the 4th Storey of the
Turnpike near the Foot thereof, oppofite to the Head of *Libertoun's*
Wynd, in the *Lawn-market*. 1725. Where Advertifements and Sub-
fcriptions are taken in.

Advertifements and *Subfcriptions* are likewife taken in at the Shop of Mr
ALEXANDER SIMMER in the Parliament Clofs ; where the *Mercury* is alfo fold,

(*Price Three Halfpence.*)

Fig. 10.7. The first advertisement for the second series of lectures by John
Thorold and Mårten Triewald in Edinburgh on the back page of the *Caledonian
Mercury* of March 25, 1725. (Photo National Library of Scotland, Edinburgh)

agement for their Pains and great Charge in carrying on their Course, they design to depart the City". The fee for the series was two guineas.[146]

The series of lectures dealt with "Natural and Experimental Philosophy", and included mathematical, optical, hydrostatic and pneumatic experiments. This means that Triewald was already dealing in 1724 with the four branches of physics that he discussed in his lectures in Stockholm in 1728. The audience received a programme of the contents of the lectures and a list of all the apparatus and instruments that were used in the experiments. For the second series they were promised, in addition to the "numerous and costly Apparatus", such as "the Air Gun", that had been used in the first series, experiments with new apparatus, which the lecturers had purchased since. The series of lectures, the lecturers promised, would be of great benefit to all:

> The whole Course will be found to be very Useful and Instructive, not only to those who have studied the Mathematicks, but even to such as are not at all acquainted with that Science; and thereby the Latter, as well as the Former, will, with a great deal of Ease and Pleasure, come to the knowledge of those Things, so absolutely requisite to be known by all Persons of any Profession or Employment whatsoever; which otherwise cannot be attained without the greatest Labour, Application and Loss of Time, as well as vast Expence in purchasing the proper Instruments.[147]

In a note on his published lectures, Triewald observed: "When I went to Edenburg to hold public Lectures, I had the good fortune to find there one or two listeners who had been to 'sGravesande's lectures in Leiden."[148] He implied, in other words, that it was an experienced audience that he encountered in Edinburgh, "the Athens of the North". But Triewald also makes it clear that even if his audience was a learned one, it was to him, Mårten Triewald, that they turned in order to have the wonders of science explained:

> When I gave my Lectures in Edenburg in the Year 1724, one of my Audience asked the Question: how does it come about, that when [...]? How does this agree with the Principles of Mechanics? I answered this Question like this, saying that when ...[149]

By 1724, therefore, Triewald had built up his collection of equipment: his earnings as an engineer in the Newcastle collieries were used to purchase books and scientific instruments. By 1725, he had given at least two, maybe three and possibly more, series of public lectures in experimental physics in Newcastle and Edinburgh. He had now established himself as a travelling scientific lecturer.

THE RETURN TO SWEDEN: REASONS AND TIMING

Daniel Triewald, Mårten's youngest brother, had taken part in the siege of Fredrikshald as an ensign in the Uppland Infantry.[150] In 1721,

Fig. 10.8. Martin Triewald, father of Mårten Triewald. Painting by unknown artist in 1736. (Photo Nordiska museet, Stockholm)

as a lieutenant, he left the regiment and, driven by the same wanderlust as his eldest brothers, sailed for the next few years on merchant vessels to the Netherlands, England and France. In 1724 he became a skipper. On May 16, 1726, his ship dropped anchor in Stockholm Harbour.[151] The shipping records of the City of Stockholm state that "Daniel Triewald comes from NeuCastell", and there follows a list of the cargo and its owners. Among the items is "1 200 barrels of coal" for Mårten Triewald.[152]

This was almost certainly the ship by which Triewald returned home, and the date fits well.[153] He had probably arranged for his youngest brother to come to Newcastle and fetch him, together with his collection of instruments, his library, and his savings invested in New-castle coal. He was also bringing with him knowledge, of subjects ranging from the Newcomen engine to the best way of establishing stocks of bees. He was bringing with him ideas, such as his plan to give public lectures in science in his native tongue and his dream of setting up an Academy of Science in Stockholm on the model of the Royal Society in London. But what was it that made him decide to return at this particular time?

In 1716, Triewald had been a penniless foreigner in London but by about 1725 he had attained a certain standing in England. He held a royal patent, he was affluent enough to buy books and a comprehensive collection of scientific instruments, and he had given public lectures in science in Newcastle and Edinburgh. While Mårten Triewald had been abroad, his family at home had also been climbing socially. His father, an immigrant from Germany in the 1680s, had at first had little status among his colleagues in the Stockholm Guild of Blacksmiths. But by now he had advanced to a high position in the Guild, and in 1727 he became Alderman.[154] In 1731 he would represent the burghers of Stockholm in Parliament.[155] Samuel Triewald, Mårten's eldest broth-er, had continued his career as a politician and a public servant. In 1723 he decided that it was time to apply for elevation to the nobility (which appears not to have been granted, however).[156] He left his position in the Chancery in 1725, and went over to the service of the Holstein faction, where he later rose to be a supreme court justice in 1732 and a minister in 1738. The Triewald family in Stockholm thus enjoyed much greater esteem in the mid-1720s than it had in 1713, when Mårten Triewald had left Sweden.

In the previous chapters we saw how de Valair obtained a Swedish patent on the Newcomen engine in 1723 and how discussion of the proposals of O'Kelly led to the patent being declared invalid in 1725. As has been mentioned earlier, Triewald was watching these develop-ments in Sweden from his vantage point in Newcastle. On September 24, 1723, Henrik Kalmeter had written to Jonas Alströmer:

> Mr Triewald goes to Sweden next spring, he is now settled here and in partnership with the son of the inventor of the fire-Engine.[157]

But Triewald did not return to Sweden in the spring of 1724, and it was probably de Valair's patent that caused him to postpone his plans. In a later letter, in which Kalmeter wondered whether it would be possible to "get a Man over to Sveden, that could erect a fire Engine there of the same kind as at NewCastle and otherwhere", he wrote that "Mr. Triewald, when he goes there, can best bring it about".[158] Triewald himself was also aware that he possessed unique knowledge of the Newcomen engine, for he later wrote:

Fig. 10.9. Samuel Triewald, eldest brother of Mårten Triewald. Painting by Georg Engelhardt Schröder. (Photo Swedish Portrait Archives)

I am the first and only foreigner to have gained free access in England to this magnificent invention.[159]

The timing of Triewald's return to Sweden in the spring of 1726 was therefore probably determined by the fact that by the autumn of 1725 it was clear that de Valair's patent was invalid. No doubt he had also received news of the fresh winds that were beginning to blow in Sweden as the Era of Liberty opened; if nowhere else, he would have been able to read in the *Newcastle Courant* of the privileges and rewards now offered in Sweden to foreigners possessing "any Talents for Arts, Manufactures and Trade".[160] As this chapter has shown, Triewald also possessed two of the essential qualifications of anyone wanting to build a Newco-men engine in Sweden: practical technological know-how, together with high social status and a proven knowledge of science. Only one qualification was lacking, namely the backing of a private group with the means of realizing the project. Just a few months after his return, however, he had succeeded in finding support for his idea from such a group, the Partners in the Dannemora Mines.

11. The Social Context: The Dannemora Mines

We will go on to describe the mines that may be counted among the richest, the most renowned and the most plentiful in Europe [...] These mines are thought to be the most extensive in the whole of Sweden; but they are unquestionably those which yield the finest iron.[1]

With these words, the French metallurgist Antoine Gabriel Jars (1732–1769) introduced his description of the Dannemora Mines in his *Voyages métallurgiques*, published in 1774. The Dannemora Mines are situated 50 kilometres north of Uppsala in the province of Uppland (Fig. 11.1).[2] The deposits are still worked, as they have been since the fifteenth century and probably much earlier. They were known to Georg Agricola, who mentions them in 1546.[3] No other Swedish iron mine contained ore of a quality equal to that of Dannemora, and bar iron made from Dannemora ore fetched a higher price on the Continent than any other Swedish iron.[4] In England, iron from the Dannemora Mines was already being used in the seventeenth century as the finest material for steel manufacture, and the importance of Dannemora iron in the expansion of the English steel industry during the eighteenth century has often been pointed out.

In the early seventeenth century, the ironworks of Uppland had been taken over by Dutch businessmen, who brought capital and technical know-how to the Swedish iron industry.[5] The ironworks were now extended and several new ones were established. During the period with which we are concerned, 1726–1736, there were seventeen ironworks producing bar iron from the ore of Dannemora.[6] The largest of them were Leufsta, Österby and Forsmark. Many of these ironworks were ravaged by the Russians in 1719, during the Great Northern War.[7] The Board of Mines put the total loss to the ironworks in Uppland at more than one and a half million *copperdaler*.[8] (The Board's report was written by Mine Inspector Göran Wallerius, whom we have met several times before and will meet again in Chapter 14.)

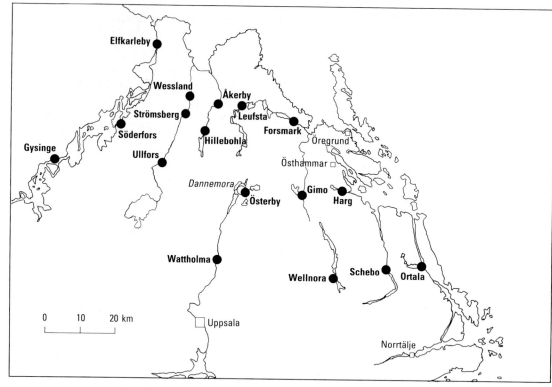

Fig. 11.1. Map of northern Uppland showing the location of the Dannemora Mines and the seventeen ironworks that were producing bar iron from the ore of Dannemora in the period 1726–1736.

Operations at the ironworks were highly centralized, a fact which is still reflected today in the strictly organized lay-out of the eighteenth-century buildings.[9] In sharp contrast to this was the decentralized operation of the Dannemora Mines, which have been described as "a monstrosity in the intricacy of their ownership and administration".[10] At this time the seventeen ironworks obtained their iron ore from twenty of the sixty or so mines that have been worked in Dannemora at one time or another.[11] Each mine was owned by several ironworks jointly, and the ironworks held differing numbers of shares in the various mines. Our main interest is in understanding the relationship between the social order and the energy supply situation of the mines. The most important aspect of this subject is the way the actual mining of the ore was organized. The ironworks were entitled to as many weeks' mining as they had shares in the mine concerned.[12] Each one had its own mine bailiff, who paid the workmen to dig ore during these weeks. The next mine bailiff then took over the work in the mine and dug ore for the duration of his employer's weekly shares. When all the Partners in the mine had worked the mine for a period corresponding to the number of their shares, the first mine bailiff began the cycle again.

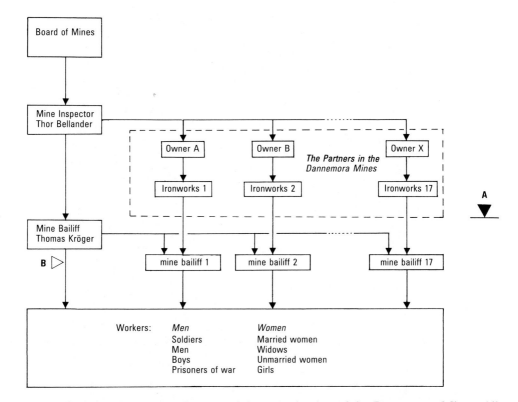

Fig. 11.2. Schematic diagram of the organization of the Dannemora Mines. All the organizational units below level *A* were located in Dannemora, while all those above level *A* were elsewhere. (*B* refers to a discussion in Chapter 15.)

A schematic diagram of the organization of the Dannemora Mines is shown in Fig. 11.2, and we will describe the different units, starting at the top. As everywhere in the mining industry, the Board of Mines was at the summit of the hierarchy. In the local organization of the Board the Dannemora Mines formed part of the 7th Mining District (*7:e Bergmästaredömet*), which covered the provinces Uppland, Gästrikland, Hälsingland and the western part of Norrland.[13] The Mine Inspector was the local representative of the Board in the district. During the period with which we are dealing, the Mine Inspector was Thor Bellander (1688–1740), then in his late thirties and early forties.[14] He lived in Gävle, a town 70 kilometres north of Dannemora, but visited the mines several times each year for the meetings of the Partners and the annual Court of Mines (*Bergsting*) in May or June. He chaired the meetings and presided at the Court. As the representative of the State, he was higher in authority than the owners of the ironworks, although the latter included both generals and privy councillors.

Bellander fits the description of a typical Board of Mines official that was given in Chapter 5. He was the son of an unusually well educated vicar of Ängsö, in the province of Västmanland, and had been to school

Fig. 11.3. Countess Eva Horn, owner of Salsta Manor and Wattholma Ironworks. Portrait by an unknown artist. (Photo Swedish Portrait Archives)

at Skara, one of Sweden's oldest gymnasiums. He studied at the
University of Uppsala from 1706 until 1714, when he was appointed
Auscultator in the Board of Mines. He rose in the hierarchy of the
Board, and became a Mine Inspector in 1723. In the same year, he
married the daughter of an Assessor in the Board, and in due course
their son became a Mine Inspector, too. The Board of Mines generally
formed a tightly-knit society, with frequent family and kinship ties of
this kind. A little episode that allows us to glimpse more of the man,
rather than the official, occurred in September 1729, on one of Bellander's visits to Dannemora to discuss Triewald's Newcomen engine

Fig. 11.4. Salsta Manor, 20 kilometres north of Uppsala and close to Wattholma Ironworks. Engraving by Wilhelm Swidde in 1693–1694 in *Suecia antiqua et hodierna*, published in 1716. (Photo Royal Library, Stockholm)

(see Chapter 13). While at Dannemora, he forgot or mislaid a treasured possession. The Mine Bailiff promised in a letter after Bellander's departure that he would keep on looking for "Pufendorf's history".[15] The well-known German political scientist Samuel von Pufendorf (1632–1694) had spent two decades in Sweden as official historian,[16] and it was one of his works that Bellander had lost at Dannemora. On his official journeys around the district the Mine Inspector read modern history in his spare time.

During the seventeenth century, the ironworks were largely owned by the noble families de Geer, de Besche and Bielke. But as a result of marriage and division of inheritances, the original shares were distributed so that by the beginning of the eighteenth century there were usually several shareholders in each ironworks. During the eighteenth century many of the ironworks were purchased by the large trading companies in Stockholm, but at the start of the century there were still a number of owners of seventeenth-century type who personally involved themselves not only in the operation of the ironworks but also in the daily work of the Dannemora Mines. One such owner was Countess Eva Horn (1653–1740, Fig. 11.3), the proprietor of Wattholma Ironworks.[17] Little is known about her, but as with so many women in history, we can get an inkling of the main features of her life from the biography of her husband.[18]

Eva Horn was the daughter of Count Gustaf Horn (1592–1657), one of Gustav II Adolf's generals during the Thirty Years' War.[19] At the age of sixteen she married Nils Bielke (1644–1716), who was then twenty-five. It has been said of Eva Horn that she lived through all the ups and downs of her husband's career with an unshakable devotion.[20] The wealth that she brought to the marriage gave Bielke the opportunity to make a career at court and in the military. He rose to the rank of lieutenant-general and became a trusted friend of Karl XI. As Swedish envoy he resided at the court of Louis XIV of France between 1679 and 1682. In 1687 he became Governor General of Pomerania, then a Swedish province. In 1698, however, Bielke's involvement in European politics led to his being accused of high treason and stripped of all office and honours. After a long trial he was sentenced in 1705 to loss of both life and property, but his life was spared although he was banished from residence in Stockholm. He spent his remaining years at Salsta Manor, which was owned by his wife and therefore not affected by the confiscatory order.[21] Nils Bielke died in 1716, and Eva Horn survived her husband by twenty-four years. During the period which concerns us she was in her seventies, and lived as a widow at Salsta (Fig. 11.4). The estate included the neighbouring Wattholma Ironworks. Many of her letters to the other Partners, the Mine Inspector and the Board of Mines have been preserved; in them she vigorously defended her rights and attempted to safeguard Wattholma's supply of iron ore.[22] Through her agents she also kept herself informed on the project to build a Newcomen engine at the Dannemora Mines.

The eighteenth century saw the emergence of a type of ironworks proprietor who was to play an important part in Sweden's cultural development.[23] This was, generally speaking, the first social class outside the aristocracy that had the time and means to pursue cultural interests. The mansions of these new ironmasters often contained impressive libraries, where the classics rubbed shoulders with modern works of science and the latest currents of thought from the Continent. Literary and musical salons were also held. The seventeenth-century type represented by Eva Horn is in marked contrast: belonging to the higher nobility and, unlike the bourgeois eighteenth-century proprietor, closely connected with the court; much travelled in connection with military campaigns and political missions, but not among the mines, commercial cities and universities of Europe. It is important to point out that this older type of ironworks proprietor was not conservative in his (or her) attitude to technological change. On the contrary, during the seventeenth century the Swedish mining industry had successfully introduced a number of new technologies from the Continent, e.g. *Stangenkunst* from Germany, French blast furnaces ("*Fransöske masugnar*") and Walloon forging from the Netherlands. These transfers of technology were however the result of rational assessments of their economic and technical value rather than technological changes based on the optimistic faith in progress and belief in science that were to be characteristic of the eighteenth century. When, shortly after his return home from England, Mårten Triewald hurried up to North Uppland with his proposal to the Partners in the Dannemora Mines, it was a meeting of two epochs.[24]

At the meetings of the Partners in the Dannemora Mines, the owners of the ironworks were normally represented by agents.[25] These were the men employed to be responsible for the running of the works, and they had the title of *Directeur* or *Inspector*. If they were absent, the Head Clerk (*Bokhållare*) might deputize, but it was equally common for another *Directeur* or *Inspector* to represent his colleague by proxy in those cases where there were matters of common interest to be dealt with and the interests of the different ironworks coincided. The composition of a meeting of the Partners could therefore vary very considerably, but, as we shall soon see, we may nevertheless treat the Partners as a single entity.

The local representative of the Mine Inspector at the Dannemora Mines was Head Mine Bailiff (*Crono Grufwefogde*) Thomas Kröger.[26] Little is known of him other than that he was appointed in 1707 and retired in 1737. He was probably in his sixties during the period we shall describe. He lived with his wife in a comparatively large house in the middle of the mine area (it is visible in the lower left corner of Fig. 11.7). All the organizational units below level *A* in Fig. 11.2 were located in Dannemora, while all those above level *A* were elsewhere. This means that Mine Bailiff Kröger was highest in rank at Dannemora, and that he was in charge of the day-to-day work of the mine. Kröger's main task was to see that work proceeded in accordance with the statutes issued by the Board of Mines (see Chapter 5). As has been mentioned, each ironworks had its own mine bailiff, who worked the mines under the weekly share system. Each mine bailiff sought only to take out his permitted quota of ore as quickly as possible, with no heed for the difficulties he might create for the one who followed. The result was, as Johan Wahlund observed in 1879 in his history of the Dannemora Mines, that the mines were worked badly and that more wood

Fig. 11.5. Working conditions in an eighteenth-century Swedish mine. Painting by Pehr Hilleström of the Great Coppermine in Falun in 1784. (Photo Stora Kopparbergs Bergslags AB)

was used on fire-setting than would have been the case if each mine had been worked by one party.[27] The mine bailiff of each ironworks had his own building at the mines,[28] and the whole area around the mines was probably much more cluttered than would appear from the idyllic view in Fig. 11.7: the homes of the officials, stores, smithies, carpenters' workshops, stables, stacks of wood, mounds of stone and ore etc. Nor can the map recreate the thronging life that must have been characteristic of the scene: workmen, horses, consignments of firewood and ore. As elsewhere, says Sven Rydberg, the bustle of activity attracted a large casual population to the place, particularly sellers of beer and spirits and other traders.[29] According to an earlier Mine Inspector there were also many cases of "fire raising, wood stealing, dealing in beer and spirits and other improprieties".[30]

Work began at five in the morning and finished at six in the evening, with an hour for breakfast and one for dinner.[31] There were prayers at the mine both morning and evening. The workmen bound themselves to work for a year at a time, from Michaelmas to Michaelmas, and had fourteen days' leave at harvest time. There are no definite details of the number of workmen, but a total of about two hundred seems probable. Soldiers and women also worked in the mines. It was not unusual for soldiers to be assigned to Swedish mines to alleviate the perpetual shortage of labour, because mining was considered of national importance. The horse whims were tended, as at other mines, by boys. At the bottom of the hierarchy in the Dannemora Mines came prisoners taken in the Great Northern War. That prisoners of war were extensively used in Swedish mines is on record from the sixteenth century.[32] There are many accounts of their grim working and living conditions, but their names and totals are seldom included in the population registers. They were long recorded as something of an expendable labour resource. They had, as King Johan III had said in 1572, "in any case forfeited their lives".[33] In 1706 the Mine Inspector wrote to the Mine Bailiff at Dannemora and told him that the Government had consented to the use in the mines of Saxon prisoners taken the same year at the Battle of Fraustadt, where the Swedish army had captured 7 600 of the enemy.[34] This was a great advantage, commented the Mine Inspector:

since it is unnecessary to expend upon them more than food and cheap working clothes. As there is always a shortage of workmen at the Mines, the Head Mine Bailiff shall immediately discuss with the mine bailiffs from the ironworks how many men may be needed and desired.[35]

The shortage of labour referred to here shows the importance of manual labour in the mines. This in turn indicates that the traditional sources of power—water, wind and horses—were insufficient. The Saxon prisoners of war were removed from the province of Uppland in 1707 and replaced by Danish and Russian prisoners in 1709, 1710 and 1715.[36] The Danes were exchanged in 1713, but the Russians remained until the end of the Great Northern War in 1721. Thus, in the 1720s, the Partners had to make the best of the domestic population, and the following section will describe an attempt to reduce the difference between potential and available resources of human muscle power.

ENERGY AND THE SOCIAL ORDER

An incident which happened to coincide exactly with the arrival at Dannemora of Mårten Triewald may serve to illustrate the importance of maintaining the prevailing social order in ensuring the energy supply of the mines. The Partners met Mårten Triewald for the first time on August 17, 1726, to discuss his suggestion that they should build a Newcomen engine to keep the mines free from water and to hoist ore.[37] They met for three days, but most of this time was devoted to a quite different matter. Indeed, before Triewald could make his appearance, the Mine Inspector gave notice that there was an important item to be dealt with. This was that the workmen had shown "recalcitrance and insubordination".[38] After listening to Triewald and giving favourable consideration to his proposal, and dealing with some minor business, on August 17, the meeting could devote August 18 and 19 to this matter.

At the Court of Mines in the spring, the Partners and the Mine Inspector had decided that ore and rock must be removed as soon as possible following a rock fall in the largest of the mines, Storrymning-en.[39] This was to be done by working two windlasses simultaneously for thirty-three days. The windlasses would have a new type of barrel, twice as large as the usual one. Eleven persons would work each windlass. But when work was due to start at six in the morning on July 1, 1726, the workmen had refused to use the new barrels. Mine Bailiff Eric Hagel, who was in charge of the work, only managed with difficulty to persuade them to attach the new barrels and start work. "But they did not continue with this longer than until half past seven, when they conferred together, became recalcitrant, and with abusive expressions left the windlasses".[40] By means of threats, Hagel induced them to resume work, but only for a short while. He was therefore compelled to call Mine Bailiff Thomas Kröger "to urge them in his official capacity to continue work".[41] Kröger emphasized the seriousness of refusing, and ordered them to resume:

to which they replied that they could not manage such hard work, and would indeed work, but no differently from in the former manner, and with the old barrels, and as they could not be brought by good words or stern admonishment to work with the new barrels, but were as insubordinate and abusive as before, the Head Mine Bailiff was forced to agree to the proposal to allow them to use the old barrels, but this on condition that there should be no more than 8 persons to each windlass.[42]

This met with new protests, but work restarted at ten in the morning. Work continued for three days, but then the protesters began to demonstrate by leaving work at three in the afternoon. On the sixth day, one windlass stood completely idle. Mine Bailiff Kröger was therefore obliged to agree to their demand to be allowed to have eleven persons to each windlass so that time "should not pass in vain".[43] When Kröger and Hagel had given an account of all this to the Partners, they asserted that the workmen must be punished for their recalcitrance and undisciplined behaviour. The meeting now called in the accused, i.e. the twenty-two people who had been working on the two windlasses. They were:

(*Miners*)
Erik Lustig
Olof And
Mats Åkare
Erik Ladman
Hans Pirgo
Johan Åhman
Pär Wäster
Anders Frisk
Erik Film

(*Soldiers*)
Pär Hagman
Noe Dandenell
Pär Broberg
Pär Lagman
Johan Lustig

(*Married women*)
Marita Hagel
Olof Bräck's wife

(*Widows*)
Johan Writz' widow
Johan Kandt's widow

(*Unmarried women*)
Marja Olsdotter Tegelberg
Brita Holm
Marja from Slacktarbo
Pär Jönsdotter Karin

These were now questioned, one by one, on the events of the days at the beginning of July. This took time, and as it was not possible to complete the questioning in a day, "it was found necessary to have these accused kept in safe custody overnight in the coffin, as they might otherwise not appear".[44] The coffin ("*Kistan*") was the everyday name for the lock-up at Swedish mines.[45] It was usually a low timber building, almost a kind of dugout; inside it a pit had been dug, in which the offenders were detained under a wooden hatch. It was a common punishment at the mines to be "put in the coffin", e.g. for drunkenness. The Partners were particularly keen to discover whether the insubordination had been caused by an agitator. Suspicion fell on the soldier Johan Lustig, who was called on the morning of August 19. He was charged with "the disturbance and the recalcitrance" for which he and the others had been responsible on the morning of July 1,[46] but he replied:

Fig. 11.6. Punishment by being flogged while suspended in manacles. Engraving by an unknown artist in 1792 (trimmed). (Photo Uppsala University Library)

that he knew of no disturbance, other than that he and his colleagues had to leave off working with the new barrels because they were too heavy, but otherwise he denied persistently that he had been recalcitrant or shown any insubordination to anyone, and although one sought to make him admit his offence, it was all in vain.[47]

The others were now called in from the coffin, and asked again who had been the troublemaker. "But they stated persistently that they had all been agreed and of one voice, and had all begun with one mind to say that they could not manage the work."[48] The blacksmith at the mines, Pär Löfbom, was then called as a witness, and he related that the miner Hans Pirgo had visited him in his smithy one morning before the insurrection. When Löfbom had left the smithy for a while to cool a glowing iron in the water trough, he had seen Pirgo fill one of the new barrels, which was standing in the smithy, with stones,

which Pirgo could not deny, but would not say whether anyone had bidden him do it, or for what reason he did it, but said that he had only done it for amusement and that he had said nothing of it to anyone else.[49]

The unity of the workmen now cracked, and some began to testify against others, while others asked for mercy. The Partners and the Mine Inspector were soon able to reach a decision ("*Resolutio*"). It was found proved that the miners Hans Pirgo, Johan Åhman, Pär Wäster, Anders Frisk, Mats Åkare and Erik Film had not only refused to work on July 1 but also shown insubordination to Mine Bailiffs Kröger and Hagel. As a result, "this work which is quite urgent to the company could not be carried out with the vigour that was appropriate".[50] For this they deserved severe punishment, but on this occasion the meeting would content itself with sentencing them under Article 27 of the Mining Statute of July 6, 1649, "as a punishment to themselves and a warning to others" to suffer 15 strokes with a pair of lashes while suspended in manacles.[51] The soldier Johan Lustig constituted a legal problem: he was evidently the one who had first sown the seeds of this defiance and thus "the prime cause",[52] but he was officially serving with the Uppland Regiment. Sentence on him was therefore deferred pending correspondence with his commanding officer, Lieutenant-Colonel Henrik Julius Voltemat, concerning how the man could be brought to justice and "have the punishment meted out to him, that his offence deserves".[53]

The girls Marja from Slacktarbo and Pär Jönsdotter Karin were also parties to the events in that they had stopped work, but as there had been an intercession of their behalf it was decided to let the matter rest with two days in the coffin on bread and water. Olof Bräck's wife had not only stopped working but had also been very argumentative and insolent when the mine bailiffs ordered her to resume. She was therefore given three days in the coffin. Marita Hagel had not protested about the work but "merely asked that the work should be as formerly was customary",[54] i.e. with eleven persons on each windlass and the usual barrels. (On Hagel's list of workmen there is a note in the margin alongside her name: "begged weepingly".)[55] She escaped punishment this time, but the others' sentences "were immediately executed".[56]

It was not the loss of output on the particular occasion that caused the Partners' to spend two days on this matter—they had daily to take decisions on technical and financial questions of far greater importance at their ironworks. But the output of the seventeen ironworks was in the last resort dependent on the manual labour performed by the miners of Dannemora. The discipline inherent in the social order had therefore to be maintained. The need for energy required every outbreak of insubordination to be forcibly suppressed, the instigator to be traced and those involved to be publicly dealt with "as a punishment to themselves and a warning to others".[57] The incident also illustrated how the Partners acted as one in matters of common interest, and they can therefore be regarded as a single entity in the events which follow. As a rule, Mine Inspector Thor Bellander can also be included, as he shared their interest in efficient hoisting and pumping work. Mine Bailiff Thomas

Fig. 11.7. The largest of the mines at Dannemora in 1715. Evidence of most of the attempts to keep the mines free from water can be seen here: the 300-metre long dam between the lake and the mines (top), the *Stangenkunst* system from the waterwheel (oblique line running across the map from the bottom to the mines), a number of horse whims and windlasses (small circles resembling dials), and the windmill ("*Wäderkonst*"). Head Mine Bailiff Thomas Kröger's house is visible in the lower left corner. Detail of a map by Johan Tobias Geisler in 1715. (Photo The National Record Office, Stockholm)

Kröger, on the other hand, occupied an intermediate position between the decisions of the Partners and the technical and social reality of the mines. In this case he had reported his failure to the Mine Inspector by July 9 ("Albeit that I have tried to obey the esteemed Mine Inspector

...).[58] But he was able to restore his prestige in the eyes of the workmen after referring the matter to his superiors, and supervising the sentences being "immediately executed". Finally, the episode also illustrated the precariousness of the energy situation in the Dannemora Mines at that time. A man can only deliver approx. 0.1 hp when working continuously for a whole day, but even this decimal was worth exploiting to the full in the Dannemora Mines in 1726.

THE STRUGGLE AGAINST WATER

The problem of keeping the workings free from water has been more difficult to overcome at Dannemora than in any other Swedish mine.[59] The mines were on the edge of a lake (Fig. 11.7), and the openings of the shafts were at a lower level than the lake surface. The lake bottom was originally covered by a watertight layer of clay, which prevented the water of the lake from permeating through the soil down into the mines. But this natural barrier was gradually destroyed by the use of the lake as a dumping place for excavated stone, which by its weight penetrated the layer of clay.[60] The problem of water in the mines then became more serious, not only as the working depth increased but also because the leakage grew as stone continued to be dumped. The latter part of the seventeenth century saw the beginning of a series of attempts to keep the mines free from water (although it was not until the nineteenth century that the water in the Dannemora Mines was brought under control).

In 1656 a substantial wall was left unexcavated between two of the mines to prevent water from filling both if it should find its way into one of them, and it was forbidden on pain of death to mine too close to this wall. In 1658 the Partners discussed lowering the level of the lake by dredging the streams that connected it with the system of lakes lower down towards Salsta. But the enterprise was deemed far too costly, especially as it would entail annual maintenance work. In 1676 the Mine Inspector proposed the building of a water-powered pumping machine. This was duly built by Olof Hindersson Trygg (1629–1699), a well-known Chief Engineer (*Konstmästare*) at the Great Coppermine in Falun,[61] and was completed in 1680. Power was transmitted from a waterwheel, 10 metres in diameter, to the pumps via a 1 500-metre *Stangenkunst* arrangement; the whole system cost 10 000 *copperdaler*. But the waterpower system proved inadequate to keep the mines free from water. An extension of the existing *Stangenkunst* system gave adequate pumping capacity, but now the problem was that the supply of water was insufficient for the waterwheel. The next solution attempted was a windmill, built in 1691–1692 to augment the power already supplied by water, men and horses. The windmill cost 9 000 *copperdaler*, but was of

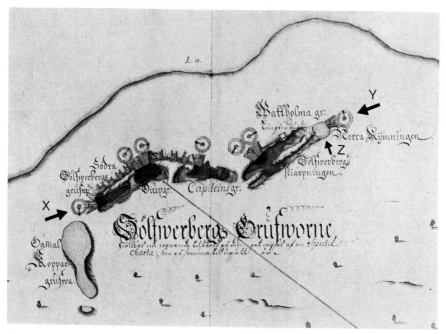

Fig. 11.8. The Silverberg mines at Dannemora, situated three hundred metres south-west (i.e. to the left) of the mines in Fig. 11.7. Z marks Norra Silverberg Mine which had been abandoned in 1709 owing to the difficulty of keeping the mines free from water with the traditional sources of power. Detail of a map by Johan Tobias Geisler in 1715. (Photo The National Record Office, Stockholm)

little use. The maintenance costs turned out to exceed the cost of doing the same amount of work with windlasses and horse whims.

In 1693 a serious cave-in occurred in one of the mines, causing several mines to fill with water and destroying the pumping machinery. The repair work was in vain since new leaks developed as the rock settled. So in 1694 efforts were made to lower the level of the lake. With great difficulty the water level was lowered by 1½ feet. This aroused protests from Uppsala University, which lost power at its flour mill further downstream, and moreover it was now realized for the first time that the leaks were from the bottom. The complete draining of the 10-metre deep lake would cost 18 000 *copperdaler*, and almost as much would be required each year to keep it dry by dredging the rivers and streams of the rest of the system. A dam, 300 metres long, was therefore built instead between the lake and the largest mines. It consisted of piles driven to a depth of 4½ metres and faced on the lakeward side with planking and compacted clay. (This dam was of great value in the fight against water in the Dannemora Mines until the mid-1960s!)[62] In 1697 most of the mines were kept free from water for the first time since the rock fall of 1693, and the total cost of repairs and construction work since the fall had by then risen to 30 000 *copperdaler*. But the same

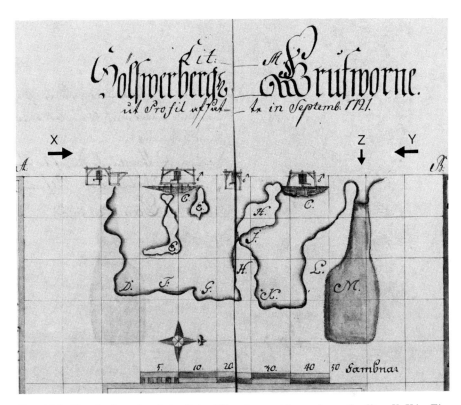

Fig. 11.9. Vertical projection of the Silverberg mines along the line *X–Y* in Fig. 11.8. *Z* marks Norra Silverberg Mine, 67 metres deep and filled with water. Map by Olof Trygg in 1721. (Photo Royal Library, Stockholm)

situation arose again: once pumps, *Stangenkunst* systems and dams were completed, the supply of energy was found to be insufficient.

The supply of waterpower and windpower had natural geographical and topographical limitations. The supply of horse power had a practical limitation in the difficulty of obtaining enough grazing land in the neighbourhood of the mines.[63] The supply of human muscle power had, as we have seen, a social and human limitation. This was the situation at the beginning of the eighteenth century. The traditional sources of power—water, wind, and muscle—had proved inadequate to cope with the problem of keeping the mines dry. In 1709 attempts to keep one of the mines free from water had to be abandoned.[64] This was Norra Silverberg (Fig. 11.8), and the 67-metre deep mine was soon filled with water (Fig. 11.9). It was at this mine that Mårten Triewald's Newcomen engine was built.

12. The Technology: The Triewald Engine

INTRODUCTION

A general description of the early Newcomen engines was given in Chapter 6, and in particular there was a discussion of the factors that were of importance to the diffusion of the new technology. But what were the distinctive characteristics of the technology that Triewald attempted to introduce in the Dannemora Mines? The description given by Triewald in his book on the Dannemora engine in 1734 might seem a natural point of departure, but his was scarcely an impartial account, as will be shown in Chapter 14. The sources cited in the following discussion have therefore been limited to the archival evidence dated before the construction work began, and the artifactual evidence that remains. The archival evidence consists mainly of the minutes of the Partners in the Dannemora Mines, as well as a set of previously unidentified drawings in the archives of the Dannemora Mines. The artifactual evidence is the engine house still standing in Dannemora: today an empty shell stripped of all remains of the engine but only apparently void of meaning. In an article in 1978, Patrick M. Malone pointed out the importance of artifacts "as sources of valuable, and often otherwise unattainable, historical information" in the study of the history of technology.[1] The following argument illustrates his point. It is a combination of archival and artifactual evidence, but its conclusiveness rests upon the artifact: the engine house.

THE DRAWINGS

In 1970, the managing director of the Dannemora Mines, Gunnar Stenberg, sent a number of drawings to Torsten Althin, who was at that

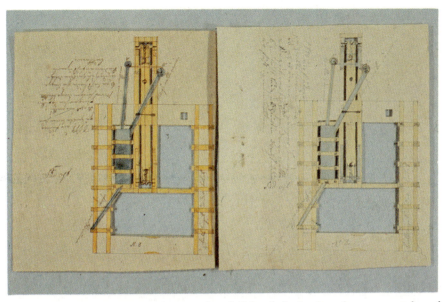

Fig. 12.1. Horizontal projection no. 8 at loft level. It depicts the water tank and the beam. The first series to the left, the second series to the right. (Photo Bengt Vängstam)

time compiling an inventory of Swedish steam engines.[2] The drawings show an early Newcomen engine, with a cylinder approx. 3 feet in diameter. The engine is driving pumps in a mine, but the remarkable thing is that it is also connected to a complex wooden hoisting machine. The drawings are neither signed nor dated. There is no mention of where the engine was, or was intended to be, built. It is clear, however, that they concern an actual steam engine project since compass directions are given for the position of the engine house in relation to the opening to the mine. Since the drawings were kept in the archives of the Dannemora Mines, it seems reasonable to assume that they had belonged to a plan to build a Newcomen engine there. Only two such projects are known. The first was Triewald's engine of 1728, and the second was a scheme in 1768 which was abandoned after a few years.[3]

The fact that the Swedish method of mapping mines is used in the drawings led Torsten Althin to assume in 1971 that the drawings had been made in 1768 by Eric Geisler (1720–1773), a well-known mine surveyor at the Great Coppermine in Falun.[4] But the archival sources prove that Geisler was only hired to survey the site for the intended engine, and moreover that this engine was designed with two boilers, which were to be in a separate boiler house.[5] This leaves us to investigate the other possibility. It may at first seem unlikely that the drawings could have anything to do with Triewald's engine, because Triewald makes no mention of any hoisting machinery in his book *Kort beskrifning, om eld- och luft-machin wid Dannemora grufwor* (A short descrip-

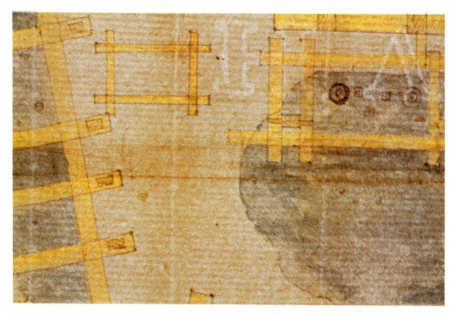

Fig. 12.2. The tiny pinholes in the corner of each detail tell us not only how such detailed drawings could be copied accurately, but also which of the two series is a copy of the other. (Photo Bengt Vängstam)

tion of the fire and air engine at the Dannemora Mines) and because of the lack of resemblance between the drawing and the well-known engraving in the book.[6]

There are fifteen sheets of drawings of various sizes up to 21×41 cm. The set consists of two duplicate series of numbered drawings, which differ only in the colour of the ink and in spelling (Fig. 12.1). The ink of the first series is brighter, and the word cylinder is spelt "Zelynder", whereas the second series is fainter in colour (or faded) and cylinder is spelt "Cylinder" and "Celynder". The first series consists of eight sheets of drawings, and the second series of seven sheets. Not only is there one sheet of drawings less in the second series, but also one sheet shows fewer vertical projections. But for these differences, the two series are almost identical.

There are tiny pinholes in the corner of each detail; for example, in the four corners of every beam (Fig. 12.2). These holes are invisible unless the drawings are held up to the light. There are more than a hundred such holes on the larger drawings. They show that one series of drawings was copied from the other by laying the first drawing over a blank piece of paper and piercing the drawing and the paper with a pin at all the intersecting points. This tells us not only how such detailed drawings could be copied accurately in those days, but also which of the two series is a copy of the other.[7] The pinholes are exactly in the corners of each detail on the drawings in the first series, but are

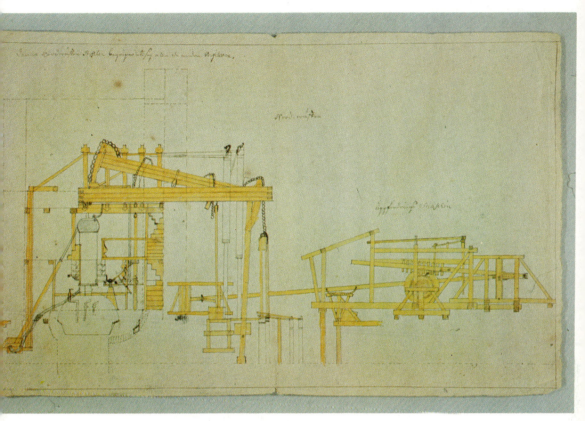

Fig. 12.3. Vertical projection no. 3. This is the main drawing of the whole series since it shows both the engine and the hoisting machinery in a cutaway view. (Photo Bengt Vängstam)

sometimes slightly out of place in the second series. If the pin was not pushed exactly vertically through the first sheet, the holes would be out of place on the second one, because the paper was quite thick (the drawings are so detailed that the distance between two adjacent holes is sometimes only a few millimetres). Had the draughtsman then merely drawn lines joining the pinholes, the beams would not have been parallel and of identical size, which they are. The draughtsman must therefore have used the holes only as guiding marks when he drew the lines. The first series is therefore the original, and the second series is a copy.

We can therefore concentrate on the first series of drawings. This is also the more complete series of the two. There are, strictly speaking, ten sheets of paper in this series, since two of the drawings (nos. 3 and 6) have smaller drawings (nos. 4 and 5) attached to them. The drawings show the engine, the engine house, the hoisting machinery and the connecting mechanism in vertical and horizontal orthogonal projections.[8] The vertical projections (*"Profilerne"*) are numbered 1–3, and

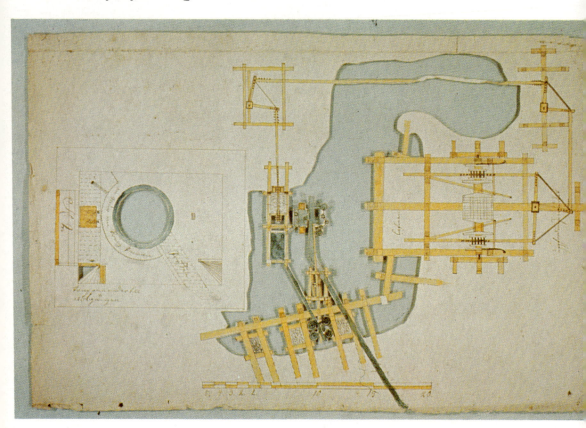

Fig. 12.4. Horizontal projection no. 2 showing the whole construction at ground level. The thin ribbon at the top represents the *Stangenkunst* linkage across the mine shaft that connects the beam of the engine to the hoisting machinery. (Photo Bengt Vängstam)

the horizontal projections 1–8. The vertical projections show the exterior of the engine house from three directions, the engine in the house from two opposite sides of the building, and a general view of the whole construction from the side. The last of these (no. 3) is the main drawing of the whole series, since it shows the entire engine together with the hoisting machinery (Fig. 12.3). It is also something of a cutaway view, as the interior of the engine house and everything below ground is shown.

The horizontal projections include two larger drawings (nos. 1 and 2, approx. 20×32 cm), showing the whole construction slightly below ground level and at ground level (Fig. 12.4). There are also six smaller drawings (nos. 3–8, approx. 15×17 cm or smaller), which show the engine and the engine house in horizontal section at different heights. The horizontal projections of the engine and the engine house have two striking features that catch the eye immediately. *First,* there are holes cut out of the paper to indicate space, or, rather, vertical passages; for example, in the boiler, the staircases, the chimney, and around some of the beams and pipes. This was in accordance with Swedish mine-surveying practice; shafts were usually cut out of the paper or parchment of the horizontal plans of the various levels in the mine.[9] *Secondly*, the back of each drawing has been used for an additional projection from below. Thus, each drawing represents a horizontal plane of projection, with projections from above and below drawn on the front and the back of the paper. This not only saves paper but also makes the construction more comprehensible. In fact, this feature, in combination with the holes cut out of the paper, makes the drawings something more than a two-dimensional representation. If the drawings are turned over, one by one in numerical order, it is easy to visualize the engine and the engine house in three dimensions. The drawings are not a model, but they fulfil a similar function in helping the mind to form a three-dimensional image of the whole construction.[10] That this was indeed the draughtsman's intention is obvious. Not only were the drawings pinned together in sequence, but on the back of drawing no. 6 there is a projection from below showing the joists supporting the floor and the huge beams supporting the cylinder. A smaller drawing, no. 5, is positioned over no. 6, and held in place by sealing wax at one end. The sealing wax works as a hinge: by turning no. 5 over one can view no. 6 (Fig. 12.5).

The drawings are remarkably detailed, considering their size. Drawing no. 5 is only 3.3×7.8 cm. There are two round holes, 0.1 and 0.2 cm, in the paper to indicate the position of the steam admission valve and the exhaust pipe for the hot condensate. The paper is also cut to leave only a thin ribbon, 0.2 cm wide, to indicate the water injection pipe from the tank on the top floor. The drawings are also true engineering drawings in the sense that they are drawn to scale. On several of the drawings there is a linear scale representing 20 Swedish ells, or "*alnar*". One "*aln*" was 2 Swedish feet, or 0.5938 m.[11] Measured on the drawings, the mean scale equivalent of 20 Swedish "*alnar*" is 128.7±0.7 mm in the first series, and 128.0±1.6 mm in the second. The larger standard deviation in the second series confirms the assumption that it is a copy of the first. This gives a scale of 1 : 92.3 in the first series.[12] The drawings bear Swedish text giving the names and functions of some of the parts of the engine.[13] This information is not very detailed, and it should be noted that it does not in any way describe the principle of the engine, the only comment being that the piston moves up and down in the cylinder. The text adds little to the visual information provided by the drawings, except for giving a basic terminology for the hardware. Many vital concepts—such as steam, condensation, vacuum, atmospheric pressure, power, and fuel consumption—are not mentioned. The text is almost identical in the two series. But the text in the second series is written in another hand, and some words are spelt differently. The handwriting in both series is in the old German style,

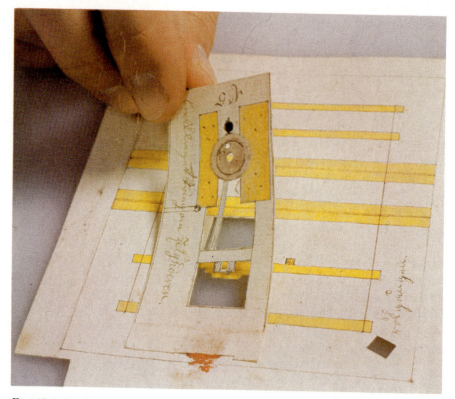

Fig. 12.5. Horizontal projection no. 5 is positioned over no. 6, and held in place by sealing wax at one end. The drawings are not signed, but the draughtsman has left his fingerprint in the sealing wax. (Photo Bengt Vängstam)

which was used in Sweden until the end of the eighteenth century. The only conclusion from the text—handwriting as well as language—is that both series were made during the eighteenth century.

Filigranology, the study of watermarks, is an auxiliary science of history which offers a method of dating archival material.[14] The paper used for the drawings is Dutch foolscap, common in Sweden in the eighteenth century.[15] There are several complete and fragmentary watermarks in the paper of the drawings (Fig. 12.6). An investigation of these suggests that the drawings were made during the period ca. 1718–1735, which is fully compatible with the assumption that they date from 1726.[16]

ARCHIVAL SOURCES

In what way do the archival sources support the suggestion that these are the construction drawings of Triewald's engine? Triewald met the Partners in the Dannemora Mines on August 17, 1726 (see Chapter 13), and at their meeting he outlined his proposal for a Newcomen engine.[17]

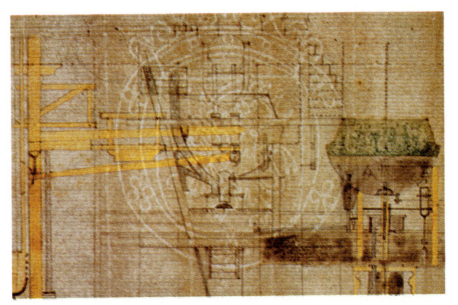

Fig. 12.6. The "*Dutch Lion*" watermark in the paper of one of the drawings. By comparing the watermarks it is possible to assign the drawings to the period ca. 1718–1735. (Photo Bengt Vängstam)

He thought that the best site for the engine would be on "the west side of the opening to the Norra Silverberg Mine",[18] The engine would "not only drain the mines of all water, but also hoist the ore in a simple and practical manner".[19] Two months later, a contract between Triewald and the Partners was signed on October 24, 1726.[20] It was agreed that Triewald should receive 5 500 *copperdaler* to order the cylinder, the regulator, the pump for the cistern, and leather for the pumps from England, as well as 2 500 *copperdaler* to order a copper boiler, copper pipes, valves, pumps and other forged goods from Stockholm. The Partners were to provide all the other materials, such as timber, bricks and lime mortar. They were also to see that workmen were available on the first day of spring in the following year to build the engine house. There are numerous archival sources of a later date that give additional information on the construction and the dimensions of the engine (see Chapter 13). But this cannot be used as evidence here, because a number of difficulties forced Triewald to change his design, and that is a different story: the story of how British technology was adapted to Swedish conditions.

The archival information quoted above should now be compared with the drawings. The most significant point is that Triewald's original suggestion on August 17 included a hoisting machine powered by the engine. The ore would be raised by this machine in two barrels,[21] which indicates that two ropes were wound in opposite directions on the same shaft, to which rotative motion was in some way imparted by the engine. This is all the archival sources tell of the design of the hoisting machine, but the point is that the drawings depict a hoisting machine ("*upfordrings Maschin*") which answers to this description: there are two winding drums on the same shaft, rotated by a ratchet mechanism powered by the engine through a *Stangenkunst*.

Triewald had suggested on August 17 that the engine should be built on the west side of the Norra Silverberg Mine. The engine on the drawings is located at the top of a mineshaft resembling that of the Norra Silverberg Mine in shape and size, but on the north-east side. He gave the dimensions of the building in which the engine was to be placed as 8.9 metres (length) by 7.1 metres (width).[22] But the engine house in the drawings is 7.5 metres long and 5.7 metres wide (inside dimensions), or 9.9 metres long and 8.1 metres wide (outside dimensions). The information about the orientation and the dimensions of the engine house proves that the drawings could not have been made before August 17, 1726. It is, in any case, unlikely that Triewald would have gone to the trouble of producing a full set of construction drawings before his proposal had been accepted.

The wooden structure supporting the engine is worked out in detail on the drawings and indicated in yellow. But vital parts of the engine, such as the steam admission and the water injection valve gears, are not shown in any detail, and there is no indication of the materials by colour. It is not possible to build a working steam engine from the drawings alone. Unlike the engravings by Barney, Beighton and Nicholls, or the one in Triewald's own book on the Dannemora engine in 1734, the drawings do not show the principle.[23] This indicates that the drawings were to serve some other purpose, and that the engine was to be built by someone who knew how to construct it and therefore did not need to have the principle shown in detail. The most reasonable conclusion is that the drawings were prepared for the benefit of those who were to build the engine house. This supports the assumption that the drawings were made by Triewald for the Partners in the Dannemora Mines. They would have needed such detailed drawings if they were to supply all the building materials needed for the house and the wooden framework supporting the engine. This would also explain how the drawings came to be kept in the archives of the company. It should be noted, though, that although the archival sources support, and in no way contradict, the assumption that the drawings are the construction drawings of Triewald's engine, they do not prove it conclusively.

THE ARTIFACT

It still remains to be shown that the drawings really are related to Triewald's engine, and this can be done by a closer look at what is left of the artifact. Triewald's engine house is still standing at Dannemora

Fig. 12.7. The engine house at Dannemora. (Photo the author, 1977)

(Fig. 12.7). The building was struck by lightning in 1736, when the chimney and the roof were damaged.[24] The remains of the engine were sold as scrap metal in 1773, when all plans to restore it had finally been abandoned.[25] Since then the building has served as a shed, a drying kiln and a smithy. The interior has been rebuilt, and a hearth and a new chimney have been installed. The building was damaged by fire in 1901, after which the present roof was built.[26] The building has been more or less officially recognized as a national industrial monument since 1932, when the National Museum of Science and Technology (*Tekniska museet*) put a memorial plaque on it.[27] Since then it has served as a storage place for various tools and machinery of historical interest associated with the Dannemora Mines. Although no parts of the Triewald engine have been preserved, it will be shown that the engine house is an artifact of significant historical value.

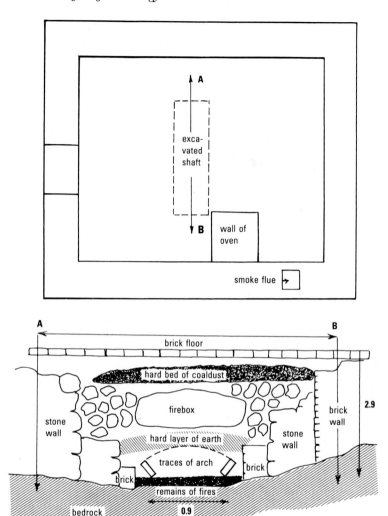

Fig. 12.8. In November 1930, the National Museum of Science and Technology (*Tekniska museet*) excavated an area in the middle of the floor in the engine house. Copy of the original drawing in the archives of the Museum. (Drawing Gunilla Johnson)

In 1931, the National Museum of Science and Technology built a model of the engine.[28] It was based mainly upon the print in Triewald's book, but supplemented by a study of the engine house and an excavation in November 1930, when a rectangular hole was dug in the middle of the floor (Fig. 12.8).[29] At a depth of 2 metres, the remains of an ash-pit, a grate and an arch were found. The span of the arch, 0.9 metres, corresponds with that of the arch on the drawings. The excavation showed that the firehole faced the door of the building (north-east), and that the chimney was in the west corner. This is also in full agreement with the drawings.

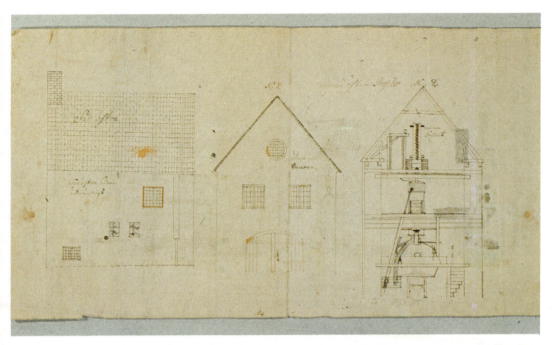

Fig. 12.9. Vertical projections nos. 1 and 2. Compare the position of the windows and the door in the gable with Fig. 12.7. (Photo Bengt Vängstam)

The engine house is situated on the north-east side of the Norra Silverberg Mine. The orientation of the building and its relation to the shaft correspond with those on the drawings, which indicate the points of the compass as "north-east", "south-west" etc. In each side of the building there are bricked-up holes in the stone wall, which correspond in size and position with the holes indicated on the drawings for the two huge beams supporting the cylinder. (This is contrary to the engraving in Triewald's book, which shows the beams running along, not across, the building.) On the south-east wall of the building is a corroded iron bar, the possible remains of a clamp supporting a pipe. It is located where the overflow pipe from the tank in the loft is indicated on the drawings. The number and positions of the windows also correspond with the ones shown on the drawings (Fig. 12.9). There is today an ordinary door to the engine house in the north-east end. But the bricks around it tell of an earlier, larger, and semi-circular opening in the stone wall, which can be seen from the inside (Fig. 12.10). This also corresponds with the door shown on the drawings. There is rectangular brickwork in the south-west gable, also visible from inside the loft, where there was a hole in the stone wall for the beam of the engine. This is placed asymmetrically, to the south-east of the centre of the building, as indicated on the drawings.

The conclusive evidence, definitely linking these drawings to the

Fig. 12.10. The original semi-circular door can be seen from the inside of the engine house. The boilers of the early Newcomen engines were often burnt out and had to be replaced, hence the shape. (Photo the author, 1981)

Triewald engine, is found in the loft. Drawing no. 8 is a horizontal projection at loft level (Fig. 12.1). It depicts the joists in the walls of the building that support the floor of the loft, the five beams in the south-east wall and the beams across the building that supported the water tank. The stone walls of the loft of the engine house at Dannemora have holes corresponding to each of these joists (Fig. 12.11).

According to the drawings, the flue was to enter the side wall in the west corner, run up through the wall, go around the corner, and end up in the west gable. The excavation in 1930 proved that the flue did indeed run towards the west corner of the side wall from the firebox. But the end of the flue, visible as a hole in the stone wall in the loft, is directly above the point where it entered the wall, and not in the adjacent gable. It is a metre out of place when compared with the drawings. This discrepancy suggests that the drawings were made before the engine house was built. Furthermore, the engine house at Dannemora is 8.1 metres wide and 9.5 metres long (outside dimensions). The engine house on the drawings is also 8.1 metres wide, but it is 9.9 metres long. This shows that the drawings were made *before* the building was erected, and not by measuring up the house and engine once completed. There were, evidently, some minor changes to the design during construction. According to archival sources, the foundation of the engine house was laid on May 2, 1727.[30] Thus the drawings must be of a prior date. This supports the assumption that they were made for the benefit of the Partners, who were to supply the building materials for the engine house. The drawings can then be dated to within six months. They must have been made prior to May 2, 1727,

Fig. 12.11. One of the holes in the stone walls of the loft that correspond exactly in size and position to the joists on horizontal projection no. 8 (Fig. 12.1). These holes are the conclusive evidence, definitely linking the drawings with the Triewald engine. (Photo the author, 1981)

but after August 17, 1726. Within this interval we can only surmise that they were made in the late autumn of 1726—as soon as the contract had been signed between Triewald and the Partners.

Thus a comparative study of archival material and artifactual evidence shows that these drawings are the original construction drawings for Mårten Triewald's Newcomen engine in Dannemora, and that they were made prior to May 2, 1727, but after August 17, 1726. It is now time to discuss the implications of this conclusion.

THE DRAWINGS: THE UNIQUENESS OF THE COMMONPLACE

The drawings are of course unique from an antiquarian point of view, since they are probably among the oldest engineering construction drawings preserved. But the question for the historian of technology is whether they were unique in their time.

In the company archives of the numerous Swedish ironworks the accounts have been kept in immaculate order for hundreds of years. It is not uncommon to find shelf after shelf of account books, one volume for each year, in consecutive order from the seventeenth century to the present day. The clerks' meticulous sense of order is reflected in the uniform parchment binding of the volumes, and the graceful handwriting on the spines. The vouchers, the minutes of the board meetings, and the correspondence are likewise kept in good order, and seldom is any volume found to be missing. But one rarely finds technical drawings in any company archive—although we know that drawings were used extensively, at least from the nineteenth century. This reflects a difference between the thinking of administrators and that of engineers: the latter saw no need, when a machine or an engine was replaced by a newer and more efficient one, to bother to keep the drawings. It was therefore an exception, rather than the rule, for any technical drawings to be kept in company archives. Triewald's engine at the Dannemora Mines was such an exception. It was a famous incident in the history of the Dannemora Mines, and its uniqueness as the first of its kind in Sweden was fully appreciated from the day when it was built. The fact that these drawings have been preserved in the archives of the Dannemora Company is hence not so surprising as it may at first seem.

These drawings were unique in Sweden, a country where another forty years were to pass before the second steam engine was built in 1766.[31] But if such drawings were used in England, a country where almost 600 Newcomen engines had been built by the 1760s,[32] they would have been anything but unique. As everyday tools of the engineers who built these engines, they would have been far too commonplace for anyone to think them worth preserving in their own right. The fact that no such drawings are known to have been preserved in England indicates that they were not unique there, but commonplace! This is a far more likely conclusion than the alternative, which would be to suggest that Mårten Triewald made a revolutionary innovation in the art of technical illustration during the short time that elapsed between his return from England and the construction of the Dannemora engine.

The first conclusion to be drawn from the drawings, then, is simply that such detailed and elaborate non-verbal technical information about a relatively new technology did indeed exist. This implies that the dissemination of the technology was not determined merely by the mobility of steam engine constructors who reconstructed engines from memory or from crude and often inaccurate engravings. No, the design of the hardware of the new technology had evidently been committed to accurate drawings, done to scale and depicting the engines in a variety of planes of projection, by 1725. This in turn implies that the technology was already well established, since the execution of such drawings presupposes a degree of standardization in the design of Newcomen engines.

THE STATE OF THE ART: ROTATIVE MOTION

It is the shortness of the time between Triewald's return in May 1726 and his proposal for the Dannemora engine in the August of the same year that makes these drawings interesting. It was too brief a period for him to have made any major improvements to the technology. We may therefore regard these drawings as a sample of the state of the art in England in ca. 1725. They indicate the level of the technology that was brought over to Sweden in 1726. As far as the technical details of the engine are concerned, the drawings confirm our knowledge of early Newcomen engines.[33] They neither add anything nor contradict anything previously known. But the remarkable feature is the hoisting machine powered by the engine.

The fact that the early Newcomen engine was used mainly for raising water has led many to believe that "it was a pumping engine only" and not a prime mover.[34] This view persists in many modern works of economic history. In *The Fontana Economic History of Europe*, for example, Carlo M. Cipolla wrote in 1973 that it was the Watt engine that "made possible the transformation of the chemical energy of coal into mechanical energy".[35] But this judgement ignores the true significance of the Newcomen engine as an engine for energy conversion, which was that it converted thermal energy into mechanical energy. The potential energy in the water raised by a Newcomen engine could also be made to produce rotative motion if used to supply a waterwheel. This was the simplest technical solution wherever there was a need for rotative motion beyond the capacity of the traditional prime movers, and where the cost of the initial investment in a Newcomen engine and the cost of maintaining it made it economical. It was not, as Samuel Lilley has claimed, "surely a sign of desperation",[36] but a perfectly rational way to achieve technological synergism. By combining two reliable and established technologies—the great pumping capacity of the Newcomen engine and the smooth and regular motion of the waterwheel—it was possible to obtain what neither could give on its own, i.e. continuous operation *and* rotative motion. There are several examples of Newcomen engines that were used in this way.[37] In the 1760s, for example, the Darby firm in Coalbrookdale installed an engine to drive several waterwheels.[38] Compared with the extent to which Newcomen engines were used to power mine pumps, this application was, admittedly, economically insignificant. But the point is that this does not allow us to regard the Newcomen engine qualitatively as anything other than an engine that "made possible the transformation of the chemical energy of coal into mechanical energy".

There were also several attempts to produce rotative motion directly from Newcomen engines. John Farey, in his *A Treatise on the Steam Engine* in 1827, mentions a number of such attempts prior to James Watt's patent in 1781 for a rotary version of his engine: Jonathan Hulls 1736, Kean Fitzgerald 1757, Joseph Oxley 1762, Dugald Clarke 1769, John Stewart 1777, Matthew Wasbrough 1779 and James Pickard 1780.[39] There was a threefold problem in producing rotative motion from a Newcomen engine: *first*, the engine only exerted power on the downward stroke; *second*, it had to be possible to change the direction of

Fig. 12.12. Working scale model (1:25) of the hoisting machinery. Built from the drawings by the author in 1975. (Photo Bengt Vängstam)

rotation if the machinery was to be used for hoisting coal out of a mine; *third*, the stroke of the beam of the Newcomen engine was irregular in length and frequency.[40] The solution in all these attempts prior to Watt seems to have been a complex mechanism of wooden ratchets, cogwheels and counterweights.[41] The Triewald drawings show such a design, with a complex wooden hoisting machine driven by the engine through a short *Stangenkunst*. It is not easy to grasp its function immediately from the drawings, and in order to understand it better I have built a working model on a scale of 1:25 (Fig. 12.12). The difficulty of building the model from the drawings convinced me that the hoisting machine was not a drawing-board design. The drawings must have been based on a working machine or model seen by Triewald in England. Such attempts must, then, have been made in England by 1725. Obviously they were not technically or economically successful, since no other records survive. But an interesting conclusion from the point of view of the history of technology is that the true nature of the Newcomen engine was realized almost immediately. To the engineers of the early 1720s, it was not "a pumping engine only", but a prime mover.

The three technical problems of rotative motion were solved in this hoisting machine in the following ways. The *first* problem was solved by two counterweights, stone-filled chests connected to the *Stangenkunst*, resetting the machinery during the non-working stroke. The *second* problem was solved by using two ratchets, one for each direction of rotation. One ratchet was lowered onto the pinion during the working stroke and lifted free during the non-working stroke by a "stirrup" on a lever. The *third* problem was solved by placing the hoisting machine opposite the engine. The operator of the hoisting machine could watch the movement of the beam, and raise or lower the ratchets accordingly. The model works very nicely, with one barrel coming up while the other is going down. Although the machine succeeds in converting the reciprocating movement of the Newcomen engine into rotative motion, it has one serious drawback. If the ratchet is not lifted free of the pinion in time, a gear tooth in the ratchet will break. The gear tooth is simple enough to replace, but it is clear that this was a *machine* rather than an *engine*: its performance depended on the skill of its operator. It will be shown in the following chapter that this human factor contributed to the failure of the hoisting machinery when it was introduced in Dannemora. Although the design solved the three technical problems involved in producing rotative motion, it needed a social setting that the Dannemora Mines did not provide.

13. Agreement, Construction and Operation, 1726–1730

AGREEMENT

Mårten Triewald had come home to Sweden on May 26, 1726,[1] and by the beginning of June he had lodged an application for a patent for the Newcomen engine.[2] This was granted on July 19,[3] and was in substance similar to de Valair's patent three years earlier (see Chapter 8). Two weeks later he wrote to the Board of Mines and requested permission to inspect various mines in order to see where his engine could be most usefully employed.[4] He had probably already decided to approach the Partners in the Dannemora Mines. Their difficulty in keeping the mines free from water would presumably have been well known among the merchants of Stockholm, and Triewald had himself visited North Uppland in 1713.[5]

At the meeting of the Partners on August 17, 1726, "a *Mechanicus* by the name of Mårten Triewald, arrived here in this country from England," announced himself and suggested that they should build an engine with the aid of which the mines could be kept free from water and the ore could be raised.[6] The ore would be hoisted out of the mine in two barrels three times as quickly as by a conventional horse whim. The pumps would restrict the water in the bottom of the mine to a depth of two or three inches. There would be no difficulty in increasing the pumping depth if the mines were excavated deeper. The engine, unlike the horse whims, could be used day and night. If the mines should happen to flood, the engine could raise more than 70 000 litres of water per hour. Triewald was convinced that the engine would be useful and economical, for there were "already thirty-three similar engines in England in full operation, demonstrating their great effect".[7]

He said that he did not yet know the prices of building materials in Dannemora, but that the cost would be about 24 000 *copperdaler*.[8] The engine would need approx. 300 *stafrum* of wood fuel per year.[9] Once erected, the engine would run for many years. It would require only minor repairs, which Triewald estimated would cost 200 *copperdaler* per year. It would be necessary to have a man "somewhat experienced in the engine and mechanical science" to supervise it,[10] so that it could be quickly repaired if needed. He would recommend a man who would be satisfied with 600 *copperdaler* per year. In addition, two "*konstknechtar*" were required who could work the engine and maintain the fire. The Partners approved the proposal. There was certainly a need for some "useful contrivance" that:

could make work in the difficult Silverberg Mines easier, and yield a good saving in the substantial cost which must annually be laid out thereon, and hitherto is said to have risen to nigh on 9 000 *copperdaler* a year, and that only for keeping the water away, and not counting the cost of drawing up the ore.[11]

It was decided to install the engine at Norra Silverberg (see Figs. 11.8–9). Since this was the deepest of the mines, the engine would keep several mines free of water. But it also meant that work in the Norra Silverberg Mine would be obstructed by the engine house at the opening of the mine shaft and by the pumps down the mine. For this reason the proposal initially met with protests from the shareholders in the Norra Silverberg Mine,[12] but on October 24 a contract was signed between the Partners in the Dannemora Mines and Mårten Triewald.[13] It was worthy of note from several points of view. First, Triewald did not contract for all the work of erecting the engine—this was divided between him and the Partners. The latter were to ensure that building materials and labour were available at the mine on the first day of spring, and they were to provide Triewald with the money to order parts for the machine from England and Stockholm. Triewald would specify what was required, and direct the work. Here there was an ambiguous division of responsibility from the outset. Was Triewald to be regarded as their employed agent or as a supplier of a finished machine? Secondly, Triewald's optimistic hopes regarding the performance of the engine and the hoisting gear were written into the contract, as was his guess that it would cost 24 000 *copperdaler* and his assumption that he would have the engine itself ready by the following June, so that he could then make a start on raising both water and ore.[14] Thirdly, the contract was unusually strictly worded in the event of the engine's performance not coming up to expectations. In that case Triewald would not only forfeit his remuneration but also indemnify the Partners for all the expenses they had incurred in connection with the erection of the engine. This contract was later to form the basis of the lawsuit between the Partners and Triewald. There its vague wording would acquire a precise legal import.

CONSTRUCTION

On April 29, 1727, Triewald arrived in Dannemora.[15] However, he had already sent his assistant, Olof Hultberg (?–1743), on ahead to supervise the building of a smithy beside the planned engine and the hewing of beams for the engine house.[16] Olof Hultberg came to play a leading role in the events that followed, but little is known of his background. He had studied at Uppsala University and appears in the university calendar for 1723 as hailing from the province of Närke.[17] Anders Gabriel Duhre had established a *Laboratorium mathematico-oecono-micum* near Uppsala in 1723, a modest attempt at a technical college, and Hultberg is mentioned there as "*mechanices studiosus*".[18] Duhre provided for him, and he studied joinery and metalwork. Duhre hoped to be able to train Hultberg until he could travel the length and breadth of Sweden building threshing machines of Duhre's design. Triewald later wrote that Hultberg was destitute when Triewald took him in hand,[19] which must have been in late 1726 or early 1727.

During the five weeks Triewald spent in Dannemora, he had the building staked out, excavated rock for the firebox, brought stone to the site and began to erect the engine house.[20] The foundation was laid on May 2, and on May 8 work on the walls began.[21] On June 6, however, Triewald had to leave Dannemora for a few weeks to return to Stockholm.[22] For the next few years he travelled between Stockholm and Dannemora several times each year, and after he became Technical Director of Wedevåg Ironworks in 1728, he travelled regularly in a triangle, the corners of which were two or three days' journey apart,[23] and he may well have spent a month on the road every year. On his return to Stockholm on this occasion he applied to the Government for a suitable appointment in keeping with the knowledge and skills he had acquired in England.[24] His application came up for discussion in a Government committee on June 19 and went the rounds between various bodies during the following weeks. On June 20, however, Triewald was back in Dannemora in time for a meeting with the Partners. There he succeeded in securing 36 *copperdaler* a month for Hultberg,[25] and Triewald was later to maintain that Hultberg, like himself, was an employee of the Partners. For their part, the Partners always maintained that Hultberg was Triewald's man. While the meeting was in progress, the large copper boiler arrived from Stockholm on June 23,[26] and on the following day the master mason could finish his work. The building was now "externally ready and the boiler walled in".[27]

On July 4 the Government decided that Triewald should be awarded the title of *Directeur* and a salary of 1 800 *copperdaler* per year. (This is equivalent to a monthly salary of 150 *copperdaler*, which may be compared with the 36 *copperdaler* Hultberg was receiving in Dannemora.[28]) Triewald's duties and his place in the administrative structure in this post are unclear. The Government recorded that he would assist the Society of Science in Uppsala, and be responsible to the Board of Mines. None of this appears to have been applied in practice, and the post is probably to be regarded as an honorary one granted to a deserving citizen on whom the Government pinned great hopes. It was emphasized, however, that for his salary Triewald should get his engine "into proper order so that the public may have the anticipated utility thereof".[29] Like Triewald himself, the Government saw the engine as of public benefit. But this was not a view shared by the Partners in the Dannemora Mines. They were concerned with their own rather than the public utility.

"*Mechanicus*" Triewald was now "*Directeur*" Triewald, or, as he henceforth styled himself, "*Director Mechanicus*". Work on the machine continued: the huge beam was hoisted into position at the top of the building, and on August 1 work began on the boring of the elm logs that were going to be used for the pump pipes in the mine.[30] These pipes were to prove a complication that delayed and raised the cost of the work. Originally Triewald had meant to cast them in iron, which he was used to employing in England, but it turned out that none of the Partners' ironworks was equal to the job. So he had to go over to the traditional Swedish method of using hollowed-out elm logs. This was a difference in traditional materials and technical capacity between England and Sweden that had a bearing on the failure.

On August 21 Triewald left Dannemora again for Stockholm,[31] this time to prepare his public lectures in experimental physics, "the new natural philosophy". He had brought a collection of instruments back from England (no doubt the one he had used for his lectures with John Thorold in Newcastle and Edinburgh), and now he was going to repeat his success in Stockholm. He applied to hold the lectures at the Palace of Nobility (*Riddarhuset*), and this established a link between the Swedish nobility and science that was to play an important part in the foundation of the Royal Swedish Academy of Sciences.[32]

On September 25 a ship from London arrived in Stockholm, carrying the parts for the Dannemora engine that had been ordered by Triewald and shipped over by the agency of Jonas Alströmer (Fig. 13.1).[33] These "Two Casks Two Cases four Boxes Sixteen parcels of Merchandize and fourteen Pipes of Lead" comprised the parts for the engine which Triewald had considered himself compelled to buy from England: the regulating mechanism, leather for sealing the cylinder and the pumps, and lead pipes. Triewald probably supervised the unloading and the forwarding of the goods to Dannemora, but he also had other matters to attend to. On October 2 the first announcement of his planned series of lectures appeared in a Stockholm newspaper.[34] On October 18, Hultberg left Dannemora, being summoned by Triewald to Stockholm.[35] Hultberg also served as Triewald's instrument maker, and there were doubtless many instruments to be repaired or manufactured before the

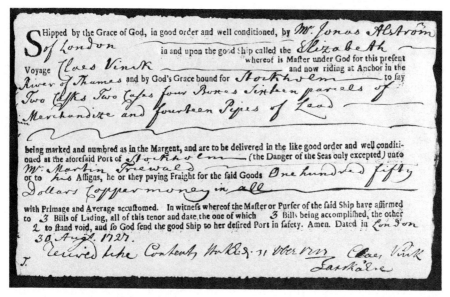

Fig. 13.1. The bill of lading for the parts for the Dannemora engine which were ordered from England. They had been shipped by the agency of Jonas Alströmer, and arrived in Stockholm on the *Elizabeth* on September 25, 1727. (Photo Stora Kopparbergs Bergslags AB)

lectures. During these years Hultberg, just like Triewald, was torn between his two roles in science and technology. Triewald had to reconcile his capacities as scientific lecturer and introducer of the steam engine, and Hultberg the more humble positions of maker of scientific instruments and engineer at Dannemora.

A problem had arisen, however, that obliged Triewald to devote himself to the project in Dannemora. On November 20 he wrote to Mine Inspector Bellander and reported that unfortunately the cylinder that he had ordered in England could not be delivered.[36] The first two attempts to cast it, Triewald said, had been failures, and when finally the third had been successful, it was too late in the year for him to dare ship the cylinder from England to Stockholm. This is an example of the way the project was delayed by Swedish climatic conditions. The cylinder in England had been cast in iron, but Triewald had now been promised by Gerhard Meijer (1704–1784) that in two months he could cast a bronze cylinder at the Royal Gun Foundry in Stockholm. Like Triewald, Gerhard Meijer came from a German family of artisans in Stockholm,[37] and as has been mentioned earlier (Chapter 10), the Triewald and Meijer families rented adjacent pews in the German Church. Meijer was a fourth-generation gun founder, and at the age of only twenty-four he took over responsibility for the Royal Gun Foundry (which was in effect a family concern) at the beginning of 1728. In the autumn of 1727 he had returned from a study trip of several years that

Fig. 13.2. The Palace of Nobility (*Riddarhuset*) in Stockholm, where Triewald gave his lectures in experimental physics. The lectures in 1728 were held in one of the rooms on the ground floor (the two windows to the far left). The statue of King Gustav Vasa in front of the Palace was cast in 1770 by Gerhard Meijer, who in 1728 had cast the cylinder for the Dannemora engine. (Photo the author, 1984)

had taken him to Holland, England, France, Austria and Germany. He had done most of the journey on foot, and had taken employment at various foundries in order to learn their trade secrets. While abroad, Meijer had, Triewald wrote, himself seen steam-engine cylinders being cast and made a drawing. According to Triewald, the whole business was almost to be regarded as an advantage, and at the end of his letter to Bellander he added:

P.S. I forgot to say that brass remains brass of the same value even after hundreds of years of use.[38]

But it was an advantage that counted for little in the eyes of the Partners, whose financial calculations were not based on such a long-term view. What mattered to them was that it should not involve increased cost or delay the project. The matter was discussed at a stormy meeting in Dannemora on December 12 and 13, in which Triewald took part.[39] They decided that they would let Triewald have the cylinder cast at Meijer's in Stockholm, but that in that case he was to assume financial liability for the cylinder in England. He would have to try to sell it where best he could, but the Partners were to be indemnified. This decision was one for which Triewald was later to criticize the Partners many times during the lawsuit. Here for the first time open distrust seems to have revealed itself.

A few days later, Triewald was in Stockholm again, this time because the instruments for his lectures were to be put on public display for a week.[40] On January 29, 1728, Mårten Triewald commenced his series of lectures at the Palace of Nobility.[41] Here he began to realize his view of the social role of science: the view that he was to expound in 1729 in his criticism of the Society of Science in Uppsala and to put into practice in 1739 by founding the Royal Swedish Academy of Sciences. First, science should not be pursued only in the universities, and it was for this reason that he was giving his lectures in the capital, which had no university. Secondly, science should not be pursued for its own sake but should be of practical service. This seems to have been a recurring theme, judging from his published lectures. Thirdly, science should be taught in the native language in order to fulfil its role as a public utility. Just as he had lectured in English in Newcastle and Edinburgh, he now lectured in Swedish in Stockholm and had his lectures published in that language.

The famous physicist Anders Celsius (1701–1744), then a student at Uppsala University, travelled down to Stockholm to be present at the whole series of Triewald's lectures. He stayed there from the end of January until April, and on April 2 he wrote to Erik Benzelius, telling him about the lectures.[42] Celsius mentioned Hultberg's activities as a maker of scientific instruments:

> The *Directeur* [Triewald] has improved the Englishmen's *Antlia*; so that I can fairly say that we have here in Stockholm the best and most useful air pump that has yet been made in Europe. There is a student here, named Hultberg, who is under Triewald's instruction. He has now progressed so far in mechanical practice, that he is now making an *Antlia* invented by the *Directeur*. Privy Councillor Gyllenborg and Baron Saack have ordered from him, and several others. He has also made many hydrostatic balances, which are quite as good as the Englishmen's.[43]

It is well known that Triewald's lectures drew many distinguished listeners: privy councillors, generals, foreign ambassadors etc.[44] At this time, as was pointed out in Chapter 9, science was not a profession but an interest shared by the higher strata of society. Hildebrand demonstrated in 1939 that among the subscribers to the published lectures there were many who later came to be elected to the Royal Swedish Academy of Sciences.[45] But it has not previously been remarked that one of the largest groups among Triewald's subscribers, and therefore probably one of the largest groups in the audience, was that of the merchants. The title *Mercator* is given for 18 of the 216 subscribers, and the list includes names representing many of Stockholm's big trading companies.[46] It is not likely that they could convert Triewald's elementary experiments into practical improvements in their own daily duties, but this shows that interest in science was now also supported by other social groups.

During the cold winter months Triewald could devote all his interest to science, but as soon as spring came, his duties as an engineer called. Thus the division of his life between science and technology assumed a natural rhythm regulated by the changing seasons. To Triewald, science became a winter activity and technology a summer activity. On May 1, 1728, he was back in Dannemora, and he stayed there until the end of the summer.[47] Two days later Hultberg arrived from Stockholm and brought the cylinder, which had been cast and bored by Meijer, and it was installed in the building on May 7.[48] As his assistant during the lectures, Triewald had employed Daniel Menlös (1699–1743), who had studied mathematics at Uppsala University.[49] Menlös also came out to help him at Dannemora that summer, arriving there on May 10.[50] (Menlös later purchased Triewald's instruments and succeeded in obtaining a professorship at Lund University after promising to donate the collection. Some 70 items from the original collection of 327 instruments are today preserved at the Malmö Museum of Science and Technology.[51])

Mine Bailiff Thomas Kröger reported regularly to Mine Inspector Thor Bellander on the progress of work on the engine. At first his reports were neutral in tone, as in this example, written on June 17, 1728:

> And as far as Mr *Directeur* Triewald's engine is concerned, work proceeds with it daily, and it is intended to be ready except for the hoisting machinery for the next Court of Mines. He had some trouble this week with the tank or water cistern, as this had not been made watertight as he desired, but this difficulty has finally been overcome, so that the tank and the boiler are now full of water. He has also begun to assemble the large pumps, but must first make a hand pump, with which he wishes to pump part of the water out of the mine before he can lower the large pumps.[52]

On July 4, 1728, the machine could be started up for the first time.[53] One year and eight months had then elapsed since the contract was signed. The timing was not determined solely by technical factors, however, i.e. by the time it had taken Triewald to get the machine ready once the cylinder had been fitted. No, July 4 was the name day of Queen Ulrika Eleonora,[54] and it was surely no coincidence that this particular day was chosen for the engine to be "set in motion [...] for the first time by Mr *Directeuren*".[55] With this, an epoch-making date in the history of Swedish technology was linked to the society of eighteenth-century Sweden, and the connection was Triewald's own ambition: his striving for notice and esteem.

OPERATION

A week later, the annual Court of Mines was held in Dannemora, and at the same time the Partners held a meeting. They seem to have been impressed by the engine, as they decided to give Triewald 3 000 *copper-daler* "of the promised remuneration on account".[56] The engine must indeed have presented a remarkable sight to those who had never seen anything similar: a large stone building with a belching chimney, and projecting from it a great wooden beam, which was driven up and down by an invisible force so violently that the ground trembled. At the Court of Mines Hultberg was also appointed *Konstmästare* (Chief Engineer), and granted a monthly salary of 50 *copperdaler* instead of his previous 36.[57] Triewald later wrote that he himself had trained and examined Hultberg, and this nomination of an outsider to the traditional community of the Dannemora Mines must have vexed those who had served at the mines for years and risen only slowly through the ranks. Hultberg, an erstwhile student from Uppsala, had been admitted to their circle on the strength of Triewald's recommendation. And now the tone of Mine Bailiff Kröger's reports changed.

In August Triewald encountered problems with the elm mine pumps, which kept breaking, and Kröger recounted the difficulties with a certain malicious delight.[58] He also wrote that no preparations had yet been made for building the hoisting gear, "which would therefore seem to have been postponed for this year".[59] It is clear that Kröger had now dissociated himself from the whole project, and regarded it as Triewald's responsibility alone. By the end of August the engine had only managed to remove 12 fathoms of water from the mine,[60] and on August 24 Kröger reported:

As far as the engine is concerned, it is still not complete, and God knows when it will properly come about, for always there is something breaking or wanting, and it is also so with the pumps; it has already wasted 57 *stafrum* of wood.[61]

Kröger had previously declared that the engine consumed 10 *stafrum* of wood per day ("no little loss for the company"),[62] but Triewald was later to maintain that it used only 6 *stafrum*.[63] At trials arranged by the Board of Mines in 1734, it was shown that the engine consumed 8.9 *stafrum* of wood per day,[64] which indicates that Triewald tried to overstate his engine's performance while Kröger sought to understate it. Kröger's note of the total wood consumption suggests that the engine had been working for altogether a week since it was first started up two weeks earlier. The use of the word "wasted" implies that Kröger made no allowance for the modification and adjustment that the engine needed and was now decidedly hostile to it.

But in Triewald's eyes it was a machine to be proud of. He had had a visitors' book placed in the engine house, and inscribed in it an introductory quotation from Ovid: "I begin a great work, but glory gives me strength" (*magnum opus aggredior, sed dat mihi gloria vires*).[65] This was related by Gustaf Benzelstierna in a letter to his brother Erik Benzelius on August 30:

While I was in Upland I went to Dannemora and there viewed Mr Triewald's new Fire-engine, which is of very great effect and draws many spectators thither, and is indeed worth seeing. He has them also write their names in a book which is there expressly for that purpose [...] Count Freytag was the first to sign it. This engine has already pumped out one shaft that has been under water ever since 1701.[66]

In fact the mine was not entirely drained, but the statement shows that visitors who were uninvolved found the engine more impressive than did Mine Bailiff Kröger. It was presumably thanks to Benzelius that a notice appeared in *Acta literaria Sveciae*, the journal of the Society of Science in Uppsala, about "this engine, admirable both in its elegance of form and in its great success".[67] Triewald was aware of the unique nature of his engine, for he had had the following inscription, framed by two palm branches, engraved on the cylinder:

Cylinder of
the first Fire-Engine
in Sweden built by Mart.
Triewald: cast by ger.
Meyer in Stockholm A.D.
MDCCXXVIII[68]

The greatest problem was, as we have seen, that the elm pump logs proved too weak. Triewald had arranged two pump assemblies, each 18 fathoms in length, in the mine, which was 36 fathoms deep. But this design had been based on the original assumption that he would be pumping through cast iron pipes. The whole construction had now to be rebuilt, and instead three pump assemblies, each 12 fathoms long, were installed, and the logs were fitted with iron rings. Now even Kröger had to admit in his report to Bellander that since this modification the pumps had "more or less held", and that the engine had reduced the depth of water in the mine to less than 20 fathoms.[69] He added, however, that the engine was "very unsteady in its action, for there is always something breaking that has at once to be repaired, and it has still not yet worked for three days at a time".[70] Nor could he refrain from adding his own gloomy prophecy: "I fear that when winter comes the whole works will freeze solid".[71]

On September 20, 1728, Triewald left Dannemora.[72] Winter was approaching, and with it the time for his return to his other life as a scientific lecturer in Stockholm. A second series of thirty lectures was due to begin on October 15, and he not only announced these but also had a sixteen-page brochure printed.[73] In Dannemora his engine was silent, for the third pump assembly needed a new piston, which had to be cast by Meijer in Stockholm.[74] But Triewald did not concern himself only with his lectures, he also wrote the first of his many letters to the Royal Society in London. On November 20 he wrote to John T. Desaguliers, on whose example he had modelled his activity as a scientific lecturer (see Chapter 10), and told him about an interesting experiment in cohesion that he had conducted "last Winter before a great & noble Assembly at my Lectures".[75] A possible explanation of this had occurred to him, he said, when he observed how the workmen at Dannemora split stone with fire. This letter, like many that followed, was published in the *Philosophical Transactions*.[76]

During the first few years after his return, Triewald had also been busy applying other knowledge acquired while he was in England. One such interest was his diving bell (see below), while another was bee-keeping. Honey was much more important in the eighteenth century than it is today, being one of the commonest sweeteners. Triewald therefore took an interest in it for purely practical reasons and not merely for the sake of the pleasant pottering sometimes associated with apiary. In September 1728 his book *Nödig tractat om bij, deras natur, egenskaper, skiötzel och nytta* (A needful treatise on bees, their nature, character, care and usefulness) was published.[77] In the book, Triewald introduced a new type of beehive, and a model of this was put on sale with the book in the shop of Wedevåg Ironworks in Stockholm.[78]

Triewald had been engaged as Technical Director at Wedevåg Iron-works, about 150 kilometres west of Stockholm, in 1728.[79] This company was untypical of Swedish ironworks of the period, in that it had a very wide range of iron and steel products. About 150 people worked there in some fifty different workshops. A third of them were foreigners, mainly from Germany but also from England, France and the Netherlands. A slitting mill had been installed at the works in 1726, the first of its kind in Sweden. In 1728, Baron Daniel Niclas von Höpken (1669–1741) had become the largest shareholder in the works. He was an eminent politician, and one of the leading figures in the Holstein party, to which Mårten Triewald's elder brother, Samuel, also belonged (see Chapter 10). Like the Triewald family, the von Höpken family were from Germany, and were members of the German congregation in Stockholm. It was no doubt due to these connections that Triewald came to be employed at the ironworks. Concerning Triewald's achievements at Wedevåg Ironworks, Sixten Rönnow said in 1944:

> By the start of the 1730s Wedevåg had overtaken its rival, Eskilstuna, and was the foremost of its kind in Sweden. The technical progress of the late 1720s can largely be attributed to one man, who became celebrated as one of the most versatile and ingenious Swedish engineers of the eighteenth century: Mårten Triewald [...] According to the company's records, he introduced a number of prototypes of new products direct from England [...] he appears to have resided in several places, chiefly Stockholm and Uppsala, and to have had other duties there in addition to his business in Wedevåg.[80]

On January 15, 1729, the piston for the third pump reached Danne-mora, but Kröger reported that nothing was being done to install it.[81]

The pump that supplied water to the tank above the cylinder was frozen, and Kröger speculated that it may have been burst by the cold. It was not Hultberg's fault that so little could be done while Triewald was in Stockholm, but that of the winter, and Hultberg does not seem to have received any assistance from Kröger. Triewald arrived back in Dannemora on March 14,[82] and four days later Henrik Kalmeter told Jonas Alströmer in a letter: "Now he is gone to Danne' to finish his engine & in the month of June he is to settle at Wedewog".[83] But Triewald did not have time to remain in Dannemora to finish his engine. The next week he had to be in Stockholm again to perform optical experiments for a couple of days between 7 and 10 o'clock in the morning.[84] "A want of fine days" had obliged him to postpone these experiments from his series of lectures the previous December,[85] and he had now to make use of the morning light of the first days of spring. It was not quite possible for him to confine his scientific activities in their entirety to the winter half of the year in a northern country.

He was soon back in Dannemora, however, and was able to attend the first day of the annual Court of Mines on May 7.[86] At the meeting two machine minders were engaged, Hultberg was given better accommodation, and it was decided that the Partners' ironworks would supply materials for the hoist which was now to be built. On the following day "the *Directeur* departed for his wedding".[87] On May 29 Triewald married the 23-year-old Anna Margaretha Brandt (1706–1738).[88] It was a marriage within his own circle. Her father was originally from Holstein, and the family were members of the German congregation in Stockholm. Hildebrand (1939) has shown how by his marriage Triewald forged kinship ties with the group that came to form the nucleus of the Royal Swedish Academy of Sciences.[89]

But it may not have been only his imminent marriage that called Triewald away from Dannemora. On May 19 he was granted a 20-year patent for his diving bell, and together with his partners he was given a 10-year licence to salvage wrecks and cargo along the coast.[90] He already had a shorter patent, obtained a year earlier, for his diving bell,[91] which was a modification of the bell he had seen demonstrated in the Thames by Edmund Halley in 1716, his first year in London. Triewald and three other persons had now formed a diving and salvage company, *Norra Dykeri- och Bergnings-Societeten.*[92]

In Dannemora, Hultberg was trying to keep the engine working, and at the same time directing work on the hoist. The problem of the pumps appears now to have been resolved, but the difficulty was to keep the engine in continuous operation. Kröger reported on June 13 that there was still 4¼ fathoms of water in the mine,[93] which means that the engine had managed to remove most of the 36 fathoms of water that had filled the mine to begin with. At this time Dannemora was visited by the only impartial witness who has left a description of the first Swedish Newcomen engine in operation. This was none other than

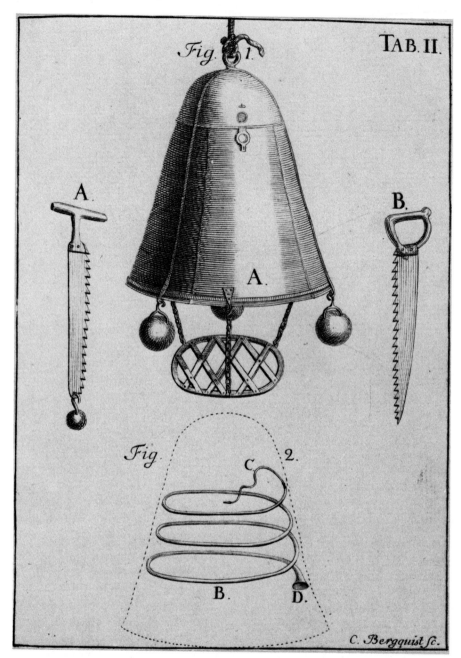

Fig. 13.3. Triewald's diving bell. Engraving by Carl Bergquist in Triewald's book *Konsten at lefwa under watn* (The art of living under water), published in 1734. (Photo Bengt Vängstam)

Linnaeus, who was still a young student. On June 23 he wrote in a letter to his Professor, Kilian Stobæus, in Lund:

At Dannemora I continuously watched Trivald's fire Engine. Whose power is so great that the whole mine, the building and all its appurtenances shake. The beam on which the shaft is mounted, which is of 3 squared timbers, bends under the weight. I could say more about this engine, for I believe I quite understand how she is driven, but I assume she is wellknown to the Professor already.[94]

(Linnaeus is the only person to refer to the Dannemora engine by a feminine pronoun. In all other sources the machine is referred to as "he" and the mine as "she".)[95] Linnaeus does not say anything about the engine having stopped, and it evidently worked often enough for a casual visitor to be able to see it in action. He also wrote that during his visit (May 26–June 10) he watched it "continuously", which confirms that it seems to have functioned reasonably well during the summer of 1729.

On September 15 Triewald was back in Dannemora, having received a letter from the Mine Inspector asking him to return.[96] The tone of his letter was still polite, but it is clear that he, too, had begun to have his doubts about the engine and the hoisting gear. Triewald attended a meeting of the Partners where they criticized him for his failure to keep the engine in operation, and for the delay in completing the hoisting machinery.[97] The main problem with the machine seems to have been the difficulty of keeping water in the tank above the cylinder in sufficient quantities to condense the steam. (Kröger had reported the previous June that the tank had been found to leak even when it was newly built.)[98] It was now decided to clad it in lead sheeting. A later proposal for repairing the engine, undated and unsigned, states that, after careful inspection of all the parts:

nothing has been found on the engine itself that requires improvement or alteration, except for the injection valve, which cannot have been well made and finished to start with and has always leaked much.[99]

These two leaks—in the tank and the injection valve—must have contributed to the difficulty of supplying enough cold water to condense the steam. But the crucial factor was probably to be sought elsewhere. Triewald cast a 36-inch cylinder instead of the one of approx. 26 inches that was the average in England when he left.[100] He thereby increased not only the power of the engine by over 90 % but also the volume of steam to be condensed. This meant that almost twice as much water was now required in order to condense the steam with the same rapidity. It is unlikely that Triewald had increased this parameter to the same extent as the area of the cylinder. The attempts to seal the tank and the injection valve were probably nothing more than symptoms of a major defect in his original design—a design which was inspired by Triewald's ambition to build not only the first Newcomen engine in Sweden, but also the biggest ever. In the letter to Desaguliers that was referred to earlier, he wrote:

I have erected the first & largest Fire Engine for drawing Water & Oare in this kingdom, the Cylinder being 2 lines more than 36 Inches in Diameter.[101]

Even a sixth of an inch was worth mentioning when Triewald wanted to boast of his achievement to Desaguliers and the Royal Society! The reasons for the operational unreliability of the Dannemora engine are therefore ultimately to be found in Triewald's personal ambition. And this in turn may stem from his background (see Chapter 10): the blacksmith's son who had emigrated to England for better or worse as a failed businessman wanted to show when he returned that he was second to nobody; the colliery engineer in Newcastle, who had worked for many years in a subordinate position, wanted to show now that he could do better than anyone in England.

At the meeting of September 16, the Partners upbraided Triewald for not yet having completed the hoist.[102] They took the view that he had neglected his obligations by not spending more time in Dannemora, and that Hultberg was quite inadequately trained to be able to complete the construction work himself. Seven workmen had now been working under Hultberg's direction since the beginning of the summer in order to complete the project. Triewald protested that the Partners had themselves to blame for having treated Hultberg so badly. Hultberg was then called to make his own complaint:

Hultberg was called, and upon being questioned, would charge Mine Bailiff Kröger that he had vented upon him harsh words and opinions and in particular that Kröger had said that when Hultberg came under the disposition of the Dannemora Mines, Kröger would have him dance to a different tune or teach him better ways, and more. Which Kröger being present steadfastly denied, stating this to be nothing more than an unfounded gossip that can never be proved.[103]

Kröger suggested that Hultberg should summon him and try to prove his allegation before a court. He himself asseverated "on his salvation" that he had placed no obstacles in the way of Hultberg but had procured for him all that Hultberg had requisitioned for the engine.[104] Mine Inspector Bellander pointed out to Hultberg that the meeting was not concerned with legal matters of this nature, which should be brought before the regular Court of Mines in the spring. "Should he feel himself to have cause", Hultberg ought to follow Kröger's advice and summon him to appear before the Court.[105] Kröger and other mine bailiffs were however warned not to treat Hultberg unjustly—which suggests that the Mine Inspector was aware that there was some substance in the accusations. But all replied:

that they had neither themselves given him the slightest affront nor heard any other do so, but had shown him every courtesy and obtained for him all that he had required for the building.[106]

A glance at Fig. 11.2 in Chapter 11 gives us reason to be sceptical of their unanimity. They were all subordinate to Kröger, and his responsibility for organizing the daily routine in the mines made them dependent on his favour. Their main concern was to maintain the supply of ore to their respective ironworks. What Kröger and the other mine bailiffs had to say to Hultberg privately after the meeting can only be left to the imagination, but in the middle of the next day's meeting:

> *Konstmästare* Olof Hultberg entered and told the Partners that he wanted to resign from his position. But he was referred to Mr *Directeur* Triewald, in order first to notify him, under whose disposition he stands.[107]

The incident shows that the Partners, just like Kröger, now regarded the machine as entirely Triewald's responsibility. They were not prepared to adopt a helpful attitude to overcome the difficulties that had arisen between them. They paid Hultberg's salary, it is true, and made accommodation available to him, but they were careful to point out that he was considered Triewald's employee. Olof Hultberg was the only person, other than Triewald, who could operate the engine and supervise the work on the hoist. If Hultberg could not be persuaded to remain in Dannemora, Triewald would be obliged to abandon his scientific ambitions in the capital and stay in Dannemora indefinitely. The winter was again approaching, and with it his programme of scientific activities. Triewald concluded an agreement with Hultberg the following day, September 18 [108] If Hultberg remained at his post and completed the hoist, he would receive 100 *copperdaler*, and he was to report to Triewald on the progress of the work every mail day. Triewald had been paid 3 000 *copperdaler* the previous summer as a remuneration on account.[109] He ought therefore to have been able to afford to promise Hultberg a much higher reward for enduring the harassment of the mine bailiffs, but he evidently found two months' extra salary sufficient for his hired assistant. Theoretical technological knowledge and practical work commanded very different scales of pay in early eighteenth-century Sweden.

Triewald left Dannemora the same day, this time bound for Wedevåg Ironworks,[110] and two weeks later he was in Stockholm to arrange a public demonstration of his diving bell in Lake Mälaren for members of the parliament.[111] An enthusiastic newspaper paragraph, probably composed by Triewald himself, reported the event:

> When the diver had thus understood the excellence and importance of this invention, he pronounced himself willing to live for a whole day in this manner on the lake bottom.[112]

A less happy state of affairs prevailed in Dannemora, however, and on October 11 Kröger reported to the Mine Inspector on events since the meeting the previous month. After Hultberg's public accusation of him, Kröger appears to have done his utmost to present the engine in as poor a light as possible.

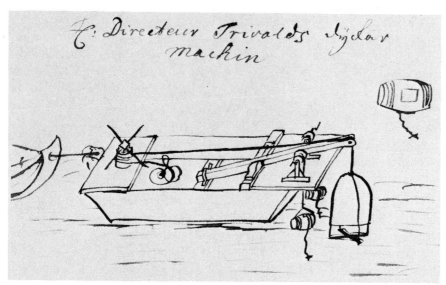

Fig. 13.4. Triewald's diving barge with the diving bell and the barrels supplying the diver with an extra reservoir of air. Drawing by Carl Johan Cronstedt
in 1729 in his sketchbook. (Photo National Museum of Science and Technology, Stockholm)

The machine has worked 3 times since the meeting, the first time on October
1 it went for 6 minutes, the next day from 2–6 o'clock in the afternoon and the
3rd time on October 8 from 2–7 in the aft. noon, and broke down every time
after the water in the tank had been used up.[113]

Hultberg worked hard on the modification of all the weak points in
the design, however, and before long he succeeded in making it deliver
18 strokes per minute (the average for the early Newcomen engines was
14).[114] He was also so near to finishing the hoist that on October 26,
1729 he could write to Triewald and ask for his 100 *copperdaler:*

The hoisting gear has already been tried and shown more promise than I
could have imagined, as Mr *Directeur* I'm sure will find [...] giving me good
reason to request 100 *copperdaler* as promised, but I do not think I dare yet put
barrels on the ropes to test them better. The foe has now to say of the engine
and hoisting gear: 'Well, bless my soul etc: don't tell me it is working now'
[...] but at the end had time to take a pipe of tobacco, while I sat and enjoyed
its motion [...] The big winding drum turned on the cogs as easily as the
bobbin of a spinning wheel.[115]

When Hultberg tried to describe the performance of the steam-
powered hoist, a new technology in Sweden, he used a comparison
borrowed from a technology traditional in rural Sweden, just as Kalmeter had likened the cylinder of the first Newcomen engine he saw in
England to a churn when he wanted to describe it to the Board of
Mines. But Hultberg's adversary did not allow himself to be impressed
with this test of the unloaded hoisting gear. The engine had not yet
succeeded in emptying the mine of water, and in his reports during the

autumn of 1729, Kröger related how unreliable it was.[116] At the end of November, however, Hultberg managed to keep the engine working for a few days so that the mine was once more drained to within 3 fathoms of the bottom, but then it proved impossible to get it working again for several weeks.[117] Meanwhile, Triewald had other preoccupations in Wedevåg and Stockholm. In the middle of December he made an interesting observation in Stockholm, which he described in a letter to Sir Hans Sloane, President of the Royal Society:

> The 15th December last coming into the Hall where my Apparatus is plac'd in the Pallace of the Nobility here, The Weather being very cold, I fear'd that the Glas (for showing the Experimt with the Cartesian Devils [...]), would be in Danger if the Water should frese in the same, I took it down from the shelf & was pleas'd to see the Water in a fluid state ...[118]

But when Triewald was about to demonstrate the experiment with the Cartesian divers to some friends who were with him, the water suddenly turned to ice and the little glass figures stopped in their movement in the middle of the glass tube. Here was yet another strange phenomenon to report to the Royal Society! And his letter was published in the *Philosophical Transactions* under the title "An Extraordinary Instance of the Almost Instantaneous Freezing of Water".[119] We shall return to this episode when we discuss Triewald's scientific method in Chapter 15. Here it may serve just to illustrate the degree to which he was preoccupied with science at the expense of technology. The freezing water in the small glass tube in his lecture theatre was more important to him than the freezing water in the huge wooden mine pumps of his engine in Dannemora.

A few days later, he received an unexpected letter from Hultberg. The latter had had enough harassment in Dannemora, and again wished to resign. Triewald wrote an indignant reply, a mixture of threats and promises.[120] Was the trouble and expense to which he had gone in order to help Hultberg to be repaid with untrustworthiness and ingratitude? If he did not fulfil his obligations then he would certainly find out what it meant to take on a whole mining company! He need not imagine that he was the only person who could do the job. But if he stayed on and carried out his duties until the next meeting of the Partners, Triewald would help him to secure the desired departure with "some sort of honour".[121] The next meeting was not due until early the following spring, and Triewald was presumably reluctant to leave Stockholm and his scientific activities before then. Hultberg was indeed the only person who could do the job, and it was therefore very important to Triewald for Hultberg to remain at his post until the engine and the hoisting gear were functioning and their operation could be transferred to other hands.

But Triewald had more important matters on his mind. A week later he was once more at Wedevåg, from where he wrote on December 27, 1729 to the Society of Science in Uppsala.[122] When Triewald had been

Fig.1.

Fig. 13.5. Triewald's glass tube with the Cartesian divers. The divine hierarchy was preserved even in a scientific instrument: man hovers between heaven and hell. Detail of an engraving probably by Carl Bergquist in the second volume of Triewald's *Föreläsningar öfwer nya naturkunnigheten* (Lectures on the new natural philosophy), published in 1736. (Photo Bengt Vängstam)

appointed *Directeur*, the Government had recorded that he was to assist the Society in Uppsala. His election did not take place for over a year, however,[123] and it is likely that the Society, like the Board of Mines, was not particularly pleased to be saddled with outsiders by the Government. Triewald attached his own interpretation to his task of "assisting" the Society, and the new member's first act after being elected was to dispatch a long letter proposing a comprehensive reorganization of the Society and its activities. Hildebrand discussed the contents of this letter thoroughly in 1939, for here Triewald anticipates everything that he was to put into effect ten years later with the foundation of the Royal Swedish Academy of Sciences.[124] What Triewald advocated, in a nutshell, was a society modelled on the Royal Society in London. Triewald considered that the Society of Science should renounce the pedantic atmosphere that was characteristic of the university city and move to the capital, that merchants and engineers should be elected as members, and that the Society's Proceedings should be published in Swedish instead of Latin. (The Society of Science did not mention his letter in its minutes.) Two days after writing this letter, Triewald made a new observation, one which again was reported to the Royal Society:

on the 29th of December 1729 [...] I was travelling in a sledge on the Ice, being then on an inled from the Botnick Gulf, my Course being Northerly I perceiv'd in the Evening about 8 o Clock, directly before me a very fine Lumen Boreale [...] I caus'd the Postillion, who drove my sledge, imediatly to stop, which he was unwilling to do.[125]

In March 1716 there had been a rare and magnificent display of the northern lights over almost the whole of Europe, and this had done much to arouse interest in the phenomenon in England and on the Continent.[126] After 1716 the Royal Society gladly published articles on the subject in the *Philosophical Transactions*. Triewald made use of his northern position as a scientific resource—here the aurora borealis was not unusual. Sweden afforded a "laboratory environment" that gave him an advantage over English observers and increased his chance of attracting notice in the Royal Society. Just as he had written about supercooled water in his glass tube with the Cartesian divers, so he also wrote about the northern lights. In the cold climate of Sweden it was natural to seek a connection between ice and the northern lights, whereas in the milder climate of England it was natural to search for a link with other phenomena (e.g. variations in the magnetic field). Triewald also believed, like many others at this time, that the aurora was caused by the reflection of moonlight in ice crystals in the atmosphere.[127] After inducing the driver to stop the horse, he made his observations while the moon was rising and the fog was "being turn'd by a severe Cold into Icy particles."[128]

In Dannemora, the Partners were now beginning to grow restless. Eighteen months had elapsed since the engine was first started up, and it had still not succeeded in draining all the water from the mine. Nor had any convincing tests of the hoist been carried out, and the hoist was in any case dependent on the continuous functioning of the engine. During the autumn of 1729 it had worked sporadically—for anything from six minutes to three days at a time—punctuated with stoppages of up to three weeks. Ever since Hultberg's accusations in September, the focus of Kröger's criticism had been the engine's unreliability:

sometimes it breaks down, sometimes it draws no water, sometimes the piston seizes in the cylinder so that the whole machine stops [...] so that one does not know what to think of this engine.[129]

This was not a situation that augured well for the continuous working of the mine, which was a necessary requirement in view of the way that work was organized in the Dannemora Mines. Each ironworks, as we have seen earlier, owned shares in the different mines, and the shares gave an entitlement to a corresponding number of weeks of extraction each year from the mine concerned. Had the mines had one common owner, the situation would have been very different, because the engine had in fact shown that it could empty large amounts of water in a short time. But an engine that could not guarantee continuous operation was quite unpractical in Dannemora. It would involve unfair advantages and disadvantages to the different ironworks in that the mines might be kept free from water for certain weeks but be waterlogged during others. It thus threatened the whole system of weekly shares, which was the basis of the social organization of the Dannemora Mines.

At the beginning of March 1730, Mine Inspector Bellander called the Partners to a meeting in Dannemora on March 18 to discuss the engine.[130] Triewald had also been summoned to the meeting—but he did not come.[131] He later claimed that he had received the letter much too late for him to be able to get there in time.[132] It is more likely, however, that he was loath to be called to account, and possible also that he thought it was too early in the year. Spring (technology) had not yet arrived; it was still winter (science). The Partners' patience was now at an end, and they wrote to the Board of Mines and asked it to instruct Triewald to go to Dannemora and repair his engine.[133] The letter reached the Board on April 7,[134] and Triewald received it a few days after he had written the above-mentioned letter to Sir Hans Sloane on "an extraordinary instance of the almost instantaneous freezing of water" and on another interesting phenomenon concerning the flowering of tulips in a cold climate.[135] The Partners' complaint appears to have been a shock to him, but rather than set off immediately for Dannemora, he wrote an agitated rejoinder, no less than 170 folio pages long and full of countercharges.[136] With this began the correspondence that was to lead to a lawsuit between Triewald and the Partners, in which the new technology was brought to trial.

14. Technology on Trial, 1730–1736

It would take far too much space to discuss the contents of the letters exchanged by Mårten Triewald and the Partners after their differences erupted into open conflict. The value of the letters as source material is in any case doubtful, and it is clear that they were often written in a state of great emotion. We shall concentrate instead on what the parties stated during the hearing in 1731. There they had to prove every allegation, and at the same time the court heard a large number of sworn witnesses. During the autumn of 1730 the Board of Mines made several attempts at conciliation,[1] but in December it gave up and convened an Extraordinary Court of Mines in Dannemora to decide the issue.[2] Facing the threat of a court case, Triewald himself tried to ward off the action by writing to the Partners.[3] He begged them not to hold his long plea of defence against him: it had been written by a scatter-brained clerk, and his own fault had only been in hastily signing it. Surely they would not hold this trifle against him? "They are but immature thoughts that have flown from the rude mind of another to the paper that I submitted and deserve thus rather to be overlooked than denounced."[4] If they nevertheless persisted with their suit they could not hope to gain much, for it was generally known that he was a poor man. Their only satisfaction would be in knowing that they had ruined a poor but well-meaning patriot. Triewald said that he had believed that the Partners wished to support him in his endeavour to introduce the engine to Sweden "to serve my fatherland and all honest patriots".[5] What he had intended was a "work of utility to the public" and to themselves,[6] and there was a risk that a lawsuit would have significantly more serious consequences than the downfall of his humble self:

this noble knowledge which has already given and daily gives such indescrib-able benefit to foreign nations would futilely be stifled and buried with me, leaving as its only obituary the engine built by the Partners at such cost, on which posterity would probably write the epitaph: A laudable intention ("*Laudanda Voluntas*").[7]

The case was heard at Dannemora on September 13–28, 1731.[8] Triewald had objected to Mine Inspector Thor Bellander as judge,[9] and the Board of Mines had instead appointed one of its Assessors, Göran Wallerius.[10] We have met him several times before. He was the young Stipendiary in Mechanics at the Board who assisted Polhem with his experiments in hydrodynamics in Falun in 1703–1704, and in 1705 he wrote the report on these experiments for the Board (see Chapter 4). In 1723, after he had risen to the position of Mine Inspector, he was instructed by the Board to assist de Valair in the event of the latter's trying to build an engine at Öster Silvberg (see Chapter 8). In every respect, Wallerius' career resembles that of the typical Board of Mines official sketched in Chapter 5. When he sat as judge at Dannemora in 1731 he had been an Assessor for the Board for just a year.[11] (He was elected a fellow of the Royal Swedish Academy of Sciences in 1742.)

The Plaintiffs were the Partners in the Dannemora Mines, represented by Johan Åhman, who was himself a district court judge.[12] The Defendant was Mårten Triewald, a newly elected F.R.S.,[13] and his counsel was Judge Advocate Isac Fritz. The complaint was one of "negligence and deficiencies" in fulfilling the agreement signed by Triewald and the Partners in 1726.[14] The Plaintiffs claimed that under the agreement Triewald should reimburse them for all their expenses and costs in connection with the engine, together with the remuneration he had received in 1728 after the first demonstration of the engine. Triewald contested the claim, and alleged that the Partners had:

directly and indirectly broken the contract, and by causing divers hindrances had to their own disadvantage and the disparagement of the project defeated the good intentions of the contract, together with the costs, effort and trouble applied.[15]

The formal procedure for the hearing was as follows. First the Partners made their accusations, after which Triewald stated his defence and adduced his counteraccusations, and this was followed by the Partners' rejoinder to his defence and reply to his counteraccusations. Finally the parties' accusations and defences were summarized and considered in the judgment. The following account follows this pattern and includes the main arguments of the parties in their pleadings. As we shall see, there were three major issues to be discussed. Whose fault was the delay in the completion of the project? Why had it cost so much? Had the engine been of the promised "utility"?

The Partners opened by asserting that they for their part had done all that was required of them under the contract. They had supplied Triewald with funds to order parts from England and Stockholm, and they had delivered all the necessary building materials promptly. They had hoped that Triewald for his part would show himself to be sufficiently diligent and conscientious to have the engine ready in June 1727, that the costs would not exceed the sum which he had stipulated and that the engine would have "the result promised and guaranteed in the contract".[16] Triewald had assured the Partners that they would make a substantial saving in their costs for pumping and ore-raising.[17] He was in breach of all these promises. The engine had not been completed until July 1728, and then only for pumping. Work on the much-vaunted hoisting machinery had at that time not even started. As far as the effect promised in the contract was concerned, the engine's capacity to pump water had proved insignificant and unreliable, and the hoisting machine had not yielded any result at all. Triewald had now procrastinated its completion into the fifth year. With regard to costs, Triewald had promised that these would not exceed 24 000 *copperdaler*, and it was only on the basis of this "acceptable sum" that they had allowed themselves to be talked into the project.[18] The total cost now stood at 52 254 *copperdaler* and 28 *öre*, and if they had had the faintest suspicion that such a figure might be involved, they would never have signed the contract with him. He ought now to repay the whole amount, for there existed between the parties:

a clear contract, which is to be the very basis and foundation by which alone the contracting parties on both sides should abide.[19]

It was then Triewald's turn to speak. It was true, said Triewald, that he had undertaken to install at the Partners' expense an engine that could drain the water from the adjoining mines, and "afterwards" to build a hoisting machine.[20] The Partners had however bound themselves to pay an annual royalty "from the date when the engine first started working",[21] and when it had proved its reliability for a year to remember him with "a decent recompense consistent with the engine's dignity".[22] Triewald therefore considered that he had fulfilled his undertakings on July 4, 1728, when he started up the engine for the first time. With that, he had shown that the engine had the promised effect. The matter of continuous operation was not fundamental, he argued, but something to be rewarded separately and later. But, he continued:

as all gentlemen's agreements are based on equality, so the mutual obligations must balance, it must have been the duty of the Partners to remove all obstacles that could arise and assist him in all that might be required for the fulfilment of the contract.[23]

He did not accept that the Partners had done this. He related how the delivery of the iron cylinder from England had been delayed when the first two attempts to cast it had failed, and how he had not dared have it shipped to Sweden so late in the year when it did not reach London until October 1727. At the meeting in December the Partners had consented to his having a brass cylinder cast in Stockholm, but had at the same time been so vindictive as to decide that he must himself meet the cost of the cylinder ordered in England. In so doing, reasoned Triewald, they had broken the terms of the contract, because he was to be regarded as their *Commissionair*, whose proposals they had voluntarily accepted. There was no reason why he should bear the financial liability for a mishap that he was not guilty of causing. It was because of the delayed delivery of the cylinder that he had not managed to have the engine ready in 1727, but he had worked hard at Dannemora throughout the spring. As soon as the cylinder had been ready in April 1728 he had hurried to Dannemora to finish the engine. On July 4, 1728, he had been able to begin emptying water from the mine. After that, several obstacles had arisen to prevent the engine "from being quickly brought to completion and put into a condition to enable it to be kept in regular and steady operation".[24] These obstacles, too, asserted Triewald, were no fault of his. First it had been necessary to chip off all the ice that had formed in winter on the walls of the mine before the pumps could be lowered. Then it had been necessary to caulk a dam between the lake and the mine because too much water was leaking through. Time after time he had reminded the Partners of their duty to have the dam repaired, yet this had not finally been done until the autumn of 1729. During the work on the dam he had been compelled to remove the pumps, so the engine had stood idle for no less than five months, doing not the slightest good. The mine pumps were a third difficulty. As pump assemblies in England were made to a length of 20 fathoms, he had first had two assemblies made for the 36-fathom depth of the mine, and therefore ordered two pistons and valves to be cast. "But as there had not been timber strong enough in this country",[25] one log after another had split when subjected to the pressure of such a high column of water. He had not only been obliged to make three pump assemblies instead, but had also had to wait until a third bronze piston and valve could be procured. Triewald took the view that he could not be blamed for these delays and that he had displayed all possible diligence, and also that he had not guaranteed the date of completion anywhere in the contract. He had merely "assumed, as the words state, that the engine could be ready the following June".[26] Moreover it had been quite impossible to foresee all the problems he had encountered.

As for the Partners' complaint that the cost had been higher than the first estimate, Triewald said that it could not "be proved with the contract or any other evidence" that he had bound himself to a sum of 24 000 *copperdaler*.[27] No, this sum had simply been an estimate that was meant to show what such an engine cost in England, and he had then been referring to the cost of the actual engine set up for pumping work. He had not included the hoisting machinery, and certainly not all the construction costs entered by the Partners in their calculations. What was more, these seemed to him to be extremely dubious: several items were too high, furthermore much was included that had nothing to do with the engine. If the calculation was limited to what Triewald considered correct, it would be clear that the engine had cost no more than 30 000 *copperdaler* to build. If the extra cost of casting the cylinder and the pump pistons in brass instead of iron were then deducted from this sum, only 24 000 *copperdaler* would remain (i.e. the amount of Triewald's original estimate). This was a matter of general interest, thought Triewald, for the Partners had:

tried to encumber the Fire and Air Engine with far too great and excessive costs; whereby to the disservice of the public this invention is likely to be discredited here in the Realm.[28]

Triewald could not understand the Partners' claim that the engine had not demonstrated the promised effect. Had it not shown that it could drain the whole mine? And had it not made it possible to resume work in the adjacent Wattholma Mine, which had formerly been so wet that operations had had to be suspended? Triewald considered that anyone and everyone could see that the engine had achieved more than had been promised in the contract. With these savings in pumping costs their investment would pay for itself within a few years. Turning to the hoisting gear, Triewald stated that here, too, he had shown all conceivable diligence. Had he not taught *Konstmästare* Hultberg carefully how to erect it?

But notwithstanding this, Hultberg suffered every kind of indignity here and was accused of negligence, until he wearied of such work.[29]

It was only with great difficulty and at great expense that Triewald prevailed upon him to remain after the meeting in September 1729, as "nobody accustomed to the work could be found in his place".[30] But shortly afterwards, Hultberg suffered further harassment in a number of ways and once again asked to resign, which had made it impossible to complete the hoisting gear. Despite this, Triewald had planned to go to Dannemora in the spring of 1730 to complete the work on the hoisting machine, start it up and "provide the works with a new *Konstmästare* to replace the former one, who had been reduced to such apathy that he was unwilling to undertake anything".[31] But just as Triewald had been preparing for the journey, he had received a copy of the Partners' complaint to the Board of Mines:

wherewith the Partners had forestalled his journey and by lodging such an obstacle prevented his coming here. For when he was thus repaid for all his toil, trouble and work [...] then he would have thought long before putting his hand again to a project that brought loss and vexation and no reward, wherefore he had not wished to hear anything more of the operation or further installation of this engine.[32]

He concluded his plea by demanding from the Partners the payment of 6 000 *copperdaler* in damages for discrediting him by imputing the failure of the project to him, 1 200 *copperdaler* in royalties calculated from July 1728 to the present date with current interest, 5 000 *copperdaler* as recompense "if it is at all to be proportioned to the dignity of the invention",[33] and 10 200 *copperdaler* in reimbursement for his expenses. This was probably a calculation that Triewald arrived at retrospectively to match the Partners' claim, as it produced a total of just over 25 000 *copperdaler*.

It was then the Partners' turn to reply to Triewald's allegations. They began by implying that no cylinder had ever existed in England. The whole question of the cylinder was irrelevant, however, because it was obvious that Triewald:

is not to be regarded in this case as a *commissionair*, but as the other party to the contract, the one who has taken upon himself not only the work but also full responsibility for the soundness of the engine, and is thus to be held solely liable for such a loss.[34]

Concerning Triewald's assertion that they had not dealt with the leaking dam, they did not accept this as proved, nor could it be described as a breach of contract, because there was not a word about the dam in the contract. Nor were the Partners responsible for the fact that the pump pipes split and that Triewald had been obliged to build three pumps instead of the two he had originally thought sufficient:

since as a prudent *Mechanicus* he ought to have attended to all such matters before making such extravagant promises to the company.[35]

Moreover, it was his own fault that the elm mine-pumps had split. He had felled the trees at the wrong time of year, and then bored them out before laying them on the ground to season. The correct procedure would have been to season them before boring them, or to store them in water until it was time to lower them into the mine. They did not deny that he had worked hard while the engine was being built, but, "since the engine was started up, he has produced many excuses and evasions".[36] They had warned him at a meeting in 1729 not to leave the engine "until it was working steadily".[37] He had given his word, but had nevertheless departed immediately afterwards.

They found unreasonable Triewald's claim that the calculation of costs should only relate to the engine set up for pumping work and not include the costs of the hoisting machine. From the "plain words" of the contract it was clear that the one intention of the engine could not be separated from the other.[38] His claim that the expenses were too high was contested: nothing had been entered at a higher amount than was shown in the ironworks' ledgers. They had not derived any benefit from the engine in working the nearby Wattholma Mine, and it was not true that it had previously been difficult to keep this mine free of water: it was only the annual spring flood that had caused trouble. Concerning Triewald's claim that the engine had drained the mine, anyone could see that it was one-third full of water. The engine had worked in vain for three years trying to empty it, and the result in no way matched the lavish promises in the contract. The Partners considered:

that at Mr Triewald's persuasion they had as good as thrown over 52 000 *copperdaler* into the lake, and not yet had a stiver's use from it.[39]

Triewald's comments on Hultberg's resignation did not concern them, and they would "restrict themselves only and solely to Mr Triewald, who is the company's man".[40] The Partners had paid Hultberg's subsistence and wages, although this was not their contractual responsibility. Otherwise they had not concerned themselves in the least with Hultberg, and had let Triewald have free use of him.

They could not understand how Triewald had regarded himself as hindered from putting the engine in continuous operation when he had received their complaint to the Board of Mines. They had merely asked the Board "most humbly" to urge Triewald to fulfil his commitments.[41] If Triewald had really been intending to come to Dannemora, then the only reply that their letter would have needed was that he was ready to come "at the Partners' pleasure".[42] But instead he had unleashed a barrage of intolerable allegations against them in his statement of defence, thereby drawing them into this lawsuit. As he had now stated during the hearing that he did not want any more to do with the engine, he ought to forfeit both his annual royalty and the advance payment that he had received earlier. The contract stated clearly that he was not entitled to them:

until the engine has achieved the promised result and saving, which it has not yet done.[43]

Following these pleadings, the court had to decide upon the three main issues: the delay, the rise in costs and the "utility". First it adjudicated upon the question of whether anyone could be blamed for the fact that the construction work had been delayed. The court could not find that the Partners had neglected their obligations, and Triewald had blamed the delay on the late delivery of the cylinder and other causes. That a cylinder ordered in England really had existed was confirmed to the court by an account from Jonas Alströmer. The court considered that Triewald could not reasonably be blamed for this delay to the work. Concerning the other obstacles mentioned by Triewald (ice on the shaft, the leaking dam etc.), these had been substantiated by several witnesses. Therefore Triewald also should be seen as blameless for the fact that the engine had not been completed on time, especially as sworn witnesses had testified "that he had been diligent during the work, and kept a sharp eye on the workmen".[44]

The second question concerned the high costs compared with Triewald's original estimate. The court did not consider that Triewald had expressly guaranteed that the engine would cost no more than

24 000 *copperdaler*. That was an estimate, and it was quite excusable if he had erred, because he had "recently come to the country and was unaware of the cost of materials and labour here on the site".[45] So Triewald could not be regarded as being in breach of contract on this point, either:

It being not unusual that on building works, costs often exceed estimates, which is all the more to be pardoned in moving machinery, as one cannot foresee all the difficulties encountered.[46]

The judgment states explicitly that the court had made special efforts to inform itself concerning the third question: "the effect and result that the Fire and Air Engine has been able to achieve".[47] It was primarily the technology that was on trial as far as the Board of Mines was concerned. The court had therefore made a thorough inspection of the engine and all its parts during the hearing, and had also watched it working for short periods. According to witnesses' evidence, the longest period for which it had ever worked continuously was nine or ten days. On one occasion it had worked for about five days and during that time reduced the depth of water in the mine from 27 or 28 fathoms to 5 fathoms. The court regarded this as a "good and reasonable effect", as it could be seen from the mine plans of 1699 and 1715 that at that depth the mine was at least 15½ by 5 fathoms in area.[48]

In order to assess the "utility" of the engine, therefore, the problem was clarified by differentiating between "effect" and "result". When Triewald had talked of the "utility" of the engine, he had been referring throughout to the fact that he had been able to demonstrate its great "effect". When, on the other hand, the Partners complained that the engine had not had the desired "utility", they had been talking about its "result". By "effect", Triewald meant high power alone, whereas by "result" the Partners meant high power *and* continuous operation. That both requirements should be satisfied was, as was pointed out in Chapter 4, the demand made on mechanical energy resources in industry. All the questions that the court had to consider had thus been reduced to one—did the engine give a good "result"? That is, could it also maintain the continuous operation that was essential to its "utility"? All the witnesses heard during the hearing had been unanimous that:

the engine had worked erratically and uncertainly in that it had sometimes worked for a day or a day and a half, sometimes for an hour and a half, but often no more than a few strokes, after which it had at times suddenly stopped and been impossible to restart, no matter how one worked on the fire; And the court has found the working of the engine to be similar during the current session of the Court of Mines.[49]

The court found it proved that Triewald had neglected his duty to put the engine in "continuous and reliable operation" after building it and demonstrating its effect.[50] He had not been seen in Dannemora since the meeting in September 1729, and had not made any improvements since, despite the Partners' complaint. The court therefore considered that he could not claim the annual royalty, as the contract clearly stated that the engine must first have given "the promised utility and saving in costs at the Silverberg mines".[51] He had also failed to appear at the meeting in March 1730, and he had not come to Dannemora or tried in any other way to do anything about the engine after the Partners complained to the Board of Mines. On the contrary, he had now said during the hearing that he wanted to have nothing more to do with the engine. Out of respect for the Board, he ought to have done his best to rectify the defects and put the engine in operation. He had after all guaranteed in the contract that it would be able to keep all the Silverberg mines free from water.

Against which all that the *Directeur* has achieved up to today and in the 4th year after the engine was first started up is no more than that the mine has once been emptied of water, but is now largely under water again, which it has since been impossible to draw off because of the engine's unreliable operation and the *Directeur*'s failure to repair it, so that the Partners have not had the compensation or comfort of the utility and saving that the *Directeur* promised.[52]

Concerning the "handsome promises" that the hoisting machine would be able to raise three times as much as an ordinary horse whim,[53] it had emerged during the hearing that it had never worked. It was hardly possible to expect it to be of any practical use as it was constructed, especially as the strokes of the engine were so irregular. Triewald had not therefore fulfilled his promises "by a long way".[54] As there was an explicit contract to follow, which governed and guided the relationship between the parties, the court found that it had only to adjudicate in accordance with the contract. So Triewald was ordered to repay the Partners' costs for the engine. The court did not consider determination of the size of the sum to be its business. The parties could try to come to an agreement in the presence of an arbitrator, but it should be borne in mind that Triewald had already admitted that it had cost at least 24 000 *copperdaler*. Triewald was also to pay court costs of 2 534 *copperdaler*. In its judgment the court did praise Triewald on one count, however, namely for his:

good intentions, in bringing an invention so noble and hidden in the laws of nature into the country.[55]

Mårten Triewald's father, the blacksmith Martin Triewald, provided surety for the costs of the hearing.[56] However, neither party was satisfied with the decision, and they both appealed to the Board of Mines.[57] The Board did not hurry itself over the matter, which dragged on for many years (see Chapter 16).

Early in 1734, two and a half years after the hearing, the Board tried to induce the Partners to make another attempt with the engine, but they declined: it would be sure to prove as unreliable as before, and would only cost them large sums for wood fuel unnecessarily.[58] But if the Board wished to try to make it work "on behalf of the public", it was welcome to do so.[59] At the same time the Partners asked the Board to give a final decision on their suit against Triewald as soon as possible, because as long as the hoisting gear was there they could not begin working the Norra Silverberg Mine. The Board then asked Triewald if he would not be prepared to make a new attempt.[60] He would redeem his own reputation and if he succeeded he would avoid losing the case. Triewald declared himself willing, and the Board was eager to try, knowing the useful work that these engines were doing at mines in England and in several other countries.[61] Mine Inspector Bellander was therefore entrusted with the task of buying sufficient firewood, about 380–400 *stafrum*, through Mine Bailiff Kröger at the Board's expense. Kröger raised a number of difficulties, however, and said he was having trouble in persuading the farmers around Dannemora to supply the wood.[62] It is evident that he did what he could to prevent a new attempt with the engine from being made, for Mine Inspector Bellander reported him to the Board for refusing to carry out orders (an unusual and drastic measure).[63] Bellander also wrote a strictly formal letter to Kröger ordering him to give Triewald every assistance during the tests.[64] Triewald had asked the Board if he and Kröger might keep a joint log of the engine's wood consumption, the running times and the depth of the water in the mine.[65] But this was not done—Kröger and Triewald each kept their own record.

The tests took place between May 17 and 30, 1734. Triewald's carefully completed log was signed, not merely by Triewald himself, but also with the marks of the miners who helped him.[66] This shows that the only local assistance that he could count on after Olof Hultberg left Dannemora was that of illiterate miners. A single individual, Hultberg, had been the narrow social basis on which he had tried to build the whole project.

The foregoing log has been read out to us, who have been used at the engine from the start to the finish, and everything has taken place as related in this log, which we can on our sincere oath confirm where it should be required.

Dannemora, May 30, 1734

Andreas ⇥ Widberg
mine smith

Per Bång ⌒
machine minder

⅄ Eric Anderson Gräse
machine minder

Per ⌇ Bom
miner

Eric And: ⌁ Söderberg
miner[67]

The log notes how many hours the engine was in operation each day, by how many feet and inches it lowered the water level in the mine during the same time, and how much wood was delivered each day to the engine. To summarize, the engine had worked for 185¼ hours in a period of 14 days, i.e. for somewhat more than half the time. At the most it had worked continuously for 5½ days, and the shortest period was 3 hours. During this time it had consumed a total of 69 *stafrum* of wood fuel, which represented a consumption of 8.9 *stafrum* per day. (Triewald tried to maintain, however, as always, that conditions had been exceptional and that one ought therefore to allow a maximum of 7 *stafrum* per day.) The water level had been lowered by a total of 26 Swedish ells and 1 inch (15.5 metres). This was the last time the engine worked. It had been started for the first time on the Queen's name day, July 4, 1728, but it was shut down on an ordinary Thursday, May 30, 1734, "at half past two in the morning".[68] The reason was that the bottom pump assembly was broken and it was therefore impossible to pump any deeper. The engine had already begun to deteriorate through lack of maintenance.

On June 10, Triewald was called to the Board of Mines.[69] In the drawing room outside the conference room, he read for the first time Kröger's account and log of the test running. When the Board read Triewald's and Kröger's reports it found that "the latter's account differed from the former's".[70] The Board asked Triewald how he could claim that the engine had shown that it could pump up 20 000 barrels of water every day—a pronouncement that Triewald had made in the Stockholm journal *Then swänska Argus* after the tests.[71] But he was able to support his calculations, and the Board urged him to submit proposals for repairing the bottom pump assembly.[72] This invitation appears to have come to nothing, and the Board took no further steps to have the engine put in order. As has been pointed out in Chapters 8 and 9, the Board was primarily an advisory body and did not itself initiate technological change.

A year later, in September 1735, the Partners in the Dannemora Mines wrote to the Board and requested permission to dismantle the hoisting machine and remove the pump assemblies from the Norra Silverberg Mine.[73] They now wanted to begin working the mine again, but they could not drain it with the traditional horse whims while this equipment was in the way. The question was thoroughly discussed for a couple of days at the end of December.[74] Triewald had been called to the meetings, together with those of the Partners who were in the city. Once again the parties exchanged allegations and recriminations.

The character of the Board of Mines as an impartial authority is underlined by the fact that Mine Councillor Johan Bergenstierna left the room every time this question came up on the agenda. (Triewald was married to his stepdaughter, which meant that his objectivity was considered challengeable.) The Board realized the necessity of draining the mine in order to safeguard the ironworks' supply of ore, but on the other hand it did not wish to interfere with the state of the suit between

Triewald and the Partners. The Board's consent to the request might have been interpreted as expressing a point of view before the matter was decided. There was a clash between economic reality and the requirements of the legal situation, but the Board of Mines found a solution worthy of any bureaucracy. It gave the Partners permission to dismantle the hoisting machinery and, if necessary, the pump assemblies, and to erect in their place two horse whims—with the reservation that this "economic disposition" must not be regarded as having altered the situation.[75] This elevated the dispute between Triewald and the Partners to a more ethereal plane on which the engine and the hoisting machine remained intact. But it is the real world of the Dannemora Mines that interests us.

On March 12, 1736, Mine Inspector Bellander wrote to Mine Bailiff Kröger and ordered him to dismantle the hoisting machine: all the iron parts and the smaller wooden parts were to be laid in the engine house, the heavy timber was to be stacked against the south wall of the building and covered with boards, and a meticulous record was to be kept of all the items.[76] This was to be done as soon as possible and within eight days at the outside. Where the hoisting machinery had stood, two horse whims were to be set up, and these would be used to begin draining the Norra Silverberg Mine. As thieves had broken into the engine house and stolen lead piping and a metal cock from the engine, Kröger was to make sure that the windows were fitted with strong iron bars, that the hole in the wall for the great beam was blocked up and that the door was securely locked.

With this, steam power technology had been rejected by the Dannemora Mines, which returned to the traditional sources of power. It was no doubt with great satisfaction that Mine Bailiff Kröger obeyed his orders. Some time towards the end of March, 1736, he was able to close the door on the hated machine for the last time and symbolically turn the key on the new technology. At the same time, Triewald was in Stockholm, congratulating himself on the safe arrival of twenty white mulberry trees from Lübeck—the twelve black mulberry trees that he had ordered the previous autumn had arrived late and been killed by frost.[77] On May 22 he sent the second volume of his lectures, which had just been published, to Sir Hans Sloane.[78] He took the opportunity to ask for help in acquiring a number of scientific instruments that he needed for his experiments, and enquired: "Pray does the Royal Society still make use of Mr Hauksbee's Barometers & Thermometers? Which I am likewise unprovided with."[79] On June 17 he sent Sloane a description of "a new Invention of Bellows call'd Water Bellows".[80] A few weeks later an event occurred in Dannemora that Mine Bailiff Kröger and many others must have taken as a divine ratification of the decision to shut down the engine:

1736, 6 July somewhat before midday lightning struck the north side of the roof of the engine house and shattered a tile, leaving through the south side of the roof above the reservoir, there smashing 2/3 of the roof, so that the chimney also cracked.[81]

The emptying of Norra Silverberg Mine with horse whims had begun on May 12, and on August 18 there were only 10 metres of water left.[82] On November 11, 1736, the mine was empty of water and it was possible to start excavating at the bottom for the first time since 1709.[83] Kröger hurried to send the good news to Mine Inspector Thor Bellander at Gävle. Bellander replied by sharply rebuking Kröger for his unauthorized removal of the last pump assembly (which Kröger claimed was impeding the work). But Bellander also was pleased with the progress: "Otherwise I am delighted to hear that all is well at the mine".[84]

DANNEMORA REVISITED: THE ENGRAVING IN TRIEWALD'S BOOK

The publication of Triewald's book on the Dannemora engine was advertised in a Stockholm newspaper on December 2, 1734.[85] In other words, after the Partners in the Dannemora Mines had declared that they wished to have nothing more to do with the engine, but also after the Board had lost interest in further attempts to put the engine in order. It is against this background that the publication of this book has to be seen, and we could now discuss it page by page to compare Triewald's claims for the engine with what has emerged in this and the previous chapters. But there is an easier way to arrive at the same result—i.e. an assessment of its objectivity and purpose. We shall instead study the frequently reproduced engraving in his book (Fig. 14.1), assuming that what is true of the illustration is also true of the text. In an article in 1978, Patrick M. Malone stressed the importance of the artifact as direct evidence of technology:

Although written language, with its nuances and complexities of expression, may generally be more informative than inanimate objects, there are fewer biases in artifacts. These material products may sometimes be misleading, but they were rarely designed to deceive, to cover up the truth, or to press a particular opinion. The authority of physical evidence refutes the data found in many surviving documentary or graphic records, particularly in the area of engineering construction and machine design.[86]

This is true of the visual information conveyed by the engraving in Triewald's book *Kort beskrifning, om eld- och luft-machin* (A short description of the fire and air engine). It has been uncritically accepted that this is a faithful reproduction of how the engine *looked*. But we are now in a position to compare it with the remains of the artifact, together with the drawings and archival sources (see Chapter 12). The first thing that strikes the eye is that the engine house is shown to be much

Dannemora Eld och Luft Machin,
Kongl. Maij:ts och Riksens Höglofliga Bergs Collegio
Underdån ödmukast Dedicerad af Mårten Triewald.

higher than it actually was. Did the artist make the roof higher simply to show the cylinder and the water tank more clearly? The length of the cylinder is known from several different sources, and if we compare it with other parts we find that the engraving is not drawn to scale at all! The building is also much shorter, for example. The engine is housed in a building that is both higher and narrower than it was, and this makes the engine look more dramatic.

A closer look reveals that the perspective of the cylinder is incorrect: the upper half touches the side of the building, but the lower part comes down over the boiler in the centre. But it is soon obvious that this was not an oversight. The artist has in fact changed the alignment of most parts of the engine. All the parts have been brought as close as possible to the observer, and those concealed by the choice of perspective have been moved around quite freely. The beams supporting the engine have been turned through 90 degrees. They did not rest in the end walls, but in the side walls. The opening to the firegrate, the staircase leading down to it, the trap-door above, the hot well and the exhaust pipe—i.e. the whole bottom part of the engine—have also been turned 90 degrees to face the observer. The overflow pipe from the tank in the loft ran down the back of the building, but here it, too, has been turned 90 degrees. This not only makes it visible, but also brings it as close as possible to the observer without hiding other parts of the engine. The chimney is placed on the left gable, and the flue is seen running up the far left corner. They have both been turned around 180 degrees; otherwise the flue would have obstructed the view of the engine. The beam, the mine pumps, the top of the cylinder, and the plugrod have all been moved closer to the observer with no regard for their true position. Perspectives, dimensions, positions and orientations—all had to give way for the purpose of instruction. But the changing of the dimensions of the building to make the engine look more dramatic indicates that the artist had additional ambitions.

The three persons needed in order to run the engine are shown on the engraving: the engineer, the stoker and the machine-minder. The three figures serve the purpose of stating the personnel requirement, but their height is grossly reduced in comparison with the length of the cylinder, and this also was probably done quite deliberately to make the engine look even more impressive. All the windows, except for one, have been left out so as not to confuse the details of the engine. At the remaining window stands a man (Fig. 14.2). We assume that he is the engineer or supervisor of the engine because of his cane, which takes the form of the ceremonial pickaxe traditionally carried by mine officials.[87] He has

---

Fig. 14.1. The engraving made by Eric Geringius in 1734 in Mårten Triewald's book *Kort beskrifning, om eld- och luft-machin wid Dannemora grufwor* (A short description of the fire and air engine at the Dannemora Mines), published in the same year. (Photo Bengt Vängstam)

Fig. 14.2. Detail of Geringius' engraving in Fig. 14.1. (Photo Bengt Vängstam)

placed it in the window bay while he lights his pipe (the puff of smoke is visible over his right shoulder). This creates a psychological effect. The engine is obviously hard at work, because there is a fire burning in the grate and water is pouring out from the pump in the mine and from the overflow pipe. But he, the supervisor of the engine, can peacefully turn away from the engine to attend to his pipe. It conveys the impression that the new technology is fully reliable. This effect is accomplished thanks to the remaining window: the characteristic handle of his cane and the smoke from his pipe are visible only in the light through the window.

But there are other psychological overtones. In the foreground stands a man, with his back to the observer, watching the engine in the company of a boy and a dog (Fig. 14.3). He is elegantly dressed, with a cocked hat, a pigtail and a Spanish cloak. The boy is pointing excitedly and tugging at his sleeve to draw attention to the huge cylinder (which would not in reality have been seen, since this is a cutaway view). The man is obviously a nobleman, and we infer that he is one of the Partners in the Dannemora Mines, the owner of an ironworks in the province of Uppland. The impression created is that he has come to watch the engine that is so beneficial to his mines. A profitable investment to be contemplated with satisfaction.

Down in the right corner stands an old man with a white beard, dressed in rags and leaning heavily on a thick stick (Fig. 14.4). Most likely he is a retired miner who has come to watch, in awe, the new wonder: a house that makes the ground quake, consumes enormous

Fig. 14.3. Detail of Geringius' engraving in Fig. 14.1. (Photo Bengt Vängstam)

amounts of wood, spits out smoke, sparks and steam—and yet, does the work of hundreds of horses, dozens of men, numerous horse whims, waterwheels and lengths of *Stangenkunst*. He is a representative of the old technology. The artist has put him in the corner to imply that the old technology of muscle, water and wood has already been displaced by the new technology of steam and iron.

Numerous other engravings of Newcomen engines were published during the eighteenth century. A distinct style in depicting these engines emerged early; it can be traced back to the first engraving by Henry Beighton in 1717 (see Fig. 6.3). Most of these drawings show a certain similarity. This was probably the result of a conscious attempt to adapt to a traditional style, rather than of a mechanical copying of a predecessor's work. The perspective is the same, showing the engine house from the side in a cutaway view. The angle of incidence of the light falling into the building is also the same. Most parts of the engine, moreover, are positioned and oriented the same way: the beams supporting the cylinder rest in the end walls, the firehole faces the observer etc. Certain decorative elements are the same, for example the mouldings on the chimney. The elegant gentleman with a cane watching the engine and the relaxed supervisor also appear on several different engravings.

The engraving in Triewald's book is strikingly similar to that published by Sutton Nicholls in 1726 (Fig. 14.5). Not only are the letters designating the various parts the same,[88] but if we compare the proportions of the house, the angle of the beam, and the positions of all

Fig. 14.4. Detail of Geringius' engraving in Fig. 14.1. (Photo Bengt Vängstam)

the other parts, it seems almost to have been a copy.[89] The assumption that the artist who drew the Dannemora engine based his drawings on the one by Nicholls is further strengthened by the fact that the former is a mirror image of the latter. Nicholls' engraving was said to represent a Newcomen engine at York Buildings Waterworks in the Strand. Thus, the engraving in Triewald's book shows the London engine in a Swedish setting, rather than a realistic view of the Dannemora engine.

The engraver of the Dannemora engine, Eric Geringius (1707–1747), added a naturalistic touch. He not only threw in a few extra people (to say nothing of the dog), but also placed the engine in a dramatic natural setting: a hilly foreground, a lake with three men in a boat and a sloop with flying pennants, distant hills, and dark clouds in the sky. There were indeed dark clouds over the engine at this time, and the engraving and the book should be seen in this perspective. It was a petition in favour of its author.

It has not been observed before that Triewald makes no mention of the output of the Dannemora engine in his book. Instead, his argument runs something like this: It is known that the Newcomen engine in Königsberg, Hungary, raises so-and-so many barrels of water per hour. The area of the cylinder of the Dannemora engine is so-and-so many times larger than that of the Königsberg engine. Therefore, the Dannemora engine is so-and-so many times more powerful than the Königsberg engine. Consequently, the Dannemora engine raises, or ought to raise, so-and-so many barrels of water per hour. *Quod erat demonstrandum.* The reason for this evasive argument was that the failure of the engine

The ENGINE for Raising Water by Fire

Fig. 14.5. Engraving by Sutton Nicholls of the Newcomen engine at York Buildings Waterworks in the Strand, London, in 1725. This was probably the model for Geringius' engraving. Note the similarity between the letters marking the parts of the engine and the mirror-imaged Fig. 14.1. Nicholls' engraving was published a few weeks before Triewald departed for Sweden in 1726. (Photo Science Museum, London)

to achieve its promised output was, as we have seen, the matter in dispute in the lawsuit. A nice touch is that Triewald, the defendant, dedicated his book to the Partners in the Dannemora Mines, the plaintiffs. The engraving was dedicated to the Board of Mines, the judiciary authority. Nor was there any mention of the hoisting machine in the book, but this was also a subject of dispute. From archival sources we know that the hoisting machine was in fact in place at the time when the engraving was made. But this sensitive issue was left out of the engraving altogether. This emphasizes the point made by Patrick M. Malone. The engraving was designed, if not to deceive or to cover up the truth, at least to advance a particular point of view. The authority of the physical evidence at Dannemora refutes the data found in the documentary and graphic record of Triewald's book.

The engraving is not a true image of the engine—but that was not the intention. The artist had other, far more complex ambitions than to show how it was constructed. *First and foremost,* he wished to show the principle and function of the engine, i.e. to supply visual technological rather than technical information. *Secondly,* he was trying to convey a favourable view of the new technology. To achieve these goals, he did not feel bound by the physical reality. He omitted what did not suit his purposes, and what he included, he freely turned around, moved about and modified. Neither positions nor dimensions are objectively accurate. *Thirdly,* in his choice of perspective, and of decorative and symbolic elements, he adapted to the style that had already crystallized in the depiction of Newcomen engines. This is a point of some general interest in the history of technology. We are used to modern photography, which captures a split-second of reality in a faithful two-dimensional representation, and we therefore tend to believe that every picture is the result of an honest attempt to depict reality. But the case of the Dannemora engraving shows *an attempt to convey a principle and state a point of view in a contemporary style.* Perhaps this is also the case with other graphic records of the history of technology, which are usually considered historical sources in their own right?

15. Critical Factors in Technology Transfer

The introduction of the Newcomen engine in the Dannemora Mines in 1726–1736 was a failure, in the simple sense that the objectives of those introducing it were not realized. This is a fact, not a value judgement passed with the benefit of hindsight. It is, however, all too easy to regard this as a failure simply because we know that steam power technology was successfully introduced into Swedish industry ca. 1800, but it should be remembered that by then both the social context and the technology had changed. What, then, were the reasons for the lack of success of this attempt at technology transfer? In the events described here, we can discern critical factors of a number of different kinds: technical, geographical, economic, social and cultural.

1. TECHNICAL FACTORS

Swedish mechanical technology during the eighteenth century was based mainly on wood. Strictly speaking, it was a composite wood-and-iron technology, as iron was used for shaft ends, bearings and brackets, but wood was still the main material for machine parts. Triewald had counted upon casting the pumps in iron, having been familiar with the practice in England, but no Swedish ironworks could manage this. Instead he had to build the pumps from bored-out elm logs in the traditional Swedish manner. However, the logs proved too weak for his original design, which was based on the assumption that iron would be used, and this was a problem that appears to have taken about a year to overcome. There was a difference in *technical expertise* between England and Sweden, and it necessitated a period of modification that both delayed and increased the cost of the project. In an article on the transfer of technology to America in 1800–1870, Nathan Rosenberg has said:

it required considerable technical expertise to borrow and exploit a complex foreign industrial technology. This should hardly be a surprising proposition, but it seems to be worth repeating in view of the vast number of foreign aid and economic assistance programs in the years since the Second World War which have come to grief because of the absence of the appropriate skills in the receiving country.[1]

Once the difficulty with the pumps had been resolved, the problem seems to have been to maintain a sufficient supply of water in the tank above the cylinder to condense the steam. The efforts to seal the tank with lead sheet and the complaints about the leaking injection valve were, as has previously been remarked, probably symptoms of a major defect of design. Triewald had not been content to build an engine of normal cylinder diameter when he transplanted the Newcomen engine to Sweden from its British matrix ca. 1725. He had instead increased the most important parameter of the engine—the area of the cylinder—by 90 per cent.

But one cannot increase just *one* parameter of a technological system without being confronted with a series of difficulties. The Newcomen engine, with its many connected and interdependent components, constituted such a system. In Great Britain, the average cylinder area used in the engines grew gradually and organically, along with the dimensions of all the other components of the system: the boiler capacity, the depth and diameter of the mine pumps, the robustness of the automatic valve mechanism, the capacity of the pump supplying water to the tank, the tubes and fittings, the valves etc. The sizes of all these components were mutually dependent, and they had to be chosen accordingly. This could be done only by a slow process of trial and error, step by step. The growth in the area (or diameter) of the average Newcomen engine cylinder therefore shows the same exponential character as the growth of any critical parameter in any technological system.

When Triewald quite drastically increased the cylinder diameter to 36 inches, he strained all the other components of the system to the limit. That is why he ran up against problems such as the inadequacy of the water supply for condensing the steam. This was probably also the problem underlying the engine's frequent breakdown due apparently to other causes. As Kröger's reports showed, it was often out of service during its first few years. To overcome these problems, a longer period of modification would have been needed: a technical development that would have taken time and money. But no such development ever took place, and the reasons for this were not technical but, as we have seen, economic, social and cultural.

The hoisting machinery is another example of the introduction of a level of technology that was too far ahead of its time—as yet, it was not even established in England. It is not known how much experimentation with wooden hoisting machinery was taking place in England at this time, but it can never have proceeded further than the experimental stage, as there is no evidence of such machinery ever having been used. The hoisting machinery that Triewald built in Dannemora was in other words a development which was still only in an experimental stage in England and which was later rejected. The example in Dannemora proved costly and took a long time to complete, and it was never convincingly tested under working conditions. This helped to exacerbate the Partners' dissatisfaction with the project, and was therefore another contributory cause of the failure.

There is a parallel between the transfer of technology from England to the Dannemora Mines in the 1720s and the transfer of technology from industrialized to developing countries today. The level of technological development in a nation has a great symbolic importance: it is an expression of the nation's affluence, military strength and culture in general. A developing country will therefore often acquire the most modern technology possible, and regard as patronizing any idea of introducing an older technology that is less than modern in the donor country. The recipients are not willing to put up with crumbs from the rich man's table. Also, there are usually strong reasons in the donor country for exporting the most modern technology. This is because it is principally the largest companies—with a heavy investment in modern technology—that are in a position to sell on distant and difficult markets. These companies make their profits from modern technology and have no interest in a return to earlier, obsolete technology. The attitude of the receiving country and the economic interests in the donor country coincide, and the result is that the technology introduced into developing countries is often far too sophisticated. In addition, there are the usual problems of adaptation and modification, and in consequence the projects often fail. On the other hand, there are examples of successful technology transfer when the technology exported is one that has already been partially replaced by a new, more advanced technology in the donor country. The exported technology is only "unmodern" in a relative sense, and its great advantage is that it has been well tried and tested. It has been used for a long time, and its teething troubles have long since been overcome. There is also ample knowledge of the way to use it, the commonest problems arising, and the solutions to these problems.

If Triewald had been content to build a Newcomen engine of average size it would probably have worked much better. In fact the best thing he could have done would have been to dismantle an old machine that he himself had helped to erect and operate for several years, and then bring it over and re-erect it at the Dannemora Mines. Even this would have called for modification in a new environment, but it would not have resulted in so many new technical problems if the level of the technology had been the same as, or preferably below, the average in the donor country. *The tendency to export far too advanced a technology* (N.B.

by comparison with the general level in the *donor* country) may thus be regarded as a mistake commonly made in technology transfer. The complex relationship between the donor country and the receiving country was in the case of the Dannemora engine embodied in one person: Mårten Triewald. His reason for choosing an advanced level of technology was, as suggested in Chapter 13, his personal ambition, which was in turn to be explained by his background.

2. GEOGRAPHICAL FACTORS

The severity of the winter in Sweden by comparison with that in Great Britain upset Triewald's original schedule. As construction work could only be undertaken during the summer half of the year, the building of the engine fell almost a year behind time. Once it was complete, the engine was out of action for long periods when necessary repairs were prevented by ice in the mine or frozen pumps. The delays increased the Partners' irritation, and the *difference in climate* thus contributed to the failure.

The major geographical difference was in *natural resources*. In Great Britain, Newcomen engines were coal-fired, but Sweden was to all intents and purposes devoid of fossil fuel. To run a Newcomen engine on wood required considerable quantities, because the heat content of wood is far less than that of coal. Wood was, as we saw in Chapter 3, a natural resource that was at a premium in the mining districts, where it had many important uses: fire setting, ore calcination, in blast furnaces and forges etc. Indeed, this was the reason for the first misgivings expressed by the Board of Mines when discussing the Newcomen engine in 1725 (see Chapter 9). The difficulty of maintaining an adequate supply of firewood and timber at the Dannemora Mines is plain to see from Mine Bailiff Kröger's reports, but it was not an actual unavailability of wood fuel that was the direct cause of the engine's failure. The problem lay rather in the fuel economics, which will be discussed in the next section. It should be noted, however, that at the time the Newcomen engine was being introduced, another new techno-logy was also being tried out in the Dannemora Mines. This was the use of gunpowder blasting as an alternative to the extravagant practice of fire setting, and trials began at the order of the Board of Mines in June 1729—only a few weeks before Triewald's engine ran for the first time.[2] During the years that followed, the new technology was success-fully introduced with the help of a blasting expert who had learned the art in Saxony. The minutes of the Partners' meetings are filled with as much discussion of this new technology as of the Newcomen engine during the period 1726–1736. According to the Mine Inspector's report in 1737, the new method resulted in a reduction in wood consumption from 12 000 or 13 000 *stafrum* per year before 1729 to approx. 7 000 or 8 000 *stafrum* in 1737.[3] There was, in other words, an additional delivery capacity of several thousand *stafrum* of wood fuel a year during this period which might have been used for the engine, but the decisive factor was the price.

3. ECONOMIC FACTORS

There is no comparative study of the cost-effectiveness of different systems of hoisting ore and draining water in Swedish mines. There is, however, ample information in the historical records of the Dannemora engine to give us an idea of the economics of the engine. The Partners had been receptive to Triewald's proposal to build an engine that would keep all the Silverberg mines free from water and hoist the ore. They needed such a "useful contrivance", because the draining of the mines was costing them 9 000 *copperdaler* a year. The cost of drawing up the ore is not known, nor is the profit they expected to make if Norra Silverberg, the deepest of the Silverberg mines, could be drained and re-opened.

Triewald had put the cost of building the engine at 24 000 *copperdaler*, but the cost of adapting the technology to the new environment raised the total cost to over 50 000. Despite this outlay, the engine still did not provide continuous operation, and the hoisting gear was obviously a failure. The Partners were reluctant to invest more in the project, but it is possible that the engine could have been used successfully to drain water, if only more time and money had been spent on development. However, this project was financed on a very different basis from later schemes to introduce the Newcomen engine into Sweden. The next such attempts were not made until the Swedish Ironmasters' Association, founded in 1747, agreed to underwrite them.[4] These attempts were also unsuccessful, but the costs were not borne by the individual ironworks, which would not have ventured for a second time to participate in these experimental projects without a financial guarantee. As all transfers of new technologies give rise to additional costs for adaptation to a new environment, the absence of *external financial backing in the initial stage of technology transfer* appears to have been a contributory cause of the failure of the Dannemora engine.

But the crucial question is of course whether the engine would have been deemed economic by comparison with the traditional technologies. The archives of the Dannemora Mines contain an estimate by Triewald (prepared in 1733 or 1734) of the engine's running costs,[5] which may be summarized as follows:

	Annual cost
Wood fuel (6 *stafrum* per day at a cost of 3 *copperdaler* per *stafrum*)	6 570
One engineer (*Konstmästare*, at 50 *copperdaler* per month)	600
Two machine operators (at 24 *copperdaler* per month each)	576
Repairs	1 164
Triewald's royalty under the agreement	1 200
	10 110

Presumably the Partners made similar calculations, but they would doubtless have allowed for a higher fuel consumption. The Board of Mines tests in 1734 showed that the engine consumed 8.9 *stafrum*, and the Partners ought to have known by 1730 that fuel consumption was about 9 *stafrum* per day, which would raise the first figure in Triewald's estimate to 9 855 *copperdaler*. Triewald had based his estimate on the cost of repairs to the Königsberg engine in Hungary,[6] but the Partners knew by now that any estimate by Triewald had to be doubled. Let us assume that they estimated the cost of repairs at approx. 2 000 *copperdaler*. They would, however, have been entitled to withhold any royalty, because the hoisting gear did not function as stated in the agreement. Based on these assumptions, their estimate would come to approx. 13 000 *copperdaler*. This exceeded what it was currently costing them to drain all the mines except Norra Silverberg by 4 000 *copperdaler* per year. It is clear that the Dannemora engine was less economical than the traditional sources of power. This consideration of running costs did not include the capital investment of 50 000 *copperdaler*, but the Partners had invested on such a scale before (see Chapter 11) in the hope of ridding the mines of water from the lake. It is possible that they would have considered the engine economic in the long run, had it only provided a permanent solution to the problem. A necessary condition would however have been that the engine could have been relied on to give continuous service to allow all the Silverberg mines to be worked. If Triewald's original estimate of the fuel consumption, i.e. that the engine would only consume 300 *stafrum* per year, had been correct, the Partners' calculation would only have come to 4 000 *copperdaler* per year. But he underestimated the fuel requirements by a factor of 10 when inferring the Newcomen engine's consumption of wood fuel from its consumption of coal. The Newcomen engine was economic in the context in which it had been developed, i.e. the British coal industry, but it was, as Graham J. Hollister-Short has written, "so symbiotically linked to the mining of coal that [...] it could not in any significant or lasting way break clear of the technological matrix in which it had first come to maturity".[7] The basic economic reason for the failure of the Dannemora engine was thus the *difficulty of reproducing the cost-effectiveness of the technology in a new environment.*

4. SOCIAL FACTORS

The Newcomen engine was a centralized prime mover in that it produced high power and could replace a large number of the traditional, smaller power units in the mining industry. The work in the Dannemora Mines was organized on a system of weekly shares in the different mines, in other words production was decentralized. The *adaptation of the new technology to the existing social order* at the Dannemora Mines presented a major problem. If the engine was unreliable as far as continuity of operation was concerned, it might unfairly favour or penalize the different inronworks by keeping the mines free from water in some weeks but failing to do so in others. It thus threatened to undermine the whole system that formed the basis of the social order.

The main problem of adaptation at the Dannemora Mines was thus that of guaranteeing continuous operation. It was said in Chapter 6 that one of the disadvantages of the early Newcomen engines lay in the scarcity of experienced persons to keep them in service. In a paper entitled "The Transfer of Power and Metallurgical Technology to the United States, 1800–1880", Brooke Hindle stresses the role of immigrant mechanics in the initial transfer of steam engine technology.[8] Their importance in general is confirmed by the case of the Dannemora engine, and we may add a social dimension.

There were only two men in Sweden who could keep the engine working. Triewald and Olof Hultberg. The reasons for Triewald's reluctance to spend his time in Dannemora will be reviewed below, but why did Hultberg, the "immigrant mechanic", not remain in Dannemora? The answer appears to lie in his status, or lack of it, at the Dannemora Mines. He was an outsider, a student at Uppsala University, who had suddenly been entrusted with a responsible post in a traditional mining community. His position in relation to Head Mine Bailiff Thomas Kröger and the mine bailiffs of the Partners was ambiguous. But Hultberg had to work and live within the mining community, and he was dependent on Mine Bailiff Kröger and the rest of the community for wood fuel, materials and labour for the engine, as well as for his salary, food and housing. They for their part were also dependent upon him, for his efforts to keep the engine in regular service affected the work of all who hoped to be able to exploit the Norra Silverberg Mine and the other mines with which this mine was connected. It must have been obvious to them that the reliability of the engine would determine how much ore each ironworks could mine during its allotted weeks, and they held Hultberg responsible for the engine's failings.

Hultberg had been appointed *Konstmästare*, a title usually given to a person responsible for all the mechanical contrivances used for draining water and hoisting ore from a mine. But Hultberg was *Konstmästare* "at the engine", which meant that his authority extended only to the

engine, its pumps and the hoisting machinery. However, the consequences of the engine's performance reached beyond the stone walls of
the engine house and the wooden pump pipes. This led naturally to the
friction between Hultberg and the mine bailiffs that has been portrayed
in Chapters 13 and 14. Olof Hultberg had to represent the entire social
basis of the new technology when Triewald was away, but he had not
been properly integrated in the social order in terms of authority and
prestige. A glance at Fig. 11.2 in Chapter 11 will tell us that the only
position in the hierarchy of the Dannemora Mines that Hultberg could
occupy, as the engineer resposible for a centralized power system, was
at *B*, i.e. below Mine Bailiff Thomas Kröger, but above the mine bailiffs
of the various ironworks. Kröger must have realized this, because he
told Hultberg in 1729 that he would "have him dance to a different
tune or teach him better ways", once Hultberg came under the authority of the Dannemora Mines.

It is understandable that Hultberg preferred to leave Dannemora for
what was sure to be a considerably more peaceful existence as a maker
of scientific instruments in Uppsala (see Chapter 16). After he had left,
the engine remained silent: all the hardware was there, but there was
nobody to operate it. The case of the Dannemora engine illustrates the
fact that the hardware components are only a part of a technology, and
thus only a part of a successful technology transfer. Practical know-
how, experience and skill must also be transferred, and these are all
human qualities. The engine may have worked by vacuum, but its
engineer could not work in a social vacuum. Thus, this case shows in
addition the importance of *integration of immigrant mechanics in the existing
social order*. This is not only necessary to a successful operation in the
initial stages of technology transfer, but a precondition of the diffusion
of knowledge into the new environment. The engine house that still
stands in Dannemora is a silent reminder that technology is more than
hardware, more than precision-bored cylinders, pressure-resistant boilers and so on. It is also a matter of human beings.

The problems of the Dannemora engine, and their contribution to
this failure in technology transfer, discussed so far may be summarized
by a quotation from the paper by Rosenberg mentioned above:

new techniques frequently require considerable *modification* before they can
function successfully in a new environment. This process of modification often
involves a high order of skill and ability, which is typically underestimated or
ignored. Yet the *capacity* to achieve these modifications and adaptations is
critical to the successful transfer of a technology—a transfer which is too
frequently thought of as merely a matter of transporting a piece of hardware
from one location to another.[9]

Apparently neither Triewald nor the Partners realized what this need
for modifications demanded of them. If the Partners had realized, they
would have made sure that Hultberg enjoyed the authority and prestige

appropriate to his responsibilities. They would also have shown more understanding of the delays and increased costs that the modifications entailed. They would have been wise to apply the old engineering rule: "multiply all estimated costs and time schedules by π". But as we have seen, by about 1730 they were probably not very interested in investing more time and money in attempts to make the engine run properly. As an alternative to the traditional methods of draining the mines, the engine would at best be equally expensive, but besides this there was the specific problem of its unreliability as a centralized power system and the effect of this on the decentralized system of working the mines. The calculations might have looked different if the hoisting machinery had been successful: a number of hoists of this type, all powered by the engine, might have made the engine an economic proposition. But it was clear in 1730 that the hoisting machinery was never going to be of any practical use. If Triewald, on the other hand, had realized the need for modifications, he would have spent more time at Dannemora trying to make his engine run properly than he did. Triewald's absences from Dannemora once the engine had been built call for an explanation, and in order to provide this we have to examine a further set of critical factors in technology transfer.

5. CULTURAL FACTORS

In his paper "Technology Assessment from the Stance of a Medieval Historian", written in 1970, Lynn White, Jr., showed some of the different levels of complexity that may be found in the assessment of new technologies.[10] He concluded:

> My thesis is that technology assessment, if it is not to be dangerously misleading, must be based as much, if not more, on careful discussion of the imponderables in a total situation as upon the measurable elements. *Systems analysis must become cultural analysis*, and in this historians may be helpful. (My italics)[11]

So far we have been able to distinguish a number of measurable elements: *technical* (technical expertise, the level of the technology), *geographical* (climate, natural resources), *economic* (external financial backing, reproducing cost-effectiveness) and *social* (adaptation to the existing social order, integration of the immigrant mechanics). Finally we must take into account the imponderables in a specific cultural situation.

Triewald asserted that he had introduced "this glorious invention" to his native country by building the Newcomen engine at the Dannemora Mines. This was an expression of the utilitarian ideas that came to dominate Sweden during the eighteenth century, but which did not make their breakthrough as a politicoeconomic programme until after the change of regime in 1738/39.[12] In England, the concept of science and technology as public utilities had long been institutionalized in the Royal Society: this was a line of thought that could be traced back to Francis Bacon. It was from the circle around the Royal Society that Triewald had derived the inspiration for his work, and his values had been shaped during his years in England. When Triewald claimed that he had built the engine in Dannemora for the "public utility", this was an early and private attempt to apply utilitarian ideals as a politico-economic programme. This led to something of a cultural conflict in Dannemora, for the Partners' values had been moulded by a harder reality: the technical and economic situation in a traditional Swedish mining society. When Triewald spoke of the "utility" of his engine, he used the term in a broader and more general sense than the Partners. Throughout the project, the parties applied different criteria of "utility", and this brought them into conflict. In court, the Partners argued that theirs were the criteria that should apply in that the literal wording of the contract should be observed, and here Triewald's more vaguely formulated utilitarianism came off second best.

But Triewald also had more personal ambitions. He wanted to be recognized by the Royal Society, and he succeeded in this by being elected a Fellow in 1731 and having his letters published in the *Philosophical Transactions* in the 1730s. The Dannemora engine was only one route to this goal, but he omitted neither to mention it in his first letter to the Royal Society in 1728 nor to send a copy of his book on the Dannemora engine to the President of the Society in 1734.[13]

In the course of this study the definitions of science and technology postulated in the introduction (Chapter 1) have acquired a social dimension. This is nothing new, but the social dimension has been confirmed by the historical data and set in a specific historical situation. It also goes some way towards explaining Triewald's lack of interest in keeping his engine in operation, once it had "demonstrated its great effect". His main ambition was to realize his English ideals at a time when science was not yet a profession but an interest shared by the higher strata of society. It was through technology that he succeeded in acquiring the position he sought. When he came back to Sweden in 1726 he was referred to as *Mechanicus*, a general title applied to anyone who had studied technology. But at Wedevåg, Stockholm and Dannemora he demonstrated the knowledge he had acquired in England, and in 1727 he was appointed *Directeur* of mechanics. With his books on bee-keeping, on the diving bell and on the Dannemora engine, and with his published lectures, he showed that he was able to impart this knowledge to others, and in 1735 he was made Head of Education in the Fortifications Corps.[14] As *Capitaine-Mechanicus* he had a well-paid position with high social status that gave him the opportunity to pursue the role of gentleman scientist, and by the end of the 1730s he had succeeded in attaining the goals formulated during his years in Eng-

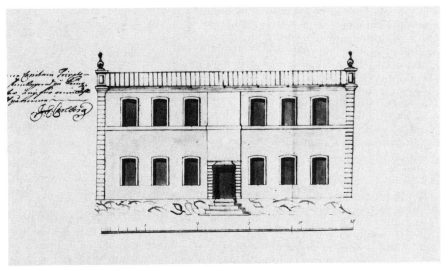

Fig. 15.1. The original drawing for Mårten Triewald's house on Kungsholmen in Stockholm, which was built in 1738. Detail of a drawing in Stockholm City Archives. (Photo Stockholms Stadsbyggnadskontor)

land. He lived in an elegant house that he had had built in Stockholm (Fig. 15.1),[15] and on the edge of the city he had a summer house with large grounds (Fig. 15.2).[16] In his gardens he cultivated exotic fruits and experimented with economic plants found in Sweden.[17] He was a member and indeed a founder of an academy of sciences enjoying royal patronage in the capital city. In its proceedings he wrote in his native language on all kinds of experiments and observations that might be of "public utility". The erstwhile Newcastle colliery engineer had come a long way.

The role of technology in realizing his ambitions is illustrated by a closer look at the dates of his letters to the Royal Society after his return to Sweden and before the founding of the Royal Swedish Academy of Sciences. During the period 1728–1738 he wrote ten letters to the Royal Society, and eight of these were written on Saturdays, or on the Sabbath or other holidays.[18] A natural rhythm was superimposed on the division of his life between science and technology, not only by the changing seasons (as we saw in Chapter 13) but also by the alternation of working week and week-end. Technology was his breadwinner, and in the long term a way to improve his social position. And for the time being science was something to which he could devote himself only when his weekly work as an engineer was over.

Finally we shall touch on yet another cultural factor that may have contributed to this failure in technology transfer, namely culturally conditioned ideas of the nature of technological development. Mårten Triewald did not employ a systematic method in his scientific work.

Fig. 15.2. Mårten Triewald's summer house at Marieberg in Stockholm. His property included approximately the area visible on the photograph. His garden, where he tried to grow exotic fruits and experimented with native economic plants, was situated on the site of the building to the far left (the present Embassy of the Soviet Union in Stockholm). (Photo the author, 1983)

Scientific discoveries were to him just that—"discoveries" in the strict sense of the word, connections that were suddenly revealed to the attentive observer. We saw several examples of this in Chapter 13, as when he was travelling on a sledge across the ice one winter evening and saw the northern lights glowing as the moon rose, or when he noticed how the workmen in Dannemora split blocks of stone, or when he entered his lecture theatre at the Palace of Nobility on a winter's day and saw the water in the glass tube containing the Cartesian divers suddenly turn to ice. Triewald himself formulated his method in a discussion in the newly established Royal Swedish Academy of Sciences in the autumn of 1739. At one meeting Jonas Alströmer had presented a divining rod, with the aid of which he believed it was possible to detect metals.[19] Linnaeus was sceptical,[20] but von Höpken thought that the Academy ought not to reject uncertain observations, for in many cases they had turned out to lead to important discoveries. He was supported by Triewald, who, according to the record, said:

Yes, indeed, most things in the world have been discovered more by chance than *a priori*.[21]

This opinion was not the expression of a superficial approach to science, but of a deep religious conviction. Like many of his contemporaries, Triewald was influenced by physicotheology.[22] The English naturalist William Derham was one of the foremost proponents of this

Fig. 15.3. Mårten Triewald. Painting by Georg Engelhardt Schröder, probably in 1735. (Photo Swedish Portrait Archives)

philosophy, which sought to show the existence of God through the purpose and beauty of Nature. The catalogue of the library left by Triewald includes several of Derham's works, in English, German and Swedish editions.[23] It is reasonable to assume that Triewald had bought and been influenced by Derham's *Physico-Theology* (1713) while he was in England, because he quotes it several times in his published lectures of 1735–36, which in all other respects reflect what he had learned in England.[24]

At the time of his death he also possessed at least four different works by Christian von Wolff, all dating from before his return to Sweden in 1726.[25] Among these was Wolff's *Vernünfftige Gedancken von Gott, der Welt und der Seele des Menschens* (Rational thoughts on God, the world and the human soul), and he seems likely to have owned at least this one in the early 1730s, as he quotes from it in 1734 (see below). Moreover, it was in about 1730 that Wolff's ideas began to gain followers in Sweden.[26] However, Tore Frängsmyr has pointed out that there were two kinds of Wolffianism, since Wolff himself:

made a distinction between learned men and common people. In his Latin work the physico-theology was not to be found at all; he rejected its value before strict scientific readers, but he accepted it for a wider public.[27]

This intellectual double-dealing by Wolff was a conscious social stratification of his philosophical arguments. Triewald's disdain for the academic world and his campaign against Latin as a language reserved for "the learned" has been mentioned earlier.[28] It is not surprising therefore that Triewald, the son of a German blacksmith, came into contact with Wolff's philosophy in its more popular form, i.e. through books in which Wolff put forward physicotheological arguments in German for the common people. (All the books by Wolff in Triewald's library were in German.)[29]

Triewald's physicotheological views had an important bearing on his views on technological development. Mechanical inventions were not "inventions" to Triewald in the true sense of the word: creations of something that did not exist before. No, they were more like discoveries: findings of something that had existed before but that had remained unknown. "To make an invention" was just another way of revealing God's ingenuity in creating this world. God was in a sense the true Inventor of all mechanical constructions, and man a mere "discoverer". Triewald expressed this view in the first chapter of his book *Konsten at lefwa under watn* (The art of living under water), in 1734:

The great God is the only and true originator of all things; whereas we wretched beings can but now and again become aware of a part of the infinite wisdom in God's creation: and turn the same to our use when we put our minds to a particular end, having previously made ourselves fully aware of all the laws of nature, which the Lord of Nature has so laid down that they can never be upset or changed by us; so we cannot credit ourselves with any more, when we discover what is useful, than having after contemplation come upon the way in which God's work has been revealed to us.[30]

For this passage, Triewald gave a reference to Wolff's *Vernünfftige Gedancken von Gott* in a footnote, and the illustration at the top of the page showed a symbolic sun, containing a motif signifying God (Fig. 15.4). Wolfgang Philipp has written that this was the symbol favoured above all others in the physicotheological books of the Enlightenment: "Its rays fill nature. They wrap men who 'contemplate' the objects of

CAP. I.

§ 1.

DEn ſtora Guden ár den endaſte och
ſanna Uphofsmannen af alt; hwar emot
wi uſla Menniſkier allenaſt då och då, kunna bli warſe något af Guds oendeliga Wis-
het i de ſkapade ting: och lempa oß de ſamme til nytta,
då når wi wåre tankar et wiſt ſyſtemål föreſåttia, ſamt
förut wel giort oß alla naturens Lager bekante, hwilka
Lager naturens HErre ſå ſtadfåſt, at de aldrig hår i tiden
af oß kunna rubbas eller åndras; wi kunna således mye-
et mindre tilſkrifwa oß något mera, når wi hwad nyt-
tigſt år påſinna, ån at wi igenom efterſinnande kommit på den wågen, hwarpå wi Guds Werk blifwit
warſe. *

A   § 2.

─────────────────────────

* Beſe Hof-Rådet Wolfs vernünfftige Gedancken von GOtt,
der Welt und die Seele des Menſchen.

Fig. 15.4. The first page of Mårten Triewald's book *Konsten at lefwa under watn*
(The art of living under water), published in 1734. (Photo Bengt Vängstam)

nature and become 'enlightened' in this contemplation".[31] The con-
tinuation on the next page of Triewald's book makes it clear that these
"discoveries" also included mechanical inventions. He states that all
new inventions "redound to the glory of God and the utility of mankind",[32] and that the inventors "have by their discoveries brought
forth the glory of God, and thus remarkably served the human race, to
alleviate the troubles that exist in this wretched life".[33]

Triewald's view of the act of invention is expressed clearly in his well-
known story of how Thomas Newcomen discovered the method of
condensing steam by the injection of cold water. This appears in his

book *Kort beskrifning, om eld- och luft-machin* (A short description of the fire and air engine), which was published in the same year as his book on the art of diving and thus written in the same spirit:

> For ten consecutive years Mr. Newcomen worked at his fire-engine which never would have exhibited the desired effect, *unless Almighty God had caused a lucky incident to take place.* It happened at the last attempt to make the model work that a more than wished-for effect was suddenly caused by the following strange event [Triewald then continued to report the story] thus convinced even the very senses of the onlookers that they had discovered an incomparably powerful force which had hitherto been entirely unknown in nature [...] Though somebody might think that this was an accident, *I for my part find it impossible to believe otherwise than what happened was caused by a special act of providence.* To this conclusion I—who knew personally the first inventors—have been brought more than ever when considering that *the Almighty then presented mankind with one of the most wonderful inventions which has ever been brought into the light of day,* and this by means of ignorant folk who had never acquired a certificate at any University or Academy.[34]

The italics are mine, but they could just as easily have been Triewald's. This passage has often been quoted by historians, as it is the only account of the moment of invention of the steam engine.[35] Unfortunately it is doubtful whether we can regard this as a historical source in the sense that it tells us anything of what actually took place in that historic Dartmouth workshop sometime during the first years of the eighteenth century. It would be more correct to regard it as simply another illustration of Triewald's religious beliefs.

This physicotheological view of the act of invention ("technotheology", we might call it) also implied that the existence of the steam engine gave us knowledge of God. At a time when the whole mining industry of England was threatened by the problem of flooded workings, God had in His wisdom revealed to mankind the steam engine to enable the riches and bounty which He had laid down in the earth to be brought up. Like the wonderful and wise arrangements of Nature, the steam engine bore witness to its divine origins and to the fact that everything was arranged in the most wonderful way in this best of all worlds. There are some celebrated lines by Linnaeus, expressing his belief in the miracle of creation as a proof of the benevolence and omnipotence of God. I saw, Linnaeus wrote:

> the eternal, all-knowing, all-powerful God from the back when he advanced, and I became giddy! I tracked his footsteps over nature's fields and found in each one, even in those I could scarcely make out, an endless wisdom and power, an unsearchable perfection.[36]

Triewald had witnessed the impressive power of the Newcomen engine and its usefulness to the mining industry of England. When he saw these engines, "which by the power of fire and air drain an incredible amount of water from the deepest shafts", he became giddy!

He saw in their mechanical design an endless wisdom and power, an unsearchable perfection. This explains Triewald's indignation at the Partners' complaint to the Board of Mines. The Partners had not only impugned his good faith ("*Laudanda Voluntas*") when he wanted to introduce the Newcomen engine into Sweden "for the public good", they had done something far worse. The Newcomen engine was, to Triewald, like natural science to Linnaeus, a proof of an almighty and benevolent God. How dared they question this? Had he not revealed to them the Glory of God in Dannemora?

16. Epilogue

The lawsuit between Mårten Triewald and the Partners in the Dannemora Mines continued for many years. It appears to have contributed
to the livelihood of a series of lawyers and clerks, for the names of the
attorneys succeed one another in a collection of documents that has
been preserved.[1] The final one is from 1746, and from this we learn that
the case had been filed with the Svea Court of Appeal. It is likely that
the lawsuit continued for as long as Triewald lived, but it was now
purely a legal matter of the size of the damages. The new technology
had already been brought to trial in Dannemora.

Triewald continued, apparently unhindered, his many other activities. He devoted himself with great enthusiasm to his very dearest
"invention", the Royal Swedish Academy of Sciences, and was its most
enthusiastic member and a frequent writer in its Proceedings—albeit
often naive and uncritical in his optimism. His first wife died in 1738,
and in 1740 he married Elisabeth Worster (1720–1789),[2] who was only
twenty years old. She was the daughter of Samuel Worster, the British
merchant in Stockholm by whom Triewald had been employed when a
young man. His first marriage had been childless, but he and Elisabeth
had five children.[3] At six in the morning on August 8, 1747, Mårten
Triewald was found "suddenly dead sitting in a privy" at his house on
Kungsholmen.[4] His friend Linnaeus wrote on the flyleaf of his copy of
Triewald's lectures:

Obiit optimus Triewald d. 8 Augusti 1747. Asphixia[5] (The excellent Triewald died
on August 8, 1747. Asphyxia)

Olof Hultberg left Dannemora for good sometime in the early 1730s.
In December 1729, when conditions for him in Dannemora were at
their most unbearable, he had taken a few days' leave from his work at
the engine and gone to Uppsala to re-enrol at the university.[6] In the
spring of 1730 he took a month off for a temporary assignment in
Stockholm to build an air-pump for Count Carl Gustaf Tessin
(1695–1770).[7] It is clear that Hultberg, just like Triewald, deliberately

abandoned a career in engineering for a comparatively peaceful existence in the scientific community. He married one Magdalena Björkgren
in October 1730,[8] and settled on Kungsängsgatan in Uppsala, where
he established himself as a scientific instrument maker.[9] Anders Celsius, who had already met Hultberg in 1728 when he attended
Triewald's lectures, engaged him in 1741 to work on the astronomical
sector that Celsius had purchased from London for Uppsala Observatory.[10] Olof Hultberg—engineer of the first Swedish steam engine
—died in Uppsala in 1743. He had kept his title of *Konstmästare*,[11] but
that was all his Dannemora experience had brought him.

The Dannemora Mines had reverted to the traditional sources of
power in 1736, but the Newcomen engine was kept in the locked and
barred building for nearly forty years. In 1768 the crown prince, the
future Gustaf III, visited the Dannemora Mines on a journey through
the mining districts of Sweden.[12] A salute of 60 blasts was given at the
mine, and while the powder smoke was dispersing, the royal company
were able to watch the work:

> Women and girls work there like the miners, and do so little shrink from
> going down on the rope to their work, that they stand on the edge of the barrel,
> grasp the chain with one arm, and then knit a stocking as they ride: and sing a
> hymn while ascending or descending, giving a weird echo.[13]

The company also viewed "the remains of the renowned Triewald's
Fire and Air Engine".[14] When His Royal Highness asked the causes
of this engine's unserviceability, He was told that it must have been
incorrectly proportioned, and that Captain Triewald had tried to use
the engine for hoisting ore before it had been properly adapted for its
rightful purpose, the pumping of water. (Gustaf III seems to have been
among the last to hear of the hoisting machinery, which later fell into
oblivion.) At the end of the 1760s, work was started on the building of
another Newcomen engine at the Dannemora Mines—the Partners had
been encouraged by the good results initially achieved by the second
Newcomen engine in Sweden, at the Persberg mines in the province of
Värmland.[15] But when eventually the Persberg engine nevertheless
proved a failure, the Dannemora project was also abandoned, and in an
attempt to cut the losses all the parts of Triewald's engine were sent to
Stockholm in 1773 to be sold.[16] The cylinder appears to have lain in
Grill's "Terra nova" shipyard (where present-day Strandvägen meets
Grev Magnigatan) until 1784, when it was sold for scrap to save the
storage costs.[17] The reason for the engine's being kept for so long at
Dannemora seems to have been that it remained a potential alternative
to the traditional sources of power. That plans to put it back into
service were never entirely given up is evident from the following
episode.

In 1751, the Vicar of Dannemora, Per Hallerström (1715–1779),
approached the Partners with the request that they donate the cylinder

of the engine to the parish.[18] If they did, it would be recast as a bell for Dannemora church. Perhaps the vicar (who is intriguingly described in the diocesan annals as "quite a rough and ill-mannered man") had entered the engine house and seen the huge bronze cylinder hanging in the dark like a great church bell.[19] But the Partners refused his request because they were "minded to have the same Fire- and Air-Engine repaired".[20] Had they acceded to it, a bell cast from the 1 700 kg bronze cylinder would have been among the largest in the diocese. Every time the Sabbath was rung in, it would have echoed the knell of steam power technology in the Dannemora Mines in 1726–1736, especially as the bell would have had a very different tone from that of an ordinary church bell. (The cylinder was probably cast in gun metal, which has a lower tin content than bell metal, a fact that would affect its timbre.)

But we may still hear sounds to remind us of the failure of the Dannemora engine if we travel twenty kilometres due north from Dannemora to Leufsta Ironworks, the largest of the Partners' ironworks in the eighteenth century. The church of the ironworks contains the only genuine preserved baroque organ of its kind in Sweden.[21] It was built by the celebrated organ builder Johan Niclas Cahman in 1725–1728,[22] i.e. while the Newcomen engine was being built in Dannemora, and it was completed in the same year. But unlike the Dannemora engine, the Leufsta organ is still working today, and it is noted for its beautiful tone. There are many technical similarities between a Newcomen engine and an organ: for to what might an organ more aptly be compared than hundreds of small Newcomen engines connected together? We find the same mechanical components: cylinders, tubes and valves, and a control mechanism. But an organ is so infinitely more sophisticated.[23] The cylinders are precisely dimensioned, not merely ground until roughly smooth on the inside with a horse-powered borer like the cylinder of the Dannemora engine. The wind channels and valves of an organ are airtight; there are no dripping valves and leaking packings as in a Newcomen engine. The control mechanism of an organ permits an exact distribution of the air to the desired pipe, and it makes the automatic valve mechanism of a Newcomen engine look positively simple. When we listen to the organ at Leufsta, we realize that it was not the advanced level of technology *per se* of the Newcomen engine that was the cause of the failure in Dannemora.[24]

Organ building, just like the building of Newcomen engines, was a technology that was for the most part transmitted as an empirical craft, a body of non-verbal knowledge. Although there were books on organ building, such as Bendeler's *Organopoeia* (ca. 1690), the first book in Swedish on the subject did not appear until 1773.[25] But organ building had a long tradition in Sweden, and there were a large number of craftsmen trained in the building, maintenance and repair of organs. This was a decisive difference, for the Newcomen engine at the Danne-

Fig. 16.1. The baroque organ in the church of Leufsta Ironworks. Built by Johan Niclas Cahman in 1728—the same year as the Newcomen engine in the Dannemora Mines. (Photo Jan Borgfelt/Ögats Glädje & Idé AB, 1980)

mora Mines stood or fell with Olof Hultberg. Michael Praetorius described the capabilities of the organ in his book *De Organographia* in 1619 as follows:

Indeed, this polyphonic and delightful mechanism contains all the music that can be conceived and composed and produces a tone and a harmony like a whole choir of musicians, in which one hears many kinds of melody, sung by the voices of young boys and grown men. The organ possesses, quite simply, and contains alone within itself all the other *instrumenta musica*, both large and small.[26]

Technologically, an organ may be described as a centralized system consisting of a large number of separate instrumental units; some of them have a timbre found only in the organ, while others serve as imitations of other musical instruments and of the human voice (*Vox humana*). Its development is to a large degree bound up with the church, and it was a centralized technology that was well adapted to the Christian liturgy. The organ in the church of Leufsta Ironworks was also suited to its environment in that there was no conflict with the social structure of the ironworks. The arrangement of the pews in the church reflects the strict hierarchy of the ironworks: a pew for the owners at the front, followed by a pew for the respected Walloon smiths, separate pews for men and women, and an enclosed pew at the back for those sentenced to church discipline. This also reflects the centralized organization of the ironworks of the Partners, which was in such marked contrast to the decentralized organization of the Danne-mora Mines. The Newcomen engine, like the organ, was a centralized technology. As a prime mover it replaced a large number of separate power units, but, unlike the organ, it was badly suited to the social environment into which it was introduced.

Finally, an organ is a representative of the symbolic function of technology mentioned in Chapter 1. Economic criteria are not applied to symbolic technologies, any more than they are to technologies that fulfil a military function. But the Dannemora engine was a different case, being intended to fulfil the third function of technology, namely production. Its output could be measured and compared with the traditional technologies in *copperdaler*, and that was why it was rejected at the Dannemora Mines. Nobody, on the other hand, would have thought of assessing the utility of a technology, however advanced and costly, whose purpose was "the glory of God and the fostering of piety in the temple of the Lord."[27]

Abbreviations

Archives

BKA	Bergskollegiums arkiv (in RA)
BM	British Museum Library and Archives, London
BMA	Bergmästarämbetets i Gävleborgs, Uppsala och Stockholms län arkiv (in LAU)
DGA	Dannemora gruvors arkiv (in SKCA)
FBA	Forsmarks bruks arkiv (in LAU)
KB	Royal Library, Stockholm
KRA	Military Archives, Stockholm
KTHB	Royal Institute of Technology Library, Stockholm
KVA	Stockholm University Library with the Library of the Royal Academy of Sciences, Stockholm
LAU	Landsarkivet i Uppsala, Uppsala
LUB	University of Lund Library, Lund
NMA	Nordiska Museets arkiv, Stockholm
RA	National Record Office, Stockholm
RS	Royal Socety Library, London
SKCA	Stora Kopparbergs Bergslags AB Centralarkiv, Falun
SSA	Stockholm City Archives, Stockholm
TM	National Museum of Science and Technology, Stockholm
UUB	University of Uppsala Library, Uppsala

Publication

KVAH *Kungl. Vetenskapsakademiens Handlingar* (Proceedings of the Royal Swedish Academy of Sciences)

N.B. the title *Daedalus* refers, unless otherwise stated, to the yearbook of the National Museum of Science and Technology, Stockholm. It has been published annually since 1931. An index covering the series to date was included in the 1982 edition.

Currency

There were two kinds of currency in Sweden during the period of this study: *daler kopparmynt* and *daler silvermynt*. All sums in *daler silvermynt* have been converted into *daler kopparmynt*, which is abbreviated *copperdaler*.

Family Names

Several of the persons referred to in this study were later ennobled, and were thus known under other family names at the time. All persons are, however, referred to by the name by which they are most commonly known today.

Notes

1. INTRODUCTION

1. Bo Carlsson et al., *Teknik och industristruktur— 70-talets ekonomiska kris i historisk belysning* (IUI-publikation; IVA-meddelande, No. 218: Stockholm, 1979).
2. Ibid., 141–144, 152–154.
3. Erland Waldenström, "Näringslivet och den teknikhistoriska forskningen", *IVA-Nytt* 1983, No. 1, 6–8.
4. Carroll W. Pursell, Jr., "History of Technology", in: Paul T. Durbin, ed., *A Guide to the Culture of Science, Technology, and Medicine* (New York, 1980), 87.
5. For references to modern historical works on technology transfer, see: ibid., 87 f., 106–120.—See also *L'acquisition des techniques par les pays non-initiateurs* (Colloques Internationaux du Centre National de la Recherche Scientifique, No. 538: Paris, 1973). This is a collection of highly interesting papers and discussions on technology transfer at a conference in Pont-à-Mousson in 1970 organized by ICOHTEC.— *Technikgeschichte* 50 (1983), No. 3, is a special issue on technology transfer which contains some of the papers presented at the conference "Technologietransfer im 19. und 20. Jahrhundert" in Düsseldorf in 1983 organized by VDI. See esp. the introduction by Ulrich Troitzsch on pp. 177–180. See also: Hans-Joachim Braun, "The National Association of German-American Technologists and Technology Transfer between Germany and the United States, 1884–1930", *History of Technology* 8 (1983), 15–35, esp. 15 and nn. 1–6.—For a Swedish historical study on technology transfer, see: Klaus Wohlert, "Svenskt yrkeskunnande och teknologi under 1800-talet. En fallstudie av förutsättningar för kunskapstransfer", *Historisk Tidskrift* 99 (1979), 398–421 (English summary on p. 421).
6. David S. Landes, *The Unbound Prometheus: Technological Change and Industrial Development in Western Europe from 1750 to the Present* (London, 1970), 100.
7. Torsten Althin, "Stationary Steam Engines in Sweden 1725–1806", *Daedalus* 1961, 95–99.
8. Mårten Triewald, *Kort beskrifning, om eld- och luft-machin wid Dannemora grufwor* (Stockholm, 1734). Also published in an English translation by the Newcomen Society, London, in 1928 as Extra Publication No. 1, see: *Mårten Triewald's Short Description of the Atmospheric Engine. Published at Stockholm, 1734. Translated from the Swedish with Foreword, Introduction and Notes* (London, 1928).
9. The daily newspaper *Upsala Nya Tidning*, for example, has published three very similar articles on Triewald and the Dannemora engine in 1927, 1977 and 1982. See: Ernst E. Areen, "Mårten Triewald och luft- och eldmaskinen vid Dannemora gruvor. Något om en märkesman inom svenskt näringsliv under förra hälften av 1700-talet", *Upsala Nya Tidning* 1927, December 17, 1; Tore Attelid, "Världspremiär i Dannemora", ibid. 1977, December 4, 34; Lennart Liljendahl, "Dannemora gruvor först med ångmaskin i Sverige", ibid. 1982, August 7, 22.
10. See, for example: Bengt Hildebrand, *Kungl. Svenska Vetenskapsakademien: Förhistoria, grundläggning och första organisation* (Stockholm, 1939), 145–148; Sten Lindroth, *Svensk lärdomshistoria*, Vol. 3 (Stockholm, 1978), 339–341.
11. Carl Sahlin, "Länshållningen i Dannemora grufvor i äldre tider", *Teknisk Tidskrift: Kemi och Bergsvetenskap* 48 (1918), 156–158.
12. Johan Wahlund, *Dannemora grufvor: Historisk skildring* (Stockholm, 1879).

13. Ibid., 32, 85, appendix 107–109.
14. Eugene S. Ferguson, "Toward a Discipline of the History of Technology", *Technology and Culture* 15 (1974), 20 n. 20.—I have previously discussed the eight most common definitions of technology, and argued for this one on the basis of historical examples, see: Svante Lindqvist, "Vad är teknik?", in: Bo Sundin, ed., *Teknik för alla: Uppsatser i teknikhistoria* (Institutionen för idéhistoria, Umeå Universitet, Skrifter No. 17: Umeå, 1983), 2–22.
15. Helmer Dahl, *Teknikk Kultur Samfunn: Om egenarten i Europas vekst* (Oslo, 1983), 47–51.
16. Pursell, 80. See also: John M. Staudenmaier, "What SHOT Hath Wrought and What SHOT Hath Not: Reflections on 25 years of the History of Technology" (paper at the Annual Meeting of the Society for the History of Technology in 1983, to be published in *Technology and Culture*).
17. It was not listed in Sahlstedt's semi-official Swedish dictionary in 1773, but *"technologie"* appears in Rinman's mining dictionary sixteen years later. See: Sven Rinman, *Bergwerks lexicon*, Vol. 2 (Stockholm, 1789), 969 f.; Abraham Sahlstedt, *Swensk ordbok* (Stockholm, 1773), 601 f.
18. See, for example: Christopher Polhem, "Theoriens och practiquens sammanfogning i mechaniquen, och särdeles i ström-wärk", *KVAH* 1741, 149 f.
19. I have discussed this previously in another context, see: Svante Lindqvist, "Discussion: An Engineer Is an Engineer Is an Engineer?", in: Carl Gustaf Bernhard et al., eds., *Science, Technology and Society in the Time of Alfred Nobel* (Nobel Symposia, No. 52: Oxford, 1982), 298–304.—For the early history of engineers as a professional group in Sweden, see: Rudolf Anderberg, *Grunddragen av det svenska tekniska undervisningsväsendets historia* (Skrifter utgivna av Ingeniörsvetenskapsakademien, Meddelanden, No. 5: Stockholm, 1921); Karl Malmsten, "Ingenjörens titel och tradition", *Med Hammare och Fackla* 11 (1940–1941), 58–101.
20. *Svenska Akademiens Ordbok*, Vol. 12 (Lund, 1933), I 484 f.
21. See, for example, Triewald 1734 (p. 6) who refers to Samuel Calley as *"ingenieur"*.
22. This definition corresponds to the present meaning of the Swedish word *"naturvetenskap"*, while the Swedish *"vetenskap"* also includes the humanities and social sciences.
23. This discussion is based on Hildebrand 1939, 74–77, 372 f.
24. Tore Frängsmyr, "Vetenskapsmannen och samhället i historisk belysning", in: *Vetenskapsmannen och samhället: Symposier vid Kungl. Vetenskapssamhället i Uppsala 1976–1977* (Acta Academiæ Regiæ Scientiarum Upsaliensis, No. 19: Stockholm, 1977), 11–32, esp. 11 and n. 2. Cf. *Oxford English Dictionary*, Vol. 9 (Oxford, 1970), 223.
25. Gunnar Eriksson, "Den nordströmska skolan", *Lychnos* 1983, 150–157 (English summary); Tore Frängsmyr, "Vetenskapens roll i historien", ibid., 171 f.; idem, "History of Science in Sweden", *Isis* 74 (1983), 465–468.—For a critique of the "Nordström tradition", see: Bo Lindberg & Ingemar Nilsson, "Sunt förnuft och inlevelse. Den nordströmska traditionen", in: Tomas Forser, ed., *Humaniora på undantag: Humanistiska forskningstraditioner i Sverige* (Stockholm, 1978), 79–107.
26. For an assessment of Torsten Althin's achievements in the history of technology and for his biography, see: "The Leonardo da Vinci Medal", *Technology and Culture* 20 (1979), 583–589.
27. Svante Lindqvist, "Teknikhistoria—motiv och mål", *Daedalus* 49 (1980), 67–72.
28. In 1976, I made a survey of the teaching in history of technology in British universities, see: Svante Lindqvist, *Teknikhistoria som läroämne vid universiteten i Storbritannien* (Stockholm Papers in History and Philosophy of Technology, TRITA-HOT-5001: Stockholm, 1976).
29. Svante Lindqvist, *The Teaching of History of Technology in USA: A Critical Survey in 1978* (Stockholm Papers in History and Philosophy of Technology, TRITA-HOT-5003: Stockholm, 1981).
30. For surveys of perspectives and themes in the discipline of history of technology in the USA, see: Thomas P. Hughes, "Emerging Themes in the History of Technology", *Technology and Culture* 20 (1979), 697–711; Pursell 1980, 70–120. See also: Merritt Roe Smith, ed., *Military Enterprise and Technological Change: Perspectives on the American Experience* (to be published by MIT Press). In his introduction, Smith identifies four different interpretative perspectives that have been applied in recent years to the study of technological change: (1) technology as expanding knowledge, (2) technology as a social force, (3) technology as a social product, and (4) technology as a social process.
31. In 1981, a Swedish National Committee for the History of Technology was founded jointly by the Royal Swedish Academy of Engineering Sciences and the Royal Swedish Academy of Sciences, and in 1983 the National Committee be-

gan publishing a Swedish journal in the history of technology, *Polhem* 1 (1983)–. This growing interest in the history of technology had previously been supported by the National Museum of Science and Technology (*Tekniska museet*) in Stockholm. The Museum, which has published its yearbook *Daedalus* since 1931, arranged international symposia in the history of technology in 1977, 1979, 1980 and 1983.

32. See, for example: Jan Glete, "Teknikhistoria —viktig i ekonomisk och historisk forskning", *Daedalus* 49 (1980), 55–65; Carl-Axel Olsson, *Teknikhistoria som vetenskaplig disciplin* (Meddelande från Ekonomisk-historiska institutionen, Lunds Universitet, No. 11: Lund, 1980); Bernt Schiller, "Technology–History–Social Change: A Methodological Comment and an Outline of a Nordic Account", *Scandinavian Journal of History* 8 (1983), 71–82.

33. In 1980, I wrote a preliminary outline of this study which I presented at seminars at the following eight departments: History of Ideas and Science in Uppsala, Gothenburg and Lund; History of Ideas in Stockholm and Umeå, Economic History in Gothenburg and Stockholm; and the Program in Technology and Social Change in Linköping. Although this was a valuable educational experience, the suggestions I received were quite contradictory even within my own discipline. It convinced me that there is as yet no consensus in Sweden on what a history of technology ought to be.

34. Steven Shapin, "Social Uses of Science", in: G. S. Rousseau & Roy Porter, eds., *The Ferment of Knowledge: Studies in the Historiography of Eighteenth-Century Science*, (Cambridge, 1980), 108–110.

35. Nathan Rosenberg, *Inside the Black Box: Technology and Economics* (Cambridge, 1982).

36. Cf. Lindqvist 1982, 300 f.

37. *Industriföretagens forsknings- och utvecklingsverksamhet 1977–1981* (Svenska Statistiska Centralbyrån, Statistiska meddelanden, series U: Stockholm, 1982). Cf. Nathan Rosenberg, *Technology and American Economic Growth* (New York, 1972), 176–185; Rosenberg 1982, 120 f., 170 f., 207–215.

2. WOODEN IMAGES OF TECHNOLOGY

In addition to the printed works cited in the footnotes, I have also made use of Marie Nisser's unpublished dissertation, "Byggnadsteknisk debatt

och utbildning i Sverige under 1600- och 1700-talen" (mimeographed fil. lic. dissertation at the Department of History of Art, University of Uppsala, 1966).

1. For an administrative history of the Royal Chamber of Models, which also lists all the earlier literature, see: Arvid Bæckström, "Kongl. Modellkammaren", *Daedalus* 1959, 55–72. Some of the foreign travellers' comments are quoted in: Torsten Althin, "Omdömen om det tekniska museet i Stockholm för 150 år sedan", *Daedalus* 1940, 103–107. For its importance to the development of technical education in Sweden, see: Rudolf Anderberg, *Grunddragen av det svenska tekniska undervisningsväsendets historia* (Skrifter utgivna av Ingeniörsvetenskapsakademien, Meddelanden No. 5: Stockholm, 1921), 23–31; Pontus Henriques, *Skildringar ur Kungl. Tekniska Högskolans historia*, Vol. 1 (Stockholm, 1917), 57–94; Wilhelm Sjöstrand, *Pedagogikens historia*, Vol. 3:1 (Malmö, 1961), 241–244; Nils G. Wollin, *Från ritskola till konstfackskola: Konstindustriell undervisning under ett sekel* (Stockholm, 1951), 12–16.

2. Johann Wilhelm Schmidt, *Reise durch einige schwedische Provinzen* (Hamburg, 1801), 86.

3. Edward Daniel Clarke, *Travels in Various Countries of Europe Asia and Africa*, Vol. 2 (London, 1824), 127.

4. Loc. cit.

5. Bæckström, 57–62.

6. There is extensive literature on the Renaissance cabinets of artificial curiosities, but it consists mostly of catalogues or descriptive accounts. They are placed in a larger context in the following works: Silvio A. Bedini, "The Evolution of Science Museums", *Technology and Culture* 6 (1965), 1–29; Eugene S. Ferguson, "Technical Museums and International Exhibitions", ibid., 30–46, esp. 30–32; Friedrich Klemm, *Geschichte der naturwissenschaftlichen und technischen Museum* (Deutsches Museum, Abhandlungen und Berichte 41 (1973), Heft 2: München, 1973); Bruce T. Moran, "German Prince-Practitioners: Aspects in the Development of Courtly Science, Technology, and Procedures in the Renaissance", *Technology and Culture* 22 (1981), 253–274.

7. Derek J. de Solla Price, "Sealing Wax and String: A Philosophy of the Experimenter's Craft and its Role in the Genesis of High Technology" (George Sarton lecture at the Annual Meeting of the American Association for the Advancement of Science in 1983).—Price died

before his paper had been footnoted, but the lecture has been published, see: idem, "Of Sealing Wax and String", *Natural History* 93 (1984), No. 1, 49–56.

8. This comment on the function of the mechanical devices in the Renaissance cabinets of curiosities is controversial. It would need a separate study to prove my case, but I believe that their utilitarian function has been overestimated by historians.

9. Frederick B. Artz, *The Development of Technical Education in France* (Cambridge, Mass., 1966), 1–18, 55–59.—For a general study of Bacon's reform of learning, see: Charles Webster, *The Great Instauration: Science, Medicine and Reform 1626–1660* (New York, 1975).

10. Artz, 8.

11. Ibid., 11.

12. Loc. cit.

13. Ibid., 56.

14. Sten Lindroth, *Svensk lärdomshistoria*, Vol. 2 (Stockholm, 1975), 168 f.

15. Rolf Lindborg, *Descartes i Uppsala: Striderna om "nya filosofien" 1663–1689* (Lychnos-Bibliotek, No. 22: Uppsala, 1965), 67, 73 f.; Lindroth 1975, 417 f., 450.

16. Johan Eenberg, *Kort berättelse af de märkwärdigste saker som för de främmande äre at besee och förnimma utt Upsala stad* (Upsala, 1704), 87–91, Lindroth 1975, 425 f.; idem, *Christopher Polhem och Stora Kopparberg* (Uppsala, 1951), 171 ff.

17. The best descriptions of Polhem's *Laboratorium mechanicum* are found in Lindroth 1951, 82–91, and Lindroth 1975, 530–552. See also: Bæckström, 56–62; Samuel E. Bring, "Bidrag till Christopher Polhems lefnadsteckning", in: *Christopher Polhem: Minnesskrift utgifven af Svenska Teknologföreningen* (Stockholm, 1911), 30–35. For Bring's work in the English edition, see: *Christopher Polhem: The Father of Swedish Technology*, trans. by W. A. Johnsson (Hartford, Conn., 1963), 33–35.

18. Bæckström, 57–62; Sten Lindroth, *Kungl. Svenska Vetenskapsakademiens historia*, Vol. 1 (Stockholm, 1967), 276 f., 356 f.

19. Carl Knutberg, *Tal om nyttan af ett Laboratorium Mechanicum* (Inträdestal KVA: Stockholm, 1754).—For Knutberg's biography, see: Rune Kjellander, "Carl Knutberg", *Svenskt Biografiskt Lexikon*, Vol. 21 (Stockholm, 1977), 389–391; Åke Meyerson, "Carl Knutberg", *Daedalus* 1937, 102; idem; "Rationaliseringssträvanden vid svenska gevärsfaktorier under 1700-talets mitt", ibid., 96.

20. KRA, Krigskollegii brevböcker, 1743, Vol. 4,

No. 571, Letter from Carl Knutberg to the War Office, dated Paris December 19/8, 1743.

21. J. B. Gough, "René-Antoine Ferchault de Réaumur", in: Charles Coulston Gillispie, ed., *Dictionary of Scientific Biography*, Vol. 11 & 12 (New York, 1981), 326. Cf. Roger Hahn, *The Anatomy of a Scientific Institution: The Paris Academy of Sciences, 1666–1803* (Berkeley, 1971), 21–24, 65–72.

22. *Descriptions des arts et métiers* (Paris, 1761–1788). The Royal Institute of Technology Library copy consists of 22 folio volumes, and is one of the few sets of this famous work in Sweden. Cf. Eugene S. Ferguson, *Bibliography of the History of Technology* (Cambridge, Mass., 1968), 58 f.

23. *Machines et inventions approuvées par l'Académie royale des sciences, depuis son établissement jusqu'à present; avec leur description*, 6 Vols. (Paris, 1735). The Royal Institute of Technology Library copy consists of the first 6 volumes in 4:o.—Ferguson has written that "In the absence of a patent office, this 42-year series served as a sort of patent digest"; see: Ferguson 1968, 57.

24. Knutberg, 13.

25. This is only relatively true, since the Swedish Academy did serve as an advisory body to the Government, Parliament and bureaucracy in technical matters; see: Lindroth 1967, Vol. 1, 155 159.

26. Bæckström, 61.

27. Knutberg, 14.

28. Ibid., 15 f.

29. Bæckström, 61 f.

30. Johann Georg Eck, *Reisen in Schweden* (Leipzig, 1806), 233.

31. Joseph Acerbi, *Travels through Sweden, Finland, and Lapland, to the North Cape in the years 1798 and 1799*, Vol. 1 (London, 1802), 133.

32. Schmidt, 85.

33. Ibid., 86.

34. Acerbi, 134.

35. Loc. cit.

36. KB, MS Depos. 69, Carl Daniel Burén, Dagbok, Vol. 1, 1790–1792, 713 f. Cf. Bertil Boëthius, "Carl Daniel Burén", in: *Svenskt Biografiskt Lexikon*, Vol. 6 (Stockholm, 1926), 720–729; *Svenska Män och Kvinnor*, Vol. 1 (Stockholm, 1942), 504 f.

37. Translation from the Swedish edition of Miranda's diary, see: Stig Rydén, ed., *Miranda i Sverige och Norge 1787: General Francisco de Mirandas dagbok från hans resa september–december 1787* (Stockholm, 1950), 132.—For the Spanish original, see: *Archivo del General Miranda*, Vol. 3 (Caracas, 1929), 40 f.

38. De Bougrenet de Latocnaye, *Promenade d'un Français en Suède et en Norvège* (Brunswick, 1801), 72.

39. Jonas Norberg, *Inventarium öfver de machiner och modeller, som finnas vid Kungl. Modell-Kammaren i Stockholm, belägen uti gamla Kongshuset på K. Riddareholmen* (Stockholm, 1779).

40. Ibid., preface.

41. Loc. cit.

42. Eugene S. Ferguson, "The Mind's Eye: Nonverbal Thought in Technology", *Science* 197 (1977), 827.

43. Ibid., 835.

44. Louis de Boisgelin, *Travels through Denmark and Sweden*, Vol. 2 (London, 1810), 81.

45. Acerbi, 133.

46. Norberg op. cit.—There is a later catalogue listing acquisitions up to 1801. It is less descriptive, and does not divide the models into subject groups. See: Jonas Adolf Norberg, "Inventarium öfwer de nyare machiner och modeller som finnas på Kongl. Modell-kammaren i Stockholm", *Magazin för swenska hushållningen och konsterne*, Vol. 1, No. 6 (Stockholm, 1801).

47. Norberg wrote that most of the models had concerned mining, building and factories when the Royal Chamber of Models was established in 1756, but that the collection had since been supplemented by several agricultural models; see: Norberg, preface.—For the mechanization of agriculture, see: Lindroth 1967, Vol. 1, 266–276; Sten Lindroth, *Svensk lärdomshistoria*, Vol. 3 (Stockholm, 1978), 109–113.

48. For the mechanization of the textile industry, see: Björn Hallerdt, "Strumpvävstolar av Christopher Polhem", *Daedalus* 1951, 51–62; Sven T. Kjellberg, *Ull och ylle: Bidrag till den svenska yllemanufakturens historia* (Lund, 1943); Folke Millqvist, "Bomullens tidiga historia och spinningens mekanisering", *Från Borås och De Sju Häraderna* 34 (1981), 7–86; Lindroth 1967, Vol. 1, 362 f.—For the attempts to create factories, see: Eli F. Heckscher, *Sveriges ekonomiska historia från Gustav Vasa*, Vol. 2:2 (Stockholm, 1949), 587–642.

49. Lindroth 1978, 116. Cf. Edvard Hubendick, "'Konstige påfund' och tekniska frågor, dryftade på Kungl. Vetenskapsakademiens sammanträden under 1700-talet", *Kungl. Svenska Vetenskapsakademiens Årsbok* 1948, 381–427.

50. See the models listed under the following subsections in Norberg: § 1, 3–7, 11, 15–17, 25.

51. It was the availability of energy conversion processes, such as the steam engine, that led to the formulation of the principle of energy conserva-tion in the 1840s. See: Thomas S. Kuhn, "Energy Conservation as an Example of Simultaneous Discovery", in: idem, *The Essential Tension: Selected Studies in Scientific Tradition and Change* (Chicago, 1977), 66–104. Originally published in: Marshall Clagett, ed., *Critical Problems in the History of Science* (Madison, 1959), 321–356.

52. Norberg, 33. Original in Swedish: "Modell på Eld- och Luft Machine til vattns upfordring efter Engelska maneret".—This model was built by Gustaf von Engeström at the request of the Board of Mines. The beam of the model is still preserved in the collections of the National Museum of Science and Technology, Stockholm (TM No. 1815). See: Torsten Althin, "Sveriges andra ångmaskin", *Daedalus* 1939, 53, 58.

53. Norberg, 31.

3. THERMAL ENERGY

An earlier version of this chapter has been published as "Natural Resources and Technology: The Debate about Energy Technology in Eighteenth-Century Sweden", *Scandinavian Journal of History* 8 (1983), 83–107. I am indebted to Einar Stridsberg for comments on that version. In addition to the printed works cited in the footnotes I have also made use of the survey of the subject which Erik Hamberg gave in his paper "Idéer kring och förslag till förbättrad skogshushållning i Sverige på 1700-talet" (mimeographed paper presented at the Department of History of Ideas and Science, University of Gothenburg, 1975).

1. The main studies of the timber shortage in Sweden during the eighteenth century are: Eli F. Heckscher, *Sveriges ekonomiska historia från Gustav Vasa*, Vol. 2:1 (Stockholm, 1949), 293–357; Gösta Wieslander, "Skogsbristen i Sverige under 1600- och 1700-talen", *Svenska Skogsvårdsföreningens Tidskrift* 34 (1936), 593–663. For a recent study of the regulation of forestry during the eighteenth century, see: Einar Stridsberg & Leif Mattsson, *Skogen genom tiderna: Dess roll för lantbruket från forntid till nutid* (Stockholm, 1980). For a study of the supply of charcoal to the iron industry, see: Gunnar Arpi, *Den svenska järnhanteringens träkolsförsörjning 1830–1950* (Jernkontorets Bergshistoriska Skriftserie, No. 14: Stockholm, 1951). The importance of the forests to

the iron industry is also discussed in: Bertil Boëthius & Åke Kromnow, *Jernkontorets historia*, 3 Vols. (Stockholm, 1947–1968); Karl-Gustaf Hildebrand, *Fagerstabrukens historia: Sexton- och sjuttonhundratalen* (Fagerstabrukens historia, Vol. 1: Uppsala, 1957); Åke Kromnow, "Övermasmästareämbetet under 1700-talet (1751–1805): Dess organisation och verksamhet samt betydelse för den svenska tackjärnstillverkningen", *Med Hammare och Fackla* 9 (1938), 15–99, and 10 (1939), 32–94.

2. Abbott Payson Usher, *A History of Mechanical Inventions* (Cambridge, Mass., 1954, rev. ed.), 2.

3. Nathan Rosenberg, *Technology and American Economic Growth* (New York, 1972), 20.

4. Loc. cit.

5. Heckscher 1949, Vol. 2:1, 398, Vol. 2:2, 49*.

6. Loc. cit.

7. Arpi, 110–112; Wieslander, 634–636; Heckscher 1949, Vol. 2:1, 308–310.

8. Leif Mattsson & Einar Stridsberg, *Det industriinriktade skogsbruket sett ur ett historiskt perspektiv* (Kulturgeografiskt seminarium, 8/79: Stockholm, 1979), 14.

9. Wieslander (pp. 633f.) estimated the annual consumption of wood for the extraction of iron ore in Sweden during the eighteenth century at 720000 m³ (stacked cubic volume). This is an approximate estimate, but if wood for the mining of all other metals and all the wood that must have been used for investigating suspected deposits and for unsuccessful mining works that did not result in the extraction of any ore are added, then it can be said that total consumption must have reached 1 million m³.

10. Heckscher 1949, Vol. 2:1, 328–334, 349, Vol. 2:2, 24*.

11. Ibid., Vol. 2:1, 322. Cf. Lindroth 1967, Vol. 1:1, 318–323.

12. Arpi, 46; Mattsson & Stridsberg, 3.—Arpi's figures include bark, while Mattsson & Stridsberg's do not. This creates a difference of 17 % but Arpi states (p. 45) that these two different systems of measurement were both in use simultaneously in earlier times.

13. Heckscher 1949, Vol. 2:1, 322–325.

14. Anonymous, *Swar på Kongl. Wettenskaps Academiens fråga: hwilka författningar äro de bästa, at underhålla tilräckelig tilgång på skog här i landet?* (Stockholm, 1768), 9. Royal Institute of Technology Library copy, call no. Tc-42.

15. Loc. cit.

16. Cf. Carl Linnaeus (von Linné), *Lapplandsresa år 1732* (Stockholm, 1975), 56.

17. Arpi, 216. Cf. Heckscher 1949, Vol. 2:1, 308–310; Hildebrand, 262f.; Wieslander, 645–647.

18. Lars Benzelstierna, "Berättelse om åtskillige nyare malm- ock mineral upfinningar i riket", *KVAH* 1741, 237–248.

19. Daniel Tilas, "Mineral-historia öfwer Osmunds-berget uti Rättwiks sochn och Öster-Dalarne", *KVAH* 1740, 202–209.

20. Boëthius & Kromnow, Vol. 3:1, 15; Kromnow 1939, 40; Carl Sahlin, "När började den svenska järnhandteringen använda torf som bränsle?", *Blad för Bergshandteringens Vänner* 14 (1913–15), 298–302.

21. Sven Rinman, *Bergwerks lexicon*, 2 Vols. (Stockholm, 1788–1789), Vol. 1, 45f.; Per Bernhard Berndes, "Försök att använda bränbar alunskiffer såsom bränsle i ställe för ved, till åtskillige hushållsbehof", *KVAH* 1802, 91–137.

22. The twentieth-century view that natural resources are limited was alien to the eighteenth century. Not only was it believed that the country contained hidden riches which only had to be discovered and recorded, but natural resources were also regarded as dynamic entities. The amount of water in the seas seemed to be declining—was it not clear from looking at the coastline that the surface of the sea was getting lower? Moreover, did not minerals grow inside rocks in the same way as plants did in the soil?

23. Sten Lindroth, *Gruvbrytning och kopparhantering vid Stora Kopparberget intill 1800-talets början*, Vol. 1 (Uppsala, 1955), 267.

24. Ibid., 521.

25. Ibid., 517.

26. Rinman 1789, Vol. 2, 1003f.

27. Lindroth 1955, Vol. 1, 517f.

28. Ibid., 496–502.

29. Ibid., 503.

30. Rinman op. cit. Cf. Anton von Swab, *Om grufvebrytning i Sverige* (Upsala, 1780), 17–21.

31. Lindroth 1955, Vol. 1, 503.

32. Samuel Sandel, "Rön angående malm- och berg-sprengning", *KVAH* 1769, 283–311; Lindroth 1955, Vol. 1, 506–517.

33. Lindroth 1955, Vol. 1, 497.

34. Rinman 1788–1789, Vol. 1, 1017–1020, Vol. 2, 26, 411.

35. Magnus Edvardi Wallner, *Kolare konsten uti Swerige, korterligen beskrifwen* (Stockholm, 1746).

36. Carl David af Uhr, *Berättelse om kolnings-försök åren 1811, 1812 och 1813. På Bruks-Societetens bekostnad anstälde* (Stockholm, 1814), 1–8. Cf. Boëthius & Kromnow, Vol. 1, 490ff., Vol 3:1, 73. This lack of clarity can also be seen in Rinman 1788–1789, Vol. 2, 26, 411.

37. af Uhr op. cit.; Boëthius & Kromnow op. cit., 512 ff.
38. Carl David af Uhr, *Handbok för kolare* (Stockholm, 1814, 2nd ed. 1814, 3rd ed. 1823).
39. Arpi, 55 f.; Sigvard Montelius, Gustaf Utterström & Ernst Söderlund, *Fagerstabrukens historia: Arbetare och arbetarförhållanden* (Fagerstabrukens historia, Vol. 5: Uppsala, 1959), 160–214.
40. Heckscher wrote: "In numerical terms, the group of workers that was quite predominant was one which is generally the least considered when the mining industry is discussed and which was not included among mining workers in the statistics of the time. This group was made up of the peasants and crofters who were engaged in transporting the ore and in charcoal burning at the same time as they farmed the soil" (Heckscher 1949, Vol. 2: 1, 454).
41. Boëthius & Kromnow, Vol. 1, 491.
42. Hjalmar Braune, "Om utvecklingen af den svenska masugnen", *Jernkontorets Annaler* n.s. 59 (1904), 1–113; Boëthius & Kromnow, Vol. 1, 463–485, Vol. 3: 1, 10–18; Hildebrand, 220–225, 289–303; Kromnow 1938–39, passim.
43. Braune, 30.
44. Heckscher 1949, Vol. 2: 1, 478–483; Hildebrand, 300 ff.; Montelius, 144–153.
45. Kromnow 1938, 27–33. Cf. Eric Thomas Svedenstierna, *Försök till hytte-ordning för Nora, Lindes och Ramsbergs Bergslager* (Stockholm, 1819), 3 f.
46. Heckscher 1949, Vol. 2: 1, 454.
47. The President of the Board of Mines, Fredrik Gyllenborg, took the initiative in securing the establishment of *Övermasmästareämbetet* (Kromnow 1938, 50–55).
48. Ibid., 16.
49. Ibid., passim.
50. Johan Carl Garney, *Handledning uti svenska masmästeriet* (Stockholm, 1791); Sven Rinman, *Försök till järnets historia, med tillämpning för slögder och handtwerk* (Stockholm, 1782); Sven Rinman 1788–1789. Cf. Boëthius & Kromnow, Vol. 3: 1, 48–66.
51. Eric Thomas Svedenstierna reported in 1796 that few miners took the trouble to read Rinman's and Garney's large works in their entirety (Kromnow 1939, 78).
52. Kromnow 1938, 27.
53. Kromnow 1939, 44; Boëthius & Kromnow, Vol. 1, 485, Vol. 3: 1, 14 f.—Hildebrand shows (p. 296) that the consumption of charcoal in one blast furnace fell steadily between 1708 and 1798 by a total of 33 %. This suggests that a slow decline in charcoal consumption also oc-curred during the eighteenth century as a result of routine improvements. Cf. ibid., 220, 263.
54. Hildebrand, 225–238; Arpi, 95–97; Boëthius & Kromnow, passim.
55. Heckscher 1936, Vol. 1: 2, 477–490, 1949, Vol. 2: 1, 386–416.
56. Eric Thomas Svedenstierna, *Ödmjukt yttrande på anmodan af Bruks-Societetens Herrar Fullmäktige inlemnadt till Jern-Kontoret* (Stockholm, 1805), 16.
57. Rinman 1782, 422 f.
58. Heckscher 1949, Vol. 2: 1, 327–337.
59. Ephraim Otto Runeberg, "Beskrivning öfver Lajhela Socken i Österbotten", *KVAH* 1758, 108–162.
60. Heckscher 1949, Vol. 2: 1, 335–337.
61. Heckscher writes (ibid., 337) that a British expert who described the process in the middle of the nineteenth century observed that it was quite identical with the process used in ancient Greece, but did not suggest that any improvements were necessary.
62. Among the exceptions were the attempts to extract tar from charcoal piles. See the appendix by Sven Rinman in Wallner, and Alexander Funck, *Beskrifning om tjäru- och kolugnars inrättande* (Stockholm, 1748, 2nd edition 1772).
63. Heckscher 1949, Vol. 2: 1, 338–351; Wilhelm Carlgren, *De norrländska skogsindustrierna intill 1800-talets mitt* (Norrländskt handbibliotek, Vol. 11: Uppsala, 1926); Bertil Boëthius, "Trävaruexportens genombrott efter det stora nordiska kriget", *Historisk Tidskrift* 49 (1929), 273–298; Oskar Johannes Näslund, *Sågar: Bidrag till kännedomen om sågarnas uppkomst och utveckling* (Stockholm, 1937); Nils Meinander, *En krönika om vattensågen* (Helsingfors, 1945); Axel Solitander, *Några anteckningar rörande träförädlingens historia i Finland* (Helsingfors, 1930); Sven-Erik Åström, "Technology and Timber Exports from the Gulf of Finland, 1661–1740", *Scandinavian Economic History Review* 23 (1975), 1–14.
64. Wilhelm Carlgren, "Norrländsk trävarurörelse genom seklen", *Svenska kulturbilder*, n.s. Vol. 4, part 7–8 (Stockholm, 1937), 322.
65. Heckscher 1949, Vol. 2: 1, 343 f.; Meinander, 62–67; Näslund, 95 f.; Solitander, 8.—The iron used in the blades was bought from hammer mills, and the village smithies then cut teeth into the blades and capped the tips with steel. The cutting edges of the blades were about 1 cm thick and the kerf wasted about half an inch of each board they sawed.
66. Loc. cit.
67. Heckscher 1949, Vol. 2: 1, 344–347; Meinander, 89–96.

68. Ulric Rudenschöld, *Tal om skogarnes nytjande och vård* (Stockholm, 1748), 29.
69. Heckscher 1949, Vol. 2: 1, 344–347.
70. Carl Knutberg, "Beskrifning, med bifogad ritning, på en finbladig såg-qvarn", *KVAH* 1769, 13–32.—The economic advantages that were to be derived from erecting the new kind of sawmill were stressed by Pehr Kalm, *Menlöse tankar om bräd-sågning* (Åbo, 1772).
71. Heckscher 1949, Vol. 2: 1, 349–351.
72. Christopher Polhem, "Theoriens och practiquens sammanfogning i mechaniquen, och särdeles i ström-wärk", *KVAH* 1741, 160.
73. Loc. cit.
74. Ibid., 343 f.; Näslund 95 f.
75. Heckscher 1949, Vol. 2: 1, 322–325.
76. Hans Hederström, "Näsby socken i Östergötland, beskrifwen år 1755", *KVAH* 1757, 285; Runeberg, 157.
77. Anders Jahan Retzius, "Berättelse om de försök som blifvit gjorda med åtskilliga utländska träd och buskarter", *KVAH* 1798, 43 f.
78. During the eighteenth century interest in the nature of lightning was to a large extent related to the question of protection against fire. Benjamin Franklin, for example, founded the first fire insurance company in Philadelphia in 1752 and proposed many urban improvements to safeguard the colonial cities in North America from devastating fires. When Torbern Bergman addressed the Royal Swedish Academy of Sciences in 1764 about the damage caused by thunder, he mentioned such examples as houses being set on fire, gutted or burned to the ground, and even the tallest trees being struck down or split; see: Torbern Bergman, *Inträdestal, om möjeligheten at förekomma åskans skadeliga verkningar* (Stockholm, 1764), 4. Cf. Lindroth 1967, Vol. 1: 1, 447, 482 f.
79. As early as 1696 Urban Hiärne published a book with suggestions for and drawings of different types of more efficient fireplace, *En lijten oeconomisk skrifft om wedsparande, huru man i desse knappa tijder med weden som efter handen begynner at tryta, bätter omgås skall, och till wärmande anwända med bättre nytta och sparsamhet, dem oförmögnom till tröst och lindring* (Stockholm, 1696). Hjärne criticized the simple fireplaces found in the homes of the common people and regarded tile stoves as superior because they forced the smoke to circulate and therefore kept it longer in the house (p. 19). This is not the place to give a complete account of the technical development of the tile stove during the eighteenth century, but proposals were made by Nordenberg 1739 (*KVAH*), Brelin 1741 (*KVAH*), Westbeck 1760, Brelin 1763, Cronstedt 1767, Schröderstierna 1767 (*KVAH*), Hindbeck 1772, Palmstedt 1775 (Published together with the 2nd edition of Cronstedt), and Clewe 1792. The literature on the history of tile stoves has focussed on the *tile* rather than the *stove*, and deals mainly with its aesthetics rather than its technical development or social and economic importance. See, however: Gösta Selling, "Den svenska kakelugnens tvåhundraårsjubileum", *Saga och Sed* 1967, 74–101.
80. Anders Johan Nordenberg, "Rön om kakelugnar och deras omslagning", *KVAH* 1739, 71–78.
81. Ibid., 74.
82. Sigurd Erixon, "Spjället, en exponent för svensk bostadsteknik", *Svenska Kulturbilder*, n.s. Vol. 5 (Stockholm, 1937), 9–50.
83. The price of timber at the place of consumption was closely related to the cost of transportation from the place of production. For the inhabitants of a town or a peasant who lived on the plains, firewood was an expensive item and it made sense to economize on it. The plains-dwellers and the poorer townsfolk would therefore probably have been interested in the many well-designed tile stoves that were available if they had been able to afford them. But the development of fuel saving technology which was described in the technical literature was primarily available to that small group of well-to-do persons to which the authors themselves belonged and had very little effect on society as a whole during the eighteenth century (I am indebted to Einar Stridsberg for this comment).
84. Lindroth 1967, Vol. 1: 1, 217–226; Sten Lindroth, *Svensk lärdomshistoria*, Vol. 3 (Stockholm, 1978), 48–63.
85. Lindroth 1967, Vol. 1: 1, 27–35, 226–230, Vol. 2, 184–188.
86. Ibid., Vol. 1: 1, 116–117.
87. Ibid., Vol. 1: 1, 229, Vol. 2, 186. Cf. Karin Johannisson, "Naturvetenskap på reträtt: En diskussion om naturvetenskapens status under svenskt 1700-tal", *Lychnos* 1979–1980, 109–154.—The Gustavian Era was characterized by an interest in aesthetics and the beautiful (*"det sköna"*) at the expense of the interest in practical and useful (*"nyttiga"*) inventions that had characterized the Era of Liberty. This, too, is reflected in the changing interest in thermal energy technology in the Proceedings. Whereas the problem had once been to find the best ways to impregnate wooden houses to stop the timber from rotting, the problem was now to find

"brighter and merrier" colours than the common red ochre. And whereas the problem had once been to scour the country for alternative building materials to save the forests, it was now to take an inventory of the natural resources that could fulfil the aesthetic requirements, such as the porphyry of Dalarna. For the *paint*, see Johan Gottlieb Gahn's report in *KVAH* 1805, 189–301. Cf. Lindroth 1967, Vol. 2, 252 f. For the *porphyry*, see: Peter Jacob Hjelm, "Minerographiske antekningar om porphyrbergen i Elfdals-socken och Öster-Dalarna, samt deras gränsor i omkringliggande socknar", *KVAH* 1805, 1–48, 75–110. Cf. Severin Solders, *Älvdalens sockens historia. Del II. Gamla porfyrverket* (Dalarnas fornminnes och hembygds förbunds skrifter, No. 8: Stockholm, 1939); Inga-Britta Sandqvist, "Elfdalsporfyr—idé och utformning: En svensk konstindustri 1788–1856" (Mimeographed fil. lic. dissertation at the Department of History of Art, University of Stockholm, 1972).

88. Lindroth 1967, Vol. 1:1, 27–35.
89. Lindroth 1978, 51.
90. Torbern Bergman, "Anledningar at tilverka varaktigt tegel", *KVAH* 1771, 211–220; Pehr Adrian Gadd, "Rön och försök med murbruk och ciment-arter", *KVAH* 1770, 189–206; Pehr Adrian Gadd, "Rön, om skiffergångarna i Finland, och takskiffer i dem", *KVAH* 1780, 294–303.
91. Johan Gottschalk Wallerius, "Undersökning om den jords beskaffenhet, som fås af vatten, vegetabilier och mineralier. Fjerde stycket, Om kalkjordens skillnad ...", *KVAH* 1760, 252.
92. Lindroth 1978, 403, 407 (ill.).
93. Bergman 1764, 5. Cf. Gunnar Eriksson, "Motiveringar för naturvetenskap: En översikt av den svenska diskussionen från 1600-talet till första världskriget", *Lychnos* 1971–1972, 121–170. Eriksson considers (p. 136) that the technological development which occurred was fairly modest despite all that was written on the subject and that progress was undoubtedly more often the result of practical experience than of profound theoretical study. See also Johannisson, 127.
94. See, for example: Thomas S. Kuhn, "The Relations between History and the History of Science", in: idem, *The Essential Tension: Selected Studies in Scientific Tradition and Change* (Chicago, 1977), 141–147. Originally published in *Daedalus*, the journal of the American Academy of Arts and Sciences, 100 (1971), 271–304.
95. See, for example: Edwin T. Layton, Jr., *The History of Technology as an Academic Discipline* (Stockholm Papers in the History and Philosophy of Technology, TRITA-HOT-1003: Stockholm, 1981), 11–13.
96. Kuhn 1977, 144.
97. Bergman 1764, 26.

4. MECHANICAL ENERGY

This chapter was, alas, originally written before the publication of Terry S. Reynolds' book *Stronger Than a Hundred Men: A History of the Vertical Water Wheel* (The Johns Hopkins University Press: Baltimore, 1983). I am therefore in a position to testify that his book is the study that has been needed in a field where an exhaustive survey with an international perspective has been lacking. I am indebted to Edwin T. Layton, Jr., for his comments on this chapter, and I regret that I have only felt able to incorporate a few of his many suggestions.

1. Christopher Polhem, *Kort berättelse om de förnämsta mekaniska inventioner, som tid efter annan af Commercierådet Christopher Polhem blifvit påfundna* (Stockholm, 1729), § 24.
2. The data on the relative strength of men and horses are conflicting, see: Eugene S. Ferguson, "The Measurement of the 'Man-Day'", *Scientific American* 225 (1971), No. 4, 96–103, esp. 96 f. In Sweden the problem was discussed in 1800 by Erik Nordwall. He said that a horse was usually considered able to do the work of 7 men, but that this was not universally accepted, and he referred to the results of Desaguliers, Bernoulli, la Hire, Bélidor and others. See: Erik Nordwall, *Afhandling rörande mechaniquen, med tillämpning i synnerhet till bruk och bergverk*, Vol. 1 (Stockholm, 1800), 95–100.
3. Sten Lindroth, *Gruvbrytning och kopparhantering vid Stora Kopparberget intill 1800-talets början*, Vol. 1 (Uppsala, 1955), 303–306, 614–622. Cf. Nordwall, 100–108; Sven Rinman, *Bergwerks lexicon*, Vol. 2 (Stockholm, 1789), 1142 ff.
4. The military was an exception. In the naval dockyard of Karlskrona, for example, human muscle power was used to operate the pumps. To dry-dock a frigate 90 sailors worked in three shifts for four days. I do not know if this was more cost-effective than animal muscle power but I would suggest that the military

was the only organisation large and strong enough to assemble and maintain such a large workforce in one place. A later example is the Göta Canal, built in 1810–1832, which was dug by a total of 60 000 soldiers, who produced 7 millions man-days of work. See: Samuel E. Bring, ed., *Göta Kanals historia*, Vol. 2: 1 (Uppsala, 1930), 454–460; Gustaf Halldin, ed., *Svenskt skeppsbyggeri: En översikt av utvecklingen genom tiderna* (Malmö, 1963), 275 verso, Fig. 245; Johan Eric Norberg, "Rön öfver den effect, som af manskap kan användas medelst handkraft, å machiner, som sättas i rörelse genom hvef", *KVAH* 1799, 51–69.

5. Lindroth 1955, Vol. 1, 616.

6. On the development of horse whims, see: ibid., 618–622. Conical winding-drums for horse whims were discussed in *KVAH* in 1796 and 1798. Olof Åkerrehn described a horse whim for pumping water in the Proceedings in 1806, and since these articles were considered worth publishing Lindroth was probably right when he wrote (ibid., 618) that the horse whim changed very little during the eighteenth century. See: Gustaf Adolf Lejonmark, "Tilläggning til föregående afhandling, eller en jämförelse imellan den coniska och den cylindriska linkorgen", *KVAH* 1796, 106–125; Gustaf Aron Lindbom, "Beskrifning på en ny häst vind vid Persberget", *KVAH* 1796, 97–106; idem, "Om coniska hästvindar, at nyttja til upfordring vid grufvor", *KVAH* 1798, 203–220; Olof Åkerrehn, "Grufve-konst, till vatten-pumpning för dragare", *KVAH* 1806, 281–284.—In an article in 1744, Pehr Elvius compared the efficiency of four treadmills operating in Stockholm at the time: one at the building of the Royal Palace, one at Kungsholmen glass factory, and two at the construction of the sluice. But the large majority of treadwheels were of course not in the capital, but in the mines. They were never discussed in the Proceedings of the Academy, and this is another example of the social limitation in the choice of problems to consider that was suggested in Chapter 3. See: Pehr Elvius, "Rön vid trampkvarnar", *KVAH* 1744, 198–208.

7. Abbot Payson Usher, *A History of Mechanical Inventions*, (Cambridge, Mass., 1954, rev. ed.), 334–336.—For an ethnological study of windmills in Sweden, see: Sven B. Ek, *Väderkvarnar och vattenmöllor: En etnologisk studie i kvarnarnas historia* (Nordiska museets handlingar, No. 58: Stockholm, 1962).

8. Carl Knutberg, "Nytt påfund, vid väder-quar-nars inrättning, at i lugnt väder malningen må kunna förrättas medelst hästvind", *KVAH* 1751, 130–132.—There was also some interest in whether wind-turbines were more effective than the common windmill, see: Sten Lindroth, *Kungl. Svenska Vetenskapsakademiens historia 1739–1818*, Vol. 1 (Stockholm, 1967), 275 f.

9. Reynolds 1983, 172–179.

10. Carl A:son Sege, "Bidrag till kännedomen om Sala silververks vattenkraftsanläggningar", *Blad för Bergshandteringens Vänner* 20 (1931–1932), 425–518; Lindroth 1955, Vol. 1, 118–122, 314–322.

11. Harald Carlborg, "Om tramphjul och andra motorer i äldre tid vid svenska malmgruvor", *Med Hammare och Fackla* 25 (1967), 7–82. The reason why the treadwheel has been underestimated is probably that its use is considered as degrading physical labour, comparable to the monotonous work of galley slaves. It is the kind of simple everyday technology, no matter how widespread or useful, that is likely to be forgotten for the sake of more efficient and spectacular technologies. But treadwheels were used in Swedish mines in the 1880s (one is even reported to have been in use in 1896). As late as in 1881 a professional society of mining engineers discussed the depths to which treadwheels could advantageously be used in mines.—For the construction of treadwheels, see: Svante Lindqvist, "Projektet 'Det medeltida tramphjulet'—en övningsuppgift i teknikhistoria på KTH", *Daedalus* 50 (1981), 59–72.

12. Lindroth 1955, Vol. 1, 290–298. Cf. Graham Hollister-Short, "Leads and Lags in late Seventeenth-Century English Technology", *History of Technology* 1 (1976), 159–183; Reynolds 1983, 141 f.

13. For example in 1679 in Dannemora (1 500 m.), in 1684 in Strossa (2 100 m.), 1698–1700 in Bispberg (1 500 m.), and Norrbärke (1 100 m.). See: Carlborg 1967, passim.

14. Lindroth 1955, Vol. 1, 331–349, 561–576; Sten Lindroth, *Christopher Polhem och Stora Kopparberget: Ett bidrag till bergsmekanikens historia* (Uppsala, 1951), 45 ff.

15. It was stated in 1892 in the journal of the Swedish Ironmasters' Association that *Stangenkunst* systems were used to transmit a power of no more than 20–30 hp. The efficiency was said to be 60–70 % "for longer systems", which may have meant about 3 kilometres. The efficiency was probably lower in the seventeenth and eighteenth centuries, say 50 %,

which over 3 kilometres would mean a loss of 20 % per kilometre. See: Jonas Wenström & Gustaf A. Granström, "Electriciteten i grufhandteringens tjenst", *Jernkontorets Annaler* n.s. 47 (1892), 221 f.

16. Sven Rinman wrote in his mining dictionary in 1789 that a *Stangenkunst* system was effective up to a distance of a few thousand Swedish *alnar* (one *aln* was 0.6 m.), which probably meant approx. 1 800–2 100 m. See: Rinman 1789, 881 f.

17. Carlborg 1967, passim; Jalmar Furuskog, *De värmländska järnbruken: Kulturgeografiska studier över den värmländska järnhanteringen under dess olika utvecklingsskeden* (Filipstad, 1924), 248–253; Kjell Kumlien, ed., *Norberg genom 600 år: Studier i en gruvbygds historia* (Uppsala, 1958), 283–288, 315–318, 340–363.

18. Quoted from the English translation: Maurice Daumas & Paul Gille, "Methods of Producing Power", in: Maurice Daumas, ed., *A History of Technology and Invention: Progress Through the Ages*, Vol. 3 (New York, 1979), 24 ff. The French original: idem, *Histoire générale des techniques*, Vol. 3 (Paris, 1968), 11 ff.

19. Lindroth 1967, Vol. 1, 218; Sten Lindroth, *Svensk lärdomshistoria*, Vol. 3 (Stockholm, 1978), 345.—Polhem's words on the need "to unite theory and practice" led Lindroth to believe that throughout the eighteenth century there was a uniform tradition of applying to technology results reached by deduction. However, that was not what Polhem meant and the tradition was far from being uniform.

20. The best description of the hydrodynamic experiments in *Laboratorium mechanicum* is found in: Lindroth 1951, 82–91. See also: Lindroth 1955, Vol. 1, 340; Sten Lindroth, *Svensk lärdomshistoria*, Vol. 2 (Uppsala, 1975), 547 ff. Cf. Friedrich Neumeyer, "Christopher Polhem och hydrodynamiken", *Arkiv för matematik, astronomi och fysik*, Band 28 A (1942), No. 15.—Polhem's method in the hydrodynamic experiments has recently been placed in a broader context of engineering methods by Boel Berner. She is the first who has related it to Walter G. Vincenti's concept "parameter variation and optimization". See: Boel Berner, *Teknikens värld: Teknisk förändring och ingenjörsarbete i svensk industri* (Arkiv avhandlingsserie, No. 11: Lund, 1981), 78–92, 250 f. (English summary); idem, "Experiment, teknikhistoria och ingenjörens födelse", *Daedalus* 51 (1982), 39–52; Walter G. Vincenti, "The Air-Propeller Tests of W. F. Durand and E. P. Lesley: A

Case Study in Technological Methodology", *Technology and Culture* 20 (1979), 712–751.
—There is an older, incorrect description of Polhem's hydrodynamic experiments, published in 1911; see: Nils Cronstedt, "Polhems arbeten på byggnadskonstens områden", in: *Christopher Polhem: Minnesskrift utgifven av Svenska Teknologföreningen* (Stockholm, 1911), 225–228. This book was translated into English in 1963 by William A. Johnson, see: *Christopher Polhem: The Father of Swedish Technology* (Hartford, Conn., 1963), 209–211. Here it also said, quite erroneously, that Polhem's "conclusions were based solely upon theory" and that Pehr Elvius "continued Polhem's theoretical work in this subject".

21. For Buschenfelt's biography, see: Lindroth 1955, Vol. 1, 683 f.; Sven Rydberg, *Svenska studieresor till England under Frihetstiden* (Lychnos-Bibliotek, No. 12: Uppsala, 1951), 147 ff.

22. The apparatus for hydrodynamic experimentation is described in Lindroth 1951, 83 ff. Lindroth's description is based upon Mårten Triewald, *Föreläsningar öfwer nya naturkunnigheten*, Vol. 2 (Stockholm, 1736), 217–275. Triewald in his turn, had based his description on the report which Göran Wallerius wrote on Polhem's instructions in 1705, and which was sent to the Board of Mines. This report seems to have been deposited with the apparatus in the Royal Chamber of Models for it is preserved today in the archive of the Royal Institute of Technology Library, Stockholm (most of the models were kept at the Institute until 1925, but the experimental apparatus was lost some time during the nineteenth century). See: KTHB, MS Pf-38, *Kårtt och ungefärlig relation med des derhoos tillhörige rijtningar, angående de fyra af Hr. Directeuren Påhlhammar Inventerade och af Hr. Markscheidern Buschenfelt förfärdigade mekaniske machiner med des experimenter och bijfogade tabeller. Hwilcka af bemelte Hr. Markscheider jemte mig undertechnad och flere äro genomgångne, och sedan efter Inventoris egen disposition och underättelse på följande sätt deducerade af Giöran Vallerius Haraldson A:o 1705*, 3–139. Wallerius' report is bound in calf in a volume of 202 folio pages together with descriptions of three other experimental machines by Polhem. The drawing (Fig. 4.3) and the diagram (Fig. 4.5) are bound in a separate volume of illustrations which is catalogued together with Wallerius' report as MS Pf-38).—There are three known drawings of Polhem's experimental apparatus. The first (Fig. 4.3) is the one in Wallerius'

report of 1705, and was probably based on an earlier drawing by Buschenfelt (Lindroth 1951, 190 n. 5). The second (Fig. 4.7) is a drawing by Carl Johan Cronstedt, who was a pupil of Polhem at Stjernsund in 1729. Although Cronstedt must have seen the original apparatus, which was kept in Stjernsund at this time, his drawing, too, seems to have been based on the one by Buschenfelt. Cronstedt's drawing is kept, together with his other drawings from Stjernsund, at Tekniska Museet, Stockholm. See: TM, Cronstedtska planschsamlingen, No. 7404. The third drawing (Fig. 4.8) is the engraving in Triewald's lectures from 1736. This is a copy of the drawing in Wallerius' report, which Triewald had borrowed from the Board of Mines in 1734 (Lindroth 1951, 190 n. 5). Wallerius' drawing (Fig. 4.3) is thus the earliest of the three preserved drawings, and probably quite accurate since Wallerius had participated for several years in the experiments with the apparatus.—There is a copy of Wallerius' report without the drawing and the diagram in: UUB, MS A 147, *Kårtt och ungefärlig relation ... af Jöran Wallerius Haraldsson.*

23. Reynolds 1983, 228–233, 235 f., Fig. 4–11. Polhem's experiments were mentioned in: Norman F. Smith, *Man and Water: A History of Hydro-Technology* (London, 1976), 155.
24. For Wallerius' biography, see: Lindroth 1955, Vol. I, 685 f.; Rydberg 1951, 151–154.
25. Lindroth 1951, 85. Cf. Reynolds 1983, 202 f.
26. Michael S. Mahoney, "Edmé Mariotte", in: Charles G. Gillispie, ed., *Dictionary of Scientific Biography*, Vol. 9 & 10 (New York, 1981), 114–122.
27. Ibid., 119.
28. See Lindroth 1951, 83 ff., and Triewald 1736, 217–275 and Fig. XXII–XXV.
29. Lindroth 1951, 83, 190 n. 3.
30. Loc. cit.
31. Ibid., 83, 190 n. 4.
32. See n. 22.
33. Wallerius' report, 52; Triewald 1736, 239.
34. Axel Liljencrantz, ed., *Christopher Polhems brev* (Lychnos-Bibliotek, No. 6: Uppsala, 1941–46), 37 f.
35. Loc. cit.
36. Loc. cit.
37. Ibid., 53.
38. Loc. cit.
39. Cf. "Interest in Technological Development" in Chapter 3.

40. Neumeyer, 16. Cf. Berner 1981, 88 ff.; Reynolds 1983, 230, 380 n. 90; Vincenti, 715 ff.
41. Liljencrantz, 123. Original in Swedish: "alla rörliga machiner eij behåla lijka pro[por]tion i stort som smått fast alla dhelar blifva lijka och till alla dhelar effter proportion giorde". For later discussions on the problem of scale, see n. 120.
42. Wallerius said, in the title of his report, that he had interpreted the data according to Polhem's instructions.
43. Liljencrantz 1941–46, 53.
44. Ibid., 37 f.
45. Pehr Lagerhjelm, Jacob Henrik af Forselles & Georg Samuel Kallstenius, *Hydrauliska försök, anställda vid Fahlu grufva, åren 1811–1815*, 2 Vols. (Stockholm, 1818–1822), Vol. 2, 3.
46. Triewald 1736, 217–239, Tab. I–LXII, Figs. XII–XV.
47. Ibid., 239.
48. Pehr Elvius, *Mathematisk tractat om effecter af vatn-drifter, efter brukliga vatn-värks art och lag* (Stockholm, 1742). Cf. Reynolds 1983, 235 f.
49. Lindroth 1967, Vol. 1, 132.
50. Lindroth 1978, 52 f., 345; Lindroth 1967, Vol. 1, passim, esp. 42–45; Bengt Hildebrand, *Kungl. Svenska Vetenskapsakademien: Förhistoria, grundläggning och första organisation* (Stockholm, 1939), 592–598; Bengt Hildebrand, "Pehr Elvius, d.y.", *Svenskt Biografiskt Lexikon*, Vol. 13 (Stockholm, 1950), 422–427.
51. Elvius 1742, preface.
52. Loc. cit.
53. Loc. cit.
54. Elvius 1742, 29.
55. For Parent's view on the utility of mathematics, see: J. Morton Briggs, "Antoine Parent", in: Gillispie 1981, Vol. 9 & 10, 319 f.—For Parent's rule, see: Reynolds 1983, Chapter 4 passim, esp. 205–208, 210–212. Cf. Donald S. L. Cardwell, *Technology, Science and History* (London, 1972), 62 f.; Daumas & Gille, 25 f.; Terry S. Reynolds, "Scientific Influences on Technology: The Case of the Overshot Waterwheel", *Technology and Culture* 20 (1979), 275 ff., Smith, 154.
56. Loc. cit.
57. Loc. cit.
58. Loc. cit.
59. See for example Polhem's dialogue between "Lady Theoria" and "Master Builder Practicus" in: Henrik Sandblad, ed., *Christopher Polhems efterlämnade skrifter*, Vol. 1 (Lychnos-Bibliotek, No. 10.1: Uppsala, 1947), 281.
60. Lagerhjelm wrote in his critique of Elvius in

1822 that "this author is clearly at issue with himself, without having even tried to resolve it. We find no sign that Elvius' work has been known outside Sweden." See: Lagerhjelm 1822, 5.—Reynolds points out that Elvius' book may have influenced Johann Euler, see: Reynolds 1983, 235 f.

61. Lindroth 1967, Vol. 1, 132, 357 f. (The review was published in *KVAH* 1742, 74–80.)
62. Loc. cit.
63. Christopher Polhem, "Fortsättning om theoriens ock practiquens sammanlämpning i mechaniquen", *KVAH* 1742, 158.
64. Ibid., 157.
65. Pehr Elvius, "Theorien om vatten-drifter jämförd med försök", *KVAH* 1743, 76–80.
66. Ibid., 77.
67. Loc. cit.
68. Loc. cit.
69. Carl Henrik König, *Inledning til mecaniken och bygnings-konsten, jämte en beskrifning öfwer åtskillige af framledne commerce-rådet och commendeuren af Kongl. Nordstierneorden hr. Polhem opfundne machiner* (Stockholm, 1752).
70. Ibid., preface.
71. Ibid., 41–50.
72. *Biographiskt lexicon öfver namnkunnige svenska män*, Vol. 23 (Örebro, 1857), 183 f.; *Svenska Män och Kvinnor*, Vol. 8 (Stockholm, 1955), 508 f.; Bertil Boëthius & Åke Kromnow, *Jernkontorets historia*, Vol. 3 (Stockholm, 1955), passim.— Åkerrehn wrote that around 1786 he began serving ("*betjena*") industrialists by building water mills, and he was called in by owners all over the mining district during the following decades, see: Olof Åkerrehn, *Svenska blåsverkens historia, från år 1786; med tillägg, om jernberedningen i allmänhet* (Stockholm, 1805), 1, passim.
73. Olof Åkerrehn, *De lege Kepleriana, comparandi distantias planetarum medias a sole* (Diss.: Uppsala, 1784). Copy in UUB.
74. Loc. cit.
75. John Hägglund, *Kinda kanals historia* (Linköping, 1966), 30–36.
76. Boëthius & Kromnow 1955, 213.
77. Olof Åkerrehn, *Utkast til en practisk afhandling om vattenverk, grundad på physiske mechaniske lagar* (Örebro, 1788).
78. Ibid., III.
79. Ibid., V.
80. Ibid., VI.
81. Loc. cit.
82. Ibid., XI ff.
83. Loc. cit.
84. Bernard Forest de Bélidor, *Architecture hydrauli-*

que, 2 Vols. (Paris, 1737–1753); Jens Kraft, *Forelæsninger over mekanik med hosføiede tillæg* (Sorøe, 1763); idem, *Forelæsninger over statik og hydrodynamik med maskin-væsenets theorier som den anden deel af forelæsningerne over mekaniken* (Sorøe, 1764); Charles Bossut, *Traité élémentaire d'hydrodynamique ouvrage dans lequel la théorie et l'expérience*, 2 Vols. (Paris, 1771).

85. *Biographiskt lexicon öfver namnkunnige svenska män*, Vol. 23 (Örebro, 1857), 184.
86. Boëthius & Kromnow 1955, 212 ff.
87. Rinman 1789.
88. Cf. "Industry" in Chapter 3.
89. Rinman 1789, 1126–1131.
90. Ibid., 1131.
91. Bossut 1771; Kraft 1763–1764; Nicolaus Poda & Ignaz Edlen von Born, *Kurzgefaßte Beschreibung der, bei dem Bergbau zu Schemnitz in Nieder-Hungarn, errichteten Maschinen* (Prag, 1771); Christoph Traugott Delius, *Anleitung zu der Bergbaukunst* (Wien, 1773); Jacob Leupold, *Theatrum Machinarum Hydraulicarum. Tomus I. Oder: Schau-Platz der Wasser-Künste, Erster Teil* (Leipzig, 1724).
92. Boëthius & Kromnow 1955, 60 ff.
93. Sven Rinman, *Afhandling rörande mechaniquen, med tillämpning i synnerhet till bruk och bergverk*, Vol. 2 (Stockholm, 1794), preface.
94. Loc. cit. Cf. Torsten Althin, "Rinman's Treatise on Mechanics", *Technology and Culture* 10 (1969), 185 f.
95. Rinman 1794, 94–138.
96. Ibid., 97: "A tilt-hammer must in general, when running at full speed, make 80 or no more than 90 strokes per minute, which means 3 strokes in 2 seconds, and since there are four lifting cogs on the axle, the wheel must make 20 or not more than 22½ revolutions per minute. Thus if the wheel is 11 feet in diameter, its circumference, which is then 34½ feet, must have such a speed, that in one minute, at 20 revolutions per minute, it runs through a distance of 690 feet, or at 22½ revolutions per minute 765¼ feet. If the running water is to gain such speed, it ought to have at least 3½ feet of water pressure in the headrace, but since the wheel under load cannot run as fast as the water, so ought the speed of the water, at least, to be 1/3 greater, and for this 4¼ or 4½ feet of water pressure in the headrace is needed, and it is still better, if it can be 5 feet high, especially with a low undershot wheel, where only the impact of the water is acting, and where one foot of ice may cover the surface during winter. It has been found by experience

that a loaded waterwheel has its highest effect or use to draw up, or subdue, the load, when in a given time, or per minute, it does not run more than half as fast as when running freely, or with no load. This applies when the friction cannot be considered to offer any heavy resistance, and when the load is an even one. Where the load is intermittent or uneven, as it can be said to be with a tilt hammer, where the swing of the wheel helps somewhat, the effect may be said to be somewhat greater. More on this may be found under Hydraulics, in the previous [*sic*] Theoretical Volume and also later in this volume on the design of waterwheels in § 65."

97. Loc. cit.
98. Ibid., 128.
99. Ibid., 133 ff.
100. See n. 96.
101. See n. 59.
102. Ibid., 106–109.—The data had in this case been collected by Salomon von Stockenström (1751–1811), Assessor in the Board of Mines, and Rinman quoted from his report.
103. Loc. cit.
104. Ibid., 108.
105. Loc. cit.
106. Ibid., 109.
107. Reynolds 1979; Reynolds 1983, 218–226, esp. 226.
108. Rinman 1794, 133.
109. Louis C. Hunter, *Waterpower: A History of Industrial Power in the United States, 1780–1930* (Charlottesville, 1979), 91. See also: Reynolds 1983, 191–195.
110. William Fairbairn, *Treatise on Mills and Millwork*, Vol. 1 (London, 1861), v–xi. See also Hunter, 91–94.
111. Fairbairn, v.
112. See n. 2. Cf. Reynolds 1983, 257.
113. *Biographiskt lexicon öfver namnkunnige svenska män*, Vol. 10 (Upsala, 1844), 87–90; Boëthius & Kromnow 1955, passim.
114. Nordwall, preface.
115. Loc. cit.
116. Boëthius & Kromnow 1955, 95–99.
117. Nordwall, preface.
118. Lagerhjelm 1822, 26.
119. Gustaf af Uhr, *Betänkande till Herrar Fullmägtige i Jern-Contoret, innehållande en öfversigt af bergsmechanikens tillstånd* (Stockholm, 1817), 3.
120. Lagerhjelm 1822, 23–35, and af Uhr 1817, 1–38, were severely critical of Nordwall's work. This is not reflected in Boëthius & Kromnow 1955, 98 f., 209, who failed to convey this criticism.—Nordwall had, however,

made one contribution which was acknowledged by later authors. He had suggested how the results obtained from a model (i.e. dimensions, weights, times and volume of water) should be adjusted for scale effects in a full-size machine. See: Nordwall, 315 f. Cf. Boëthius & Kromnow 1955, 97; Lagerhjelm 1822, 24 f.; Zacharias Nordmark, "Om slut-följdens giltighet, att, af försök med hydrotechniska modeller i smått, dömma om maschiners verkan i stort", *KVAH* 1813, 135–154; af Uhr 1817, 1 f. Cf. n. 40.
121. Boëthius & Kromnow 1955, 209.
122. Ibid., 98, 209–210; Lindroth 1955, Vol. 1, 611 f.
123. Loc. cit.
124. Loc. cit.
125. Loc. cit.; af Uhr 1817, 38; Lagerhjelm 1818, preface.
126. For Lagerhjelm's biography, see n. 143, and for af Forselles', see: Pehr Lagerhjelm, "Jacob Henrik af Forselles", in: *Lefnadsteckningar öfver Kongl. Svenska Vetenskaps Akademiens efter år 1854 aflidna ledamöter*, Vol. 1 (Stockholm, 1869–1873), 1–9.—Pehr Lagerhjelm is primarily known for his theoretical and experimental studies in strength of materials, and in particular for his hydraulic tensile testing machine. See: Folke K. G. Odqvist, "Hållfasthetsläran som förutsättning för materialprovning, särskilt i Sverige", *Daedalus* 1977, 59–112 (English summary); Stephen P. Timoshenko, *History of Strength of Materials* (New York, 1953), 102 f., 281 f.; Isaac Todhunter, *A History of the Theory of Elasticity*, Vol. 1 (Cambridge, 1886), 186–189. I am indebted to Jan Hult for these references.
127. Nordmark's earlier theoretical work in hydrodynamics had mainly concerned naval architecture, and he was only the nominal leader of the experiments in Falun; see: Lindroth 1967, Vol. 2, 368–371. He participated in the experiments in Falun in 1812–1813, but not in 1814–1815; see: Lagerhjelm 1818, preface. Berzelius visited Johan Gottlieb Gahn in Falun in the summer 1813, and wrote in a letter to Magnus af Pontin: "Nordmark is here, lively as a firebrand, but shy in the company of people, as becomes a university professor. I meet Lagerhjelm and Forselles less often, since they live at the mine and toil every day at their experiments, which will surely yield quite important results." See: Henrik Gustaf Söderbaum, *Jac. Berzelius: Lefnadsteckning*, Vol. 1 (Uppsala, 1929), 493 and n. 4.

128. Boëthius & Kromnow 1955, 210; Lagerhjelm 1818, preface.—Lagerhjelm gave a more revealing report of his own dominant position in the planning and execution of the experiments in his autobiographical notes in the 1850s. According to Lagerhjelm, Nordmark assisted only in certain parts of the experiments. He also wrote that he asked to have Kallstenius as an additional assistant in 1813, because af Forselles had proved ignorant and stubborn: "It was soon evident that this precaution had been necessary. When Forselles did not get his way, he resigned and left like a self-willed journeyman one Monday, but Kallstenius took his place. The experiments proceeded uninterrupted and Forselles soon wanted to participate, which was granted." See: KB, MS I.l. 1 b., Självbiografiska anteckningar av Pehr Lagerhjelm.

129. Lagerhjelm and af Forselles publicized the experiments in an advertisement on the first two pages of the newspaper *Stockholms Posten* on July 22, 1816. It was a rather pompous declaration by the two young men, stating that the hydraulic experiments in Falun had been successfully brought to an end after four years of work and that a new, accurate theory had been developed. The history of science shows, they wrote, that critical examination is useful wherever it may lead. The method of Hume, "the world's greatest *Scepticus*", has always preceded successful conquests in the field of science. This advertisement came as a surprise to Nordmark, and he was furious at "the conceited and sceptical style"; see: KVA, Brev till Jacob Henrik af Forselles från Pehr Lagerhjelm, letter dated August 15, 1816. (This correspondence contains more than 200 letters over a period of fifty years, and a large number of these are from the time of the Falun experiments.)—Cf. Reynolds 1983, 251.

130. Lagerhjelm 1818, 25.
131. Ibid., 26.
132. Ibid., 28.
133. Lagerhjelm 1822, 2–49.
134. Ibid., 3–5 (Elvius), 23–35 (Nordwall).
135. Ibid., 3–9, 19–23.
136. Ibid., 19.
137. Ibid., 17.
138. Ibid., 6.
139. Ibid., preface.
140. Loc. cit.
141. Loc. cit.
142. Loc. cit. Original in Swedish: "Då vattnet verkar antingen endast och allenast medelst sin

stöt, eller medelst stöt och tryckning tillika; så är ämnets behandling beroende af Erfarenhet; dock ingalunda så, som skulle endast erfarenhet, d. v. s. försök, erfordras till utredande af ämnet. Kunskapen är allmän, och skiljer sig derigenom från abstraction ur gifven erfarenhet, att denna sednare endast gäller under de omständigheter och inom de gränsor, som väsendtligen åtföljt den Class af phenomen, man erfarit. Denna allmänelighet (form), hemtar kunskapen, från sin speculativa rot; men sanningen, öfverensstämmelsen med verkliga förhållandet (innehållet), hemtas från erfarenheten. Uplösningen af striden emellan dessa yttersta motsatser inom detta ämne, ligger i hvarje sann Naturforskares egna sätt att förfara."—It is clear that it was Lagerhjelm, and not af Forselles, who wrote the preface; see: KVA, Brev till Jacob Henrik af Forselles från Pehr Lagerhjelm, letter dated May 20, 1822.

143. For Pehr Lagerhjelm's biography, see: Olle Franzén, "Pehr Lagerhjelm", *Svenskt Biografiskt Lexikon*, Vol. 22 (Stockholm, 1979), 133–136; Georg Scheutz, "Pehr Lagerhjelm. Assessor i Kongl. Bergskollegium", in: *Lefnadsteckningar öfver Kongl. Svenska Vetenskaps Akademiens efter år 1854 aflidna ledamöter*, Vol. 1 (Stockholm, 1869–1873), 71–76 (Scheutz, a publisher and well-known as the inventor of a differential calculating engine, had published the second volume of the description of the Falun experiments); *Biographiskt lexicon öfver namnkunnige svenska män*, Vol. 7 (Upsala, 1841), 259–264.

144. For Lagerhjelm's university diploma, see: Algot Kronberg, *Pehr Lagerhjelm: Vetenskapsman, näringsidkare, politiker* (Stockholm, 1960), 18 f.—Kronberg's book is a popular biography.

145. Söderbaum, Vol. 1, 315 f., 393 f., 404, 406.

146. Ibid., Vol. 2 (Uppsala, 1929), 113; Jac. Berzelius, *Tabell som utvisar vigten af större delen vid den oorganiska kemiens studium märkvärdiga enkla och sammansatta kroppars atomer, jemte deras sammansättning, räknad i procent. Bihang till tredje delen av läroboken i kemien* (Stockholm, 1818), 10.

147. Arne Holmberg, ed., "Jac. Berzelius. Brev till medlemmar av familjen Brandel om resan till England 1812 och till Frankrike 1818/19", *Kungl. Svenska Vetenskapsakademiens Årsbok 1955* (Stockholm, 1955), 328 f. Cf. Gunnar Eriksson, *Elias Fries och den romantiska biologien* (Lychnos-Bibliotek, No. 20: Uppsala, 1962), 109–111.

148. Eriksson 1962, 110 n. 4; Gudmund Frunck, *Bref rörande nya skolans historia, 1810–1811*

(Skrifter utgifna af Svenska litteratursällskapet: Upsala, 1891), 128 and n. 3; Rudolf Hjärne, ed., *Dagen före drabbningen eller nya skolan och dess män i sin uppkomst och sina förberedelser 1802–1810* (Stockholm, 1882), 385.

149. Lagerhjelm maintained his friendship with Hammarsköld during the Falun experiments. In 1813 he asked Hammarsköld, his "Loyal Friend", who was working at the Royal Library in Stockholm, to look up what was written on Greek fire. See: KB, MS Ep. H2:3 b, Letter from Pehr Lagerhjelm to Lorenzo Hammarsköld dated June 7, 1813.

150. Lagerhjelm gave a similar epistemological declaration in a speech to the Royal Swedish Academy of Sciences in 1837, in which he spoke of the two sources of human knowledge and compared the development of science with that of organic life: "First the nourishing juice is accumulated, formless until, purified against the organ's own tissue and by its own forces, it becomes useful for its purpose, is taken into alliance with it and given form. In this manner does science perceive, discriminate and assimilate every observation proved as universal, and the true scientist overlooks neither the ideal nature of science nor its real content." Original in Swedish: "Först samlas den närande saften, formlös ännu tills den, renad invid organets egen väfnad och af dess egna krafter, blir tjenlig för dess ändamål, upptages i dess förbund och får form. Så uppfattar, urskiljer och tillegnar sig vetenskapen hvarje till sin allmängiltighet pröfvad iakttagelse, och den sanna forskaren öfverser lika litet naturkunskapens ideala väsende som dess reala innehåll." In: Pehr Lagerhjelm, *Tal om hydraulikens närvarande tillstånd* (Presidii tal KVA: Stockholm, 1837), 5.

151. I have not included six articles in 1783–1785 by Henric Nicander on the mathematical theory underlying a helical pump invented by the Swiss Andreas Wirz. This was more of a fascinating mathematical problem than a suggestion for a technological improvement. Cf. Lindroth 1967, Vol. 1, 358 f.; Rinman 1789, 1147 ff.

152. Of the remaining three, two were on improved water pumps (which could be powered by any prime mover) and one on steam power technology (see n. 154).

153. Lindroth 1967, Vol. 1, 33 f.

154. Abraham Niclas Edelcrantz, "Afhandling om nödige rättelser vid mätningen af ångornes spänstighet, och determination af deras kraft,

uti Ångmachiner, samt beskrifning på en förbättrad Ångmätare", *KVAH* 1809, 128–144.—The steam engine had only been mentioned twice before and then only briefly; see: Johan Carl Wilcke, "Försök til en ny inrättning af luftpumpar, förmedelst kokande vattuångor", *KVAH* 1769, 33–42; Norberg 1799, 60.—For Edelcrantz and the steam engine, see: Torsten Althin, "Stationary Steam Engines in Sweden 1725–1806", *Daedalus* 1961, 95–99; Carl-Fredrik Corin, "A. N. Edelcrantz och hans ångmaskinsprojekt år 1809", *Daedalus* 1940, 71–81; idem, "N. A. [*sic*] Edelcrantz och Eldkvarn", *Daedalus* 1961, 39–94; Lindroth 1967, Vol. 2, 270 ff.

5. THE SOCIAL CONTEXT: THE BOARD OF MINES

1. A. Bernhard Carlsson, *Den svenska centralförvaltningen, 1521–1809* (Stockholm, 1913); Nils Edén, *Den svenska centralregeringens utveckling till kollegial organisation i början af sjuttonde århundradet (1602–1634)* (Skrifter utgivna av Humanistiska Vetenskaps-Samfundet i Uppsala, No. 8.2: Uppsala, 1902).

2. Sten Lindroth, *Gruvbrytning och kopparhantering vid Stora Kopparberget intill 1800-talets början*, Vol. 1 (Uppsala, 1955), 156.

3. Johan Axel Almquist, *Bergskollegium och bergslagsstaterna 1637–1857: Administrativa och biografiska anteckningar* (Meddelanden från Svenska Riksarkivet, n.s. No. 2.3: Stockholm, 1909). —Surveys of the industrial policy of the Board of Mines are given in two articles: Bertil Boëthius, "Hammarkommissionerna på 1680- och 1720-talet: En studie över deras ställning i bergskollegiets brukspolitiska system", in: *En bergsbok till Carl Sahlin* (Stockholm, 1921), 193–211; Eli F. Heckscher, "Den gamla svenska brukslagstiftningens betydelse", ibid., 169–186.

4. Almquist, 9–15. This had formally been decided in 1630, but the organisation of the new body was discussed throughout the 1630s. It received its name "Bergskollegium" in 1649, when its status as an independent department was confirmed.

5. *Kongl. stadgar, förordningar, privilegier och resolutioner, angående justitien och hushållningen wid bergwerken och bruken, med hwad som ther til hörer både*

inom och utom Bergslagerne uti Sweriges rike, och ther under lydande provincier, uppå Kongl. May:ts allernådigste befalning, til almän nytto och efterrättelse igenom trycket utgifne år 1736 (Stockholm, 1736), 79.—This was the first volume of mining statutes published, and it will be referred to in the following notes as *Bergsordningar 1736*. Three further volumes were published: in 1786 for the statutes issued in 1736–1756, in 1797 for 1757–1791, and in 1837 for 1792–1836. For complete titles of the volumes published in 1786 and 1797, see n. 33.

6. *Bergsordningar 1736*, preface.
7. "Konung Magni Privilegier för Kopparbergs-Männerne i Fahlun", *Bergsordningar 1736*, 1–6.
8. This count is based on *Bergsordningar 1736*.
9. Almquist, 9 ff.
10. *Bergsordningar 1736*, index (*"Straf"*).
11. Ibid., 347. And even this was of recent date. A statute issued in 1686 prescribed that all cases of homicide should be remitted to the ordinary courts for investigation and judgment and the local courts of the Board of Mines were forbidden to concern themselves with such cases "henceforth".
12. Almquist, 12–19.
13. It was decided in 1720 that there should be only four Assessors, but it is clear from the list of appointments that there were five Assessors during the period 1715–1725. See: *Bergsordningar 1736*, 599; Almquist, 19, 96.
14. Almquist excluded some groups of civil servants in the Board of Mines. The total number was probably much higher than ninety if weighmasters, ore prospectors, physicians, and bookkeepers at the mines and blast furnaces were to be included. See: Almquist, 6, 84–88.
15. For *Laboratorium mechanicum*, see the references in n. 20, Chapter 4. Almquist, writing from the point of view of the Board of Mines, described Polhem's mechanical laboratory as a "state supported private enterprise". See: Almquist, 42.
16. *Svenska Män och Kvinnor*, Vol. 6 (Stockholm, 1949), 148.
17. Sten Lindroth, *Svensk lärdomshistoria*, Vol. 2 (Stockholm, 1975), 474 f., 544 f.; Per Hebbe, "Anders Gabriel Duhres 'Laboratorium mathematico-oeconomicum'", *Kungl. Landtbruksakademiens Handlingar och Tidskrift* 72 (1933), 576–593; Bengt Hildebrand, *Kungl. Svenska Vetenskapsakademien: Förhistoria, grundläggning och första organisation* (Stockholm, 1939), 175–186.
18. Sten Lindroth, "Urban Hiärne och Laboratorium chymicum", *Lychnos* 1946–1947, 51–116

(English summary pp. 114–116); Lindroth 1975, 509–529; Hugo Olsson, *Kemiens historia i Sverige intill år 1800* (Lychnos-Bibliotek, No. 17.4: Uppsala 1971), 40–51.—The achievements of Polhem and Hiärne in the Board of Mines have traditionally been studied from a biographical point of view. Their importance appears somewhat different when viewed from an institutional perspective.
19. Lindroth 1946–1947, 112 f.; Lindroth 1975, 528.
20. Lindroth 1946–1947, 113 f.; Sten Lindroth, *Svensk lärdomshistoria*, Vol. 3 (Stockholm, 1978), 377–386.
21. Almquist, 221 f. Cf. Lindroth 1946–1947, 61.
22. Almquist, 53–77, 130–158.
23. For the tasks of the Mine Inspectors, see: Almquist, 53 f. For a complete list of the statutes regulating their work, see: *Bergsordningar, 1736*, index (*"Bergmästare"*).
24. Almquist, 42–47, 115 ff. Cf. Staffan Högberg, ed., *Anton von Swabs berättelse om Avesta kronobruk 1723* (Jernkontorets Bergshistoriska Skriftserie, No. 19: Stockholm, 1983), 11–15.
25. Ibid., 45.—Most of the applicants referred to testimonials from the university, but these were vaguely formulated regarding the actual knowledge of the applicants. A formal university examination was instituted in 1750. It included subjects such as law, mathematics, physics and chemistry. See: Almquist, 46 f.; Sten Lindroth, *Uppsala universitet 1477–1977* (Uppsala, 1976), 98–100. Cf. Claes Annerstedt, *Upsala universitets historia*, Vol. 3:1 (Upsala, 1913), 235–262, esp. 250.
26. This discussion has been inspired by Gunnar Eriksson, "Om idealtypbegreppet i idéhistorien", *Lychnos* 1979–1980, 288–296, and Bo Sundin, *Ingenjörsvetenskapens tidevarv: Ingenjörsvetenskapsakademin, Pappersmassekontoret, Metallografiska institutet och den teknologiska forskningen i början av 1900-talet* (Acta Universitatis Umensis, Umeå Studies in the Humanities, No. 42: Umeå, 1981), 205.
27. There was a strong motive for advancement in the fact that there was a great difference in social status between high and low positions within the Board. This is clear from the differences in salary, which can probably be regarded as a measure of social status even if the salary was not always disbursed in a time when the finances of the State were strained by a long period of war. The annual salary of a Mine Councillor was 4500 *copperdaler* and that of an Assessor 3600, while a Stipendiary received only 900 and a clerk 600. Between low and high

positions the salary thus increased fivefold or sixfold. One earned prestige, at least, by being hard-working and ambitious in the Board of Mines. See: Almquist, 31–33.

28. Almquist, 17. Cf. n. 21.
29. *Bergsordningar* 1736, 599.
30. Almquist, 19 n. 3.
31. This discussion is based on the list of appointments and the biographical data in Almquist.
32. Almquist 15 f. Cf. Sten Carlsson, *Ståndssamhälle och ståndspersoner 1700–1865: Studier rörande det svenska ståndssamhällets upplösning* (Lund, 1973, rev. ed.); Ingvar Elmroth, *Nyrekryteringen till de högre ämbetena 1720–1809: En social-historisk studie* (Bibliotheca Historica Lundensis, No. 10: Lund, 1962).
33. This count is based on the two sequels to *Bergsordningar* 1736: *Kongl. stadgar, förordningar, bref och resolutioner, angående justitien och hushållningen wid bergwerken och bruken. Första fortsättningen. Från och med år 1736 til och med år 1756* (Stockholm, 1786); *Kongl. stadgar, förordningar, bref och resolutioner, angående justitien och hushållningen wid bergwerken och bruken. Andra fortsättningen. Ifrån och med år 1757 til och med år 1791* (Stockholm, 1797).
34. Heckscher 1949, Vol. 2:2, 585–593.
35. For the pre-history of *Bergsordningar* 1736 and its frontispiece, see: Kjell Kumlien, "Bergsordningarnas tillkomst—en kulturbild från frihetstiden", *Med Hammare och Fackla* 23 (1963), 109–128. Kumlien argues that the decision to publish a complete collection of all mining statutes should be seen as an expression of the historical and antiquarian interest of the time. It seems more likely, however, that a complete collection of mining statutes was published to lend weight to the rapidly growing number of statutes in the 1730s (Cf. Fig. 5.5).—I am indebted to Ulla Ehrensvärd, Allan Ellenius and John Linders for comments on my iconographical interpretation of the frontispiece.—It should be noted that *Bergsordningar* 1736 was published in 3 000 copies, which was a large edition in eighteenth-century Sweden. Several thousands of people had reason to contemplate the symbolical meaning of its frontispiece. It was published, as the title stated, for "public use and obedience".
36. Johan O. Carlberg, *Historiskt sammandrag om svenska bergverkens uppkomst och utveckling samt grufvelagstiftningen* (Stockholm, 1879), 503–554; Hjalmar Hammarskjöld, *Om grufregal och grufegendom i allmänhet enligt svensk rätt* (Upsala, 1891), 1–37; Theodor Rabenius, "Om eganderätt till grufvor", *Uppsala Universitets Årsskrift*

1863, 1–16; Emil Sommarin, "Det svenska bergsregalets ursprung", *Statsvetenskaplig Tidskrift* 13 (1910), 141–170.
37. *Bergsordningar* 1736, 629.
38. Ibid., 628–635; Carlberg, 540–545; Hammarskjöld, 4 f., 37 ff.; Rabenius, 16 f.
39. Ibid., 631.
40. One consequence was the popular guide to ore prospecting which Magnus von Bromell published in 1730, and in which he reproduced the entire statute of 1723 in the preface. See: Magnus von Bromell, *Inledning til nödig kundskap at igenkiänna och upfinna allahanda bergarter* (Stockholm, 1730). Cf. Lindroth 1975, 446.
41. Erik Salander, *Systematiske nöd-hielps tankar, eller okulstöteliga grund-saker til wälgång för höga och låga i et fattigt land, der penningen har rymt, näringen ligger öde och winsten är förswunnen; begrundade af En Som önskar allas wälmågo!* (Göteborg, 1730). Cf. Heckscher 1942, Vol. 2:2, 822, 853; Lindroth 1978, 105 f.
42. Salander, 4.
43. Ibid., 5.
44. Ibid., 6.
45. Ibid., 7 ff.
46. Ibid., 15–27.
47. Ibid., 34.
48. Ibid., 66.
49. Ibid., 27.

6. THE TECHNOLOGY: EARLY NEWCOMEN ENGINES

I am indebted to John S. Allen, Richard L. Hills and Wolfhard Weber for their comments on an earlier version of this chapter.

1. Donald S. L. Cardwell, *From Watt to Clausius: The Rise of Thermodynamics in the Early Industrial Age* (Ithaca, 1971), 15 ff.
2. Jennifer Tann, ed., *The Selected Papers of Boulton & Watt: Volume I, The Engine Partnership 1775–1825* (London, 1981), 1–19 (introduction by Tann); G. N. von Tunzelmann, *Steam Power and British Industrialization to 1860* (Oxford, 1978), 74–79.
3. John R. Harris, "The Employment of Steam Power in the Eighteenth Century", *History* 52 (1967), 133–148.
4. Lionel T. C. Rolt & John S. Allen, *The Steam Engine of Thomas Newcomen* (Hartington, 1977);

John Kanefsky & John Robey, "Steam Engines in 18th-Century Britain: A Quantitative Assessment", *Technology and Culture* 21 (1980), 161–186.

5. See, for example: Milton Kerker, "Science and the Steam Engine", in: Thomas Parke Hughes, ed., *The Development of Western Technology since 1500* (New York, 1964), 66–76 (but N.B. the Editor's note on p. 70). Originally published in *Technology and Culture* 2 (1961), 381–390.

6. Henry W. Dickinson, *A Short History of the Steam Engine* (Cambridge, 1939), 32f.

7. Eugene S. Ferguson, "The Origins of the Steam Engine", *Scientific American* 210 (1964), No. 1, 101.

8. David S. Landes, *The Unbound Prometheus: Technological Change and Industrial Development in Western Europe from 1750 to the Present* (London, 1970), 95f.; Samuel Lilley, "Technological Progress and the Industrial Revolution 1700–1914", in: Carlo M. Cipolla, ed., *The Fontana Economic History of Europe: The Industrial Revolution* (The Fontana Economic History of Europe, Vol. 3: Huntington, 1973), 203.

9. Cf. the discussion on the difference between potential and available natural resources in Chapter 3.

10. Rolt & Allen, 34–36.

11. For the pre-history of the steam engine, see: Ferguson 1964, 98–107; Graham J. Hollister-Short, "Antecedents and Anticipations of the Newcomen Engine", *Transactions of the Newcomen Society* 52 (1980–81), 103–117; Joseph Needham, "The Pre-Natal History of the Steam Engine", in: idem, *Clerks and Craftsmen in China and the West: Lectures and Addresses on the History of Science and Technology* (Cambridge, 1970), 136–202; Rolt & Allen, 14–43.—I have not listed all the literature that discusses the possible influence of the scientific community on the inventors of the steam engine; it can be found in almost any general survey of the history of technology. It should be noted, however, that most authors use expressions such as "I have no doubt that ...", "Surely he must have heard of ..." or "It seems likely therefore that ..." etc. There simply are not yet enough historical data for any statement to be made with certainty.

12. Rolt & Allen, 44–54; John S. Allen, "The 1712 and Other Newcomen Engines of the Earls of Dudley", *Transactions of the Newcomen Society* 37 (1964–65), 57–84.

13. Ferguson 1964, 102; Rolt & Allen, 90.

14. Rolt & Allen, 89–106; John S. Allen, "A Chronological List of Newcomen Society Papers on Thomas Newcomen and the Newcomen Engine", *Transactions of the Newcomen Society* 50 (1978–79), 217f.

15. Donald S. Cardwell, *Technology, Science and History* (London, 1972), 66.

16. Two parameters are needed in order to calculate the power of the engine: the area of the cylinder and the effective vacuum. A diameter of 24 inches was the average during the period 1712–1725, see: Rolt & Allen, 146–150. Cf. Kanefsky & Robey, 183 (Table 8). The effective vacuum can be assumed to have been about half an atmosphere; see: Richard L. Hills, "A One-Third Scale Working Model of the Newcomen Engine of 1712", *Transactions of the Newcomen Society* 44 (1971–72), 63–77.

17. The average length of the cylinder was about 8 feet in the period 1712–1725, see: Rolt & Allen, 146–150. The number of strokes per minute varied between 12 and 16, but 14 seems to have been an average value.

18. This was exactly the case in a French colliery where an engine was built in the 1730s. Before then 50 horses, managed by 20 men, had been working in shifts around the clock to keep the mine free from water. See: Dickinson, 57.

19. Cf. Hills 1971–72, 73–77.

20. See the discussion on this in "The State of the Art: Rotative Motion" in Chapter 12.

21. Ferguson 1964, 105.

22. The temperature of steam increases with the pressure. A pressure boiler had to be soldered together, and the low melting temperature of the solder used in those days limited the steam pressure of a boiler. It was this limitation of available manufacturing techniques that made Savery's steam engine unsuccessful in practice during the early eighteenth century since it operated at a high steam pressure.

23. Cf. Melvin H. Jackson & Carol de Beer, *Eighteenth Century Gunfounding* (Newton Abbot, 1973).—Richard L. Hills has commented: "The problem in casting steam engine cylinders was the thinness of the walls, which caused the metal to cool before it had run in properly. A gun has much thicker sides and this is not such a great problem. Even in bell founding, the wall is fairly thick" (private communication).

24. Richard L. Hills has commented: "Even today, it is not the piston which fits the bore but the rings or, in Newcomen's case, the packing. With the low-pressure steam and the weight on the pump rods, there was not much difficulty in sealing the piston on the up stroke to prevent the steam escaping. The problem was on the

down stroke and Newcomen lessened this by using a water seal. This meant that he had to seal a liquid and not a gas. James Watt had to deal with a gas at higher pressure" (private communication).

25. Note the distinction between the invention of the automatic valve mechanism and the manufacturing technique needed to build it once the problem had been solved. Most writers have concentrated on the inventive ability that this solution displays. I do not want to question that, but my point is that we must not confuse the ingenuity at the moment of invention with the manufacturing technique needed for later engines, which was very similar to that needed to build a church clock. The difficult thing was not to build the automatic valve mechanism but to adjust it properly, and that demanded practical experience and not the inventive ability of a Newcomen.

26. For a discussion of the economics of the Newcomen engine compared with other types of steam engines, see: von Tunzelmann, 46–97.

27. I am indebted to John S. Allen for this information.

28. John R. Harris, "Recent Research on the Newcomen Engine and Historical Studies", *Transactions of the Newcomen Society* 50 (1978–79), 176.

29. Alan Smith, "Steam and the City—The Committee of the Proprietors of the Invention for Raising Water by Fire", *Transactions of the Newcomen Society* 49 (1977–78), 5–20.

30. Kanefsky & Robey give the possible total as 40 up to 1720 and 100 up to 1730. An interpolation gives a possible total of 70 steam engines built by 1725 (of which some were of the Savery type). Allen has identified 50 Newcomen engines built by 1725. A reasonable estimate would be about 60, but then it should be remembered that the number of Newcomen engines known has been almost doubled during the last fifteen years. See: Kanefsky & Robey, 169 (Table 2); Rolt & Allen, 146–150.

31. Kanefsky & Robey, 169, 170 f. Their total estimate is that 2 500 steam engines were built in England during the eighteenth century, and that about two-thirds were of the Newcomen type.—The remarkable longevity of the Newcomen engines is a proof of their mechanical reliability. One engine is known to have been in operation from ca. 1750 until 1900, see: Rolt & Allen, 134.

32. For a general survey of what was known of these engines up to 1977, see Rolt & Allen, 70–81, 146–150.—For a general discussion, see: Eric

H. Robinson, "The Early Diffusion of Steam Power", *The Journal of Economic History* 34 (1974), 91–107.

33. Georges Hansotte, "L'introduction de la machine à vapeur au pays de Liège (1720)", *La Vie Wallone* 24 (1950), 47–95; Graham J. Hollister-Short, "The Introduction of the Newcomen Engine into Europe", *Transactions of the Newcomen Society* 48 (1976–77), 11–24; idem, "A New Technology and its Diffusion: Steam Engine Construction in Europe 1720–c. 1780", *Industrial Archaeology* 13 (1978), 9–41, 103–128 (the paper is in two parts).

34. Conrad Matschoss, *Die Entwicklung der Dampfmaschine*, Vol. 1 (Berlin, 1908), 145; Hollister-Short 1978, 26–28.—Wolfhard Weber has kindly provided the following references to the Kassel engine, of which little is known: E. Gerland, "Die erste in Deutschland in dauernden Betrieb genommen Dampfmaschine", *Zeitschrift des Vereins Deutscher Ingenieure* 49 (1905), 1283 f.; Conrad Matschoss, "Die ersten Dampfmaschinen ausserhalb Englands", ibid., 1971–1975; A. G. Gren, "Beschreibung der wesentlichen Einrichtung der neuern Dampf- oder Feuermaschinen nebst einer Geschichte dieser Erfindung ...", *Neues Journal der Physik* 1 (1795), 62–95, 144–191; Ludwig Beck, *Die Geschichte des Eisens in technischer und kulturgeschichtlicher Beziehung*, Vol. 3 (Braunschweig, 1897), 91–112. Cf. Wolfhard Weber, *Innovationen im frühindustriellen deutschen Bergbau und Hüttenwesen: Friedrich Anton von Heynitz* (Studien zu Naturwissenschaft, Technik und Wirtschaft im Neunzehnten Jahrhundert, No. 6: Göttingen, 1976), 58–62.—For a recent discussion on the Kassel engine, see: Mikuláš Teich, "The Early History of the Newcomen Engine at Nová Baňa (Königsberg): Isaac Potter's Negotiations with the *Hofkammer* and the Signing of the Agreement of 19 August 1721", *East-Central Europe* 9 (1982), 24–27.

35. Dietrich Hoffmann, "Die frühesten Berichte über die erste [sic] Dampfmaschine auf dem europäischen Kontinent", *Technikgeschichte* 41 (1974), 118–131; Hollister-Short 1976–77 and 1978; Mikuláš Teich, "Diffusion of Steam, Water and Air-Power to and from Slovakia during the 18th Century and the Problem of the Industrial Revolution", in: *L'acquisition des techniques par les pays non-initiateurs* (Colloques Internationaux du Centre National de la Recherche Scientifique, No. 538: Paris, 1973), 349–376, 402–406.—For the most recent discussion on this engine, see: Teich 1982, 24–38.

36. Hoffmann, Hollister-Short 1976–77 and 1978.

37. Alan Smith, "The Newcomen Engine at Passy, France, in 1725: A Transfer of Technology Which Did Not Take Place", *Transactions of the Newcomen Society* 50 (1978–79), 205–217.
38. Hollister-Short 1978, 122.
39. See Chapter 12.
40. Lewis Mumford, *Technics and Civilization* (New York, 1963), 109–112.
41. There was also a print by Thomas Barney in 1719 which will be discussed in Chapter 7 (see Fig. 7.3), and a print by Sutton Nicholls in 1725 which will be discussed in Chapter 14 (See Fig. 14.5).
42. This was illustrated by the difficulties encountered by the Department of Mechanical Engineering at UMIST in 1970 when they built a one-third scale working model of a Newcomen engine for the North Western Museum of Science & Industry in Manchester. Even the experienced staff of a technical university, with the facilities of a modern mechanical workshop, found it difficult to build an engine from one of these engravings (i.e. Thomas Barney's in 1719, see Fig. 7.3). The problem of making it run smoothly illustrated the practical experience that was needed. See: Cardwell 1972, 71 f.; Hills 1971–72.—I am indebted to Richard L. Hills for showing me this engine in operation and explaining the difficulties encountered in making it run.

7. THE TRAVELLERS, 1715–1720

An earlier version of this chapter was read at the Annual Meeting of the North Western Branch of the Newcomen Society in Manchester on October 15, 1981. I am indebted to Richard L. Hills for his comments on that version. Some of the early findings presented in this chapter were communicated to John S. Allen in 1976, and are incorporated in his revised edition of Rolt's book on Thomas Newcomen; see: Lionel T. C. Rolt & John S. Allen, *The Steam Engine of Thomas Newcomen* (Hartington, 1977), 51, 66 f. The major study of Swedish travellers in England during the eighteenth century is Sven Rydberg, *Svenska studieresor till England under frihetstiden* (Lychnos-Bibliotek, No. 12: Uppsala, 1951). Cf. Michael W. Flinn, "The Travel Diaries of Swedish Engineers of the Eighteenth Century as Sources of Technological History", *Transactions of the Newcomen Society* 31 (1957–1958 and 1958–1959), 95–109, esp. 95–100.

1. Sten Lindroth, *Svensk lärdomshistoria*, Vol. 2 (Stockholm, 1975), 556–558; Rydberg, 281–285; Emanuel Swedenborg, *Opera quaedam aut inedita aut obsoleta de rebus naturalibus*, Vol. 1 (Stockholm, 1907).
2. Swedenborg 1907, 210.
3. Ibid., 220.
4. Ibid., 209, 214, 220, 228.
5. Ibid., 224–228.
6. Ibid., 226. Original in Latin: "Machina renovata per ignem ejiciendi aquam; et construendi illas ad officinas (vulgo *Hyttor*) ubi nullus est aquae lapsus sed ubi aqua est tranquilla: Ipse ignis et Caminus satis aquae rotis suppeditare poterit."—Swedenborg mentioned this engine again in a later letter to Benzelius after his return to Sweden: "an engine to build at a blast furnace, by the side of any stagnant water, and yet the wheel will be turned by the fire and will drive the water" (Ibid., 231). Original in Swedish: "en *machine* at byggia en massung, wid hwad stilla watn man behagade, och hiulet skal doch omföras genom elden som skal drifwa watnet".
7. (Thomas Savery), "An Account of Mr. Tho. Savery's Engine for Raising Water by the Help of Fire", *Philosophical Transactions* 21 (1699), 228.
8. Swedenborg 1907, 210.
9. Henry W. Dickinson, *A Short History of the Steam Engine* (Cambridge, 1939), 24 f.
10. Ibid., 20. Savery listed six "Uses that this Engine may be applied unto" in 1702. Supplying mills with water was the *first*, and the draining of mines was the *last*. See: Thomas Savery, *The Miners Friend* ... (1702), facsimile edition (Edinburgh, 1979), 28–38.
11. Swedenborg 1907, 226 f.
12. Ibid., 230.
13. A review of the early history of the Society is given in Lindroth 1975, 552–555. For earlier, more exhaustive works, see: Samuel E. Bring, "Bidrag till Christopher Polhems lefnadsteckning", in: *Christopher Polhem: Minnesskrift utgifven af Svenska Teknologföreningen* (Stockholm, 1911), 57–80; Nils C. Dunér, *Kungliga Vetenskaps Societetens i Upsala tvåhundraårsminne* (Uppsala, 1910); Bengt Hildebrand, *Kungl. Svenska Vetenskaps Akademien: Förhistoria, grundläggning och första organisation* (Stockholm, 1939), 81–135; Axel Liljencrantz, "Polhem och grundandet av Sveriges första naturvetenskapliga samfund jämte andra anteckningar rörande Collegium curiosorum", *Lychnos* 1939, 289–308, & 1940, 21–52.
14. Swedenborg 1907, 235–288 passim. The six vol-

umes that were published are reprinted in facsimile in Dunér.

15. The title pages of *Daedalus hyperboreus* made it clear that the journal contained the works of Polhem "and other brilliant persons in Sweden".
16. Swedenborg 1907, 238.
17. For example, Swedenborg published an article on his idea of a heavier-than-air flying machine in 1716; see: *Daedalus hyperboreus* 1716, Vol. 4, 80–83.
18. *Daedalus hyperboreus* 1716, Vol. 1, 14–23.—The second volume contains a description and engraving of Polhem's most famous construction, the water-powered hoisting machinery at Blankstöten in the Great Coppermine in Falun. The author, probably Swedenborg, points out the importance of mechanical science if Sweden's potential mineral resources are to be made available, see: *Daedalus hyperboreus* 1716, Vol. 2, 25–28.
19. Rydberg, 151, 154.
20. RA, Bergskollegium till Kungl. Maj:t, June 11, 1718.
21. Loc. cit., Original in Swedish: "Men sedan de, som på förbemelte sätt giordt sig om Bergwerckens drift förfarne, hwareftter annan blifwit befordrade, så skulle på slike personer wid *Collegium* blifwa brist, der man icke i tid är omtänckt, at upmuntra fler qwicke och snälle ämnen, at på lika sätt sig widare skickelige giöra."—This is confirmed by the list of officials in the Board of Mines. During the period 1680–1710, some 20 officials of the Board travelled in Europe, and several soon left for private employment. Among those on the Board in 1718 who had travelled in England were Johan Angerstein in 1704, Lars Benzelstierna at about the same time, Göran Wallerius in 1708–1710, and Anders Svab in 1712, see: Johan A. Almquist, *Bergskollegium och bergslagsstaterna 1637–1857* (Meddelanden från Svenska Riksarkivet, n.s. 2.3: Stockholm, 1909), 161–306; Rydberg, 139–154.
22. RA, BKA, Huvudarkivet, EIV: 157, Brev och suppliker 1718, 413f.; AI: 64, Protokoll 1718, 877f.
23. For Kalmeter's biography, see: Rolf Vallerö, "Henric Kalmeter", *Svenskt Biografiskt Lexikon*, Vol. 20 (Stockholm, 1975), 574–576. For his travels in England, see: Rydberg, 154–166.
24. Anders Berch, *Åminnelsetal öfver Henric Kalmeter* (Stockholm, 1752), 8f. Original in Swedish: "Han var flitigt närvarande vid Collegii Sessioner, at utaf Ledamöternes rådslag lära, huru

ärenderne borde hanteras: de lediga stunder använde han til sin undervisning, dels uti Collegii Malm-Cabinet, dels uti Machine-Kammaren, dels uti Prober-Kammaren, dels ock uti Archivo [...] Han hade under de 4 år, han såsom ämnessvän upvaktat, haft tilfälle, at inom verket lära känna Collegii göremål, och igenom resor i bergslagerne göra sig bekant om grufvor, hyttor, smält-verk och hvad mera til deth ämne hörde."
25. RA, Bergskollegium till Kungl. Maj:t, June 11, 1718.
26. Loc. cit. Original in Swedish: "en särdeles håg, böjelse och flit, at inhämta en god kundskap om de stycken, som till berörde Bergwercks drifft och *oeconomie* höra".
27. Five volumes of Kalmeter's diaries of his journeys in Europe in 1718–1726 and 1729–1730 are in KB, MS M 249: 1–5, and a sixth volume for 1726–1727 is in RA, BKA, Huvudarkivet, EIII: 12. KB, MS M 249: 1 contains the diary of his journey in England in 1719–1721, and M 249: 3 of that in 1723–1725. His report to the Board of Mines on the mining industry in Scotland is in RA, BKA, Huvudarkivet, EIII: 1 (dated January 22, 1720), and that on the mining industry in England in EIII: 10 (dated July 5, 1725). There are a large number of letters from Kalmeter to Jonas Alströmer during this period in UUB, MS G 130. For an account of Kalmeter's years in England, see Rydberg, 154–166.
28. Rydberg, 105, 154, 402.
29. RA, Bergskollegium till Kungl. Maj:t, June 27, 1719.
30. RA, BKA, Huvudarkivet, AI: 65, Protokoll 1719, 787–790.
31. RA, Bergskollegium till Kungl. Maj:t, June 27, 1719.
32. RA, BKA, Huvudarkivet, EI: 10, Kungl. brev 1717–1721, 138f.
33. RA, BKA, Huvudarkivet, AI: 65, Protokoll 1719, 1121–1124.
34. RA, BKA, Huvudarkivet, BI: 86, Registratur 1719, 1202–1207.
35. See n. 21.
36. RA, BKA, Huvudarkivet, EIII: 1, January 22, 1720.
37. Loc. cit. Original in Swedish: "ibland hwilcka är den wid Newcastle på några ställen inrättade *Machine*, at medelst rök af warmt watn, eller rättare sagt, igenom et giordt *Vacuum* och Luftens tryckande, häfwa tyngder utur grufwor".
38. KB, MS M 249: 1, 292. Kalmeter describes the Newcastle area on pp. 292–308.

39. Ibid., 294 f. Original in Swedish: "hwarest som grufwan hade olägenhet af watn, så hade de nu under händer at bygga en *Machine* at medelst eld eller rättare dunster af röken draga up watnet utur Schacktet [...] denna invention, som den är den curieusaste som nånsin war upfinnen, så skall jag willja längre fram den samma beskrifwa, då jag fått lägenhet at se den samma i sielfwa werket".

40. Rolt & Allen, 58–88. Cf. Chapter 10 (Triewald as engineer in Newcastle) and the references in n. 103, Chapter 10.

41. KB, MS M 249:1, 295 f., 300 f.

42. Loc. cit. Original in Swedish: "des så kallade Boyler och Cylinder woro insatta".

43. Swedenborg 1907, 300. Original in Swedish: "som nyligen är kommit ifrån *Kohlmäter* som är i *Neucastel*, angående en ny *curieus* pump-*machine*".

44. Loc. cit. Original in Swedish: "Strax här utan för staden är ett nytt *inventeradt* pumpwerck opbygt för derass kohlgrufwor, som mycket af wattn äro beswärade, hwilcket är derass största *incommoditet*: detta werck är för 6 weckor sedan först fullbracht, en öfwermottan wacker *invention*, drifwes med eld och wattn; med en stor jernkittel ofwantill helt betekt, allenast ett litet hohl: i thenna kitteln kokas wattnet och hela wercket drifwes af then imen, som kommer igenom thet lilla hohlet ofwanpå, som är mechta starkt och drifwer op *ballancen* af pumpen; och som med samma, wädret förloras, suger öpningen pumpen eller *ballancen* neder igen, som på then ena ändan af *ballancen*, hwilcket dess rörelse förorsakar, går liksom en kiärna eller trumma, som man giör smör med, giord af metall, hwilcken går så tett at intet wäder kan sig på sidorna af pulssen, som går inuti trumman, intrengia. Denna konst är nestan intet at beskrifwa, ett sådant werck wore i Swerje wid Grufworna, ther intet wattufall är, högt nödigt; thet pumpar 400 oxhufwud wattn om timan, och kan än starckare drifwas; *consumerar* om dygnet ungefähr 9 tunnor stenkohl, och kan gå på hwad diup man behagar. *Secret*: *Triwaldz* broder, som tiänar här hos Herrar *Redley*, har lofwat skicka afritningen med utförlig beskrifning derom til sin broder i *Stockholm*.

Mina tanckar härom i anledning af brefwet som utaf the *modeller* som för några åhr sedan om *thylikt* äro *publicerade*, samt huru thet i Swerje står at *practiceras*, wil jag wid annat tilfelle utförligare wisa."

45. RA, BKA, Huvudarkivet, AI:65, Protokoll 1719, 787.

46. Swedenborg was in Stockholm from November 1719 to March 1720. Although he attended the meetings of the Board on only five occasions in November 1719, he associated closely with Urban Hiärne, the Vice President of the Board, and they exchanged letters they had received. See: RA, BKA, Huvudarkivet, AI:65; Swedenborg 1907, 292, 295.

47. For example: Christian Ludvig Jöransson, *Tabeller, som föreställa förhållandet emellan Sveriges och andra länders mynt, vigt och mått* (Stockholm, 1777). For Swedish units of weight and measure before the metric system, see: Sam Owen Jansson, *Måttordbok: Svenska måttstermer före metersystemet* (Stockholm, 1950).

48. Johan F. Georgii, *Jordens alla mått och vigter* (Linköping, 1842), 160 f.

49. A calculation of the pumping capacity of a Newcomen engine in Newcastle in 1747 stated "63 gallons a Hodgshead". Since a gallon in the eighteenth century equalled 3.785 litres, this also gives one hogshead as 238.5 litres. See: T. Robertson, ed., *A Pitman's Notebook: Hope and Success at Houghton, Co. Durham. The Diary of Edward Smith. Houghton Colliery Viewer 1749–1751* (Newcastle upon Tyne, 1970), 30. I am indebted to R. M. Gard, County Archivist in Northumberland, for this reference.—Hogsheads/hour seems to have been the usual unit of measure for the pumping capacity of Newcomen engines, see: Rolt & Allen, 146–147.

50. Jansson, 59.

51. Ibid., 91.

52. R. W. Raymond, ed., *Glossary of Terms used in the Coal Trade of Northumberland and Durham* (Newcastle upon Tyne and London, 1849), 8. I am indebted to R. M. Gard, for this reference.

53. In the SI-system, Kalmeter gave the pumping capacity as 94 400 litres of water per hour, and the coal consumption as 1 485 litres of coal per twenty-four hours.

54. The Swedish word "*metall*" was usually used to denominate alloys of copper, and generally for all non-ferrous metals; see: *Svenska Akademiens Ordbok*, Vol. 17 (Lund, 1945), M 859 f.

55. Use of the word "*cylinder*" for a cylindrical object is first recorded in Swedish in 1705, see: *Svenska Akademiens Ordbok*, Vol. 5 (Lund, 1925), C 277. Cf. Stig Nilsson, *Terminologi och nomenklatur: Studier över begrepp och deras uttryck inom matematik, naturvetenskap och teknik. I* (Lundastudier i Nordisk Språkvetenskap, series A No. 26: Lund, 1974) 24 ff.

56. In the letter Kalmeter used the Swedish word "*modell*". This word could be used to designate not only three-dimensional representations of

an object on a reduced scale, but also two-dimensional ones such as drawings and engravings. Kalmeter used the word in the latter sense. See: *Svenska Akademiens Ordbok*, Vol. 17 (Lund, 1945), M 1218.—Very few copies of the prints by Barney and Beighton exist today, but the Swedish travellers' familiarity with them indicates that they must have been fairly widespread at that time.

57. I have not found any evidence that Mårten Triewald ever sent any drawing or description to Sweden before his return in 1726, and it is unlikely that he did so.

58. For Alströmer's biography, see: Eli F. Heckscher, "Jonas Alströmer", *Svenskt Biografiskt Lexikon*, Vol. 20 (Stockholm, 1918), 556–564. For his industrial works in Alingsås, see: idem, *Sveriges ekonomiska historia*, Vol. 2:2 (Stockholm, 1949), 585–642; Gustaf H. Stråle, *Alingsås manufakturverk* (Stockholm, 1884). For Alströmer and the foundation of the Royal Swedish Academy of Sciences, see: Hildebrand 1939, 328–337.

59. For an account of Alströmer's journeys, see: Stråle, 40–57; Rydberg, 108–109.

60. Rydberg, 104.

61. Copies of Alströmer's diary of his journey in England in 1719–1720 are in UUB, MS X 376 (eighteenth-century copy) and KB, MS M 218 (nineteenth-century copy).

62. UUB, MS X 376, 88f. See KB, MS M 218, 170 for the date.

63. Ibid. Original in Swedish: "Litet från denna Platzen är en Eld Spruta, den första af det wärck inrättat i Ängelland, hwilken updrager watnet från några kohlgrufwor [...] Dessa wärcken äro wärd at besee. Thomas Barney filmakare, har utskurit denna Eld Machinen uti en koppar plåt med beskrifning till det samma tryckt på twenne ark, kåstade tillsammans 2 sh."

64. Richard L. Hills, "A One-Third Scale Working Model of the Newcomen Engine of 1712", *Transactions of the Newcomen Society* 44 (1971–72), 63–77; Rolt & Allen, 44–54.

65. The print portrays the engine clearly, with only a few minor errors of detail. It shows the self-acting gear, and every feature of the Newcomen engine. This technology was well established by 1719, and was to remain basically unaltered until the time of James Watt. The description accompanying the print consists of fifty single-line references to details on the drawing stating the name and function of each component and the material of which it is made. It was possible to understand the working principle and the work cycle from the print. A scale at the bottom gives the dimensions. A replica of this engine was built from the print by the Mechanical Engineering Department, UMIST, for the North Western Museum of Science and Industry in Manchester, in 1970–71. It was not found easy to construct a working engine on the basis of the print, which contains certain ambiguities. Nevertheless, it provides a detailed and informative description of the basic features of the new technology. See: Hills 1971–72. Cf. James H. Andrew, "Some Observations on the Thomas Barney Engraving of the 1712 Newcomen Engine", *Transactions of the Newcomen Society* 50 (1978–79), 202–204.

66. I have not been able to find any further references to steam engines in Alströmer's diaries or his correspondence.

67. Stråle, 40f.

68. Ibid., 38.

69. See n. 13, and Sten Lindroth, *Svensk lärdomshistoria*, vol. 3 (Stockholm, 1978), 63–65. The reports from the meetings of the Society are published in: Henrik Schück, ed., *Bokwetts gillets protokoll* (Uppsala, 1918).

70. Loc. cit. It was published quarterly from 1720, and changed its name to *Acta literaria et scientiarum Sveciae* in 1728.

71. Schück, 13. Original in Swedish: "låfwar wid tillfället communicera, hwad som kunde förefalla".

72. Swedenborg, 1907, 300. Original in Swedish: "något *curiosum*, som werdt är".

73. Schück, 17. This meeting, held on March 4, 1720, was attended by Lars Roberg, Erik Benzelius, Per Martin, Erik Burman, Johan Billmark and Jacobus Burman.

74. Ibid. Original in Swedish: "Utaf Assessor Swedenborgs bref af den 29 Feb: berättades att Auscultanten i Kongl: Bergs-Collegio Kålmeter, hade ifrån New Castel i England gifwit kundskap af en Pump-Machin som nyligen blifwit inrättad i deras kålgrufwor, huru nembl. den starcka imman ell. wädret, som går af en stor kittel med wattn uthi, och Eld under, trycker eller drifwer up och ned genom en art af wäffning 400 tunnor Wattn om timan uthi Metall pumpor med Pulsar äfwen af Metall. Denna machinen går an i all högd, och consumerar allenast 9 tl. Sten Kåhl om Dygnet."—Sven Widmalm has pointed out to me that Schück probably misread the Swedish word "*häffning*", meaning heaving or lifting, as "*wäffning*".

75. One "*oxhufwud*" was 236 litres, but one "*tunna*" of charcoal was only 165 litres, see: Jansson, 59, 91.
76. Schück, passim.
77. Carl Sahlin, "Historien om den förstenade gruvarbetaren i Falun och denna berättelses användning som diktmotiv", *Jernkontorets Annaler*, n.s. 75 (1920), 239–252; idem, "Ett nytt bidrag till historien om den förstenade gruvarbetaren i Falu gruva", *Jernkontorets Annaler*, n.s. 83 (1928), 59–64.
78. Schück, 37 f.
79. Ibid., 41 f.
80. Ibid., 43.
81. Ibid., 45.
82. Ibid., 51.
83. Ibid., 53 f., 61.
84. Adam Leijel, "Narratio accurata de cadavere humano in fodina Cuprimontana ante duos annos reperto", *Acta literaria Sveciae* 1 (1720–1724), 1722, 250–254.
85. Schück, 60.
86. Sahlin 1920, 240–242.

8. THE ADVENTURERS, 1723–1725

The Gregorian calendar (New Date) was not adopted in Sweden until 1753, but it had been adopted in the Netherlands in 1701. The Gregorian calendar was 11 days ahead of the Julian calendar (Old Date). The Swedish ambassador in The Hague used double dates in his correspondence, but all dates here are given as Old Dates to allow sequential comparison with Sweden. This is of some importance to the discussion in n. 96 and Chapter 9 on time lag in the process of technology diffusion.

1. RA, BKA, Huvudarkivet, EIV:165, Brev och suppliker 1723:I, 307–315.—That de Valair was of French origin is mentioned in the minutes of the Board, where he is referred to as the "*fransöske*" colonel; see: RA, BKA, Huvudarkivet, AI:71, Protokoll 1725, 435.
2. RA, BKA, Huvudarkivet, EIV:165, Brev och suppliker 1723:I, 312. Original in Swedish: "skulle äga särdeles wettenskaper om åtskilliga sådana konster och manufacturer".
3. Loc. cit. Original in Swedish: "uprätta och låta förfärdiga en machine, hwarigenom en otrolig myckenhet af wattn utur de djupaste Schackter och grufwor kan utdragas".
4. Loc. cit. Original in Swedish: "genom en mycken lättare och genare wäg".
5. Loc. cit. Original in Swedish: "Landets skatt och naturlige Rikedomar, som mäst och förnämligast består i metaller, ädlare och ringa, skulle til en ansenlig myckenhet ökas och förkåfras."
6. Ibid., 313. Original in Swedish: "förvandla stångjärn, utan minskning av sielva materien till det finaste stål til den myckenhet som någonsin åstundas".
7. Ibid., 312.
8. RA, BKA, Huvudarkivet, AI:69, Protokoll 1723, 644–663, 677–698, 728–756.
9. Ibid., 678. Original in Swedish: "de konster ock wettenskaper, som nu för tiden här i Riket brukas ock bekanta äro".
10. Ibid., 756. Original in Swedish: "en stång af godt fint järn".
11. Sven Rinman, *Bergwerks lexicon*, Vol. 1 (Stockholm, 1788), 4–7. See also: idem, *Försök till järnets historia, med tillämpning för slögder och handtwerk*, Vol. 1 (Stockholm, 1782), 204–220.
12. RA, BKA, Huvudarkivet, AI:69, Protokoll 1723, 747. Original in Swedish: "och hwad angår det begiärda Privilegium uppå den wattu machin, som han tänker inrätta, så finner Collegium der wid intet at påminna om, utan må den som sig af samma machin tänker betiena, bäst som gitter med Societeten öfwerens komma".
13. RA, Sammansatta kollegier till Kungl. Maj:t, Vol. 246 b, May 10, 1723.
14. Sten Lindroth, *Svensk lärdomshistoria*, Vol. 3 (Stockholm, 1978), 377, 411–412, 643.
15. Ibid., 643.
16. Gustaf Bonde was to take an active part in a later project initiated by de Valair in 1727–1730, this time to transmute copper into gold by means of a secret powder (see below).—The Board of Mines had been engaged in the late seventeenth century in another project to transmute metals, see: Nils Zenzén, "Från den tid, då vi skulle transmutera järn till koppar och få lika mycket silver i Sverige som gråberg", *Med Hammare och Fackla* 7 (1936), 88–151.
17. De Valair's steel-making project is discussed in the minutes of the Board of Mines for 1723–1726, see: RA, BKA, Huvudarkivet, AI:69–72, Protokoll 1723–1726, passim.
18. UUB, MS G 130, Alströmerska brevsamlingen, Letter from Henrik Kalmeter to Jonas Alströmer dated September 24, 1723.
19. RA, Registratur Inrikes Civil Expeditionen,

1723, 963–966.—A draft of the patent had been referred to the Boards of Mines and Commerce in September, see: RA, BKA, Huvudarkivet, AI:69, Protokoll 1723, 1267 f.—The Society of Science in Uppsala heard on October 31, 1723, about de Valair's proposal for making steel, and read the patents for his two projects on December 6; see: Henrik Schück, ed., *Bokwetts gillets protokoll* (Uppsala, 1918), 94, 97.

20. RA, Registratur Inrikes Civil Expeditionen, 1723, 964. Original in Swedish: "ny och här tillförende obekant Wattumachine".

21. Ibid. 963 f. Original in Swedish: "Såsom Wij nogsamt besinna, hwad stor lindring och fördehl Bärgwercken här i Riket tillskyndas skulle, om en sådan machine kunde inventeras, och med den effect brukas till wattnets uppdragande som Öfwersten Valair utlofwat ..."

22. RA, BKA, Huvudarkivet, EIV:168, Brev och suppliker 1724:II, 649–650.

23. Ibid. Original in Swedish: "åtskillige förnehma Män jempte andra erfarne persohner".

24. RA, BKA, Huvudarkivet, AI:70, Protokoll 1724, 894.

25. Johan O. Carlberg, *Historiskt sammandrag om svenska bergverkens uppkomst och utveckling samt grufvelagstiftningen* (Stockholm, 1879), 1–20; Gunnar Ekström, "När upptäcktes Östra Silverberget?", *Historisk Tidskrift* 68 (1948), 126–134; Abraham Hülphers, *Dagbok öfwer en resa igenom de, under Stora Kopparbergs höfdingedöme lydande lähn och Dalarne år 1757* (Wästerås, 1762), 91 f.; Ulf Qvarfort, *Sulfidmalmshanteringens början vid Garpenberg och Öster Silvberg* (Jernkontorets Bergshistoriska Utskott, No. H 20: Stockholm, 1981); Carl Sahlin, "Föremål av guld och silver, förfärdigade av metall från svenska bergverk", *Med Hammare och Fackla* 6 (1935), 26–34; Ingemar Tunander & Sigurd Wallin, eds., *Anders Tidströms resa genom Dalarna 1754* (Falun, 1954), 20 f.

26. Sahlin 1935; Sten Lindroth, *Gruvbrytning och kopparhantering vid Stora Kopparberget intill 1800-talets början*, Vol. 2 (Uppsala, 1955), 325.

27. Bror Emil Hildebrand, *Sveriges och svenska konungahusets minnespenningar, praktmynt och belöningsmedaljer beskrifna*, Vol. 1 (Stockholm, 1874), 449.

28. Sahlin 1935, 32. Original in Swedish: "Fast än mitt Magra Bergh af alla war förskutit,/Så har doch Gripenhielm min malm medh flijt uthbrutit/och Låter Werden see, att eij förachtas böör,/Hvadh Öster Silfverberg i Dalars Tuna föör."

29. RA, BKA, Huvudarkivet, AI:71, Protokoll 1725, 1813–1816.

30. *Svenska Män och Kvinnor*, Vol. 7 (Stockholm, 1954), 289.

31. See n. 22.

32. That de Valair had no command of Swedish is noted in the minutes of the Board of Mines, see: RA, BKA, Huvudarkivet, AI:69, Protokoll 1723, 1271.

33. UUB, MS G 130, Alströmerska brevsamlingen, Letter from Henrik Kalmeter to Jonas Alströmer dated October 12, 1723.

34. Rydberg, 160; RA, BKA, Huvudarkivet, AI:71, Protokoll 1725, 1813–1816.

35. Rydberg, 160; *Svenska Män och Kvinnor*, Vol. 8 (Stockholm, 1955), 69.

36. Rydberg, 160.

37. This is also confirmed by later sources.

38. See n. 9.

39. See n. 3.

40. See n. 33.

41. RA, BKA, Huvudarkivet, EIV:168, Brev och suppliker 1724:II, 651.

42. Loc. cit. Original in Swedish: "at gifwa invention på denna nya machinen".

43. Only de Valair's patent has been known, see: Björkbom 1936.

44. RA, BKA, Huvudarkivet, AI:70, Protokoll 1724, 2052.

45. RA, BKA, Huvudarkivet, EIV:169, Brev och suppliker 1725:I, 748–752, Letter from Joakim Fredrik Preiss to the Government dated March 2, 1725. See also: RA, Hollandica, Vol. 364.

46. Loc. cit. Original in Swedish: "I går afton kom till mig en Catholisk Irländsk Adelsman och Öfwerste wid namn Johan OKelly Esquier, Seigneur d'Aghrim, och sade sig besitta Wetenskapen at giöra eldmachinen till att utpumpa de med wattu öfwerströmda miner. At han en sådan Machine i Luyk hade förfärdigat, ock wore Baronen Wansul och andra förnäma herrar däri delaktige ock hans associerade. At wärket woro istånd ock draga om dygnen omtränt 14 000 Tunnor Watn utur Miner."

47. Carl Björkbom, "Ett projekt att bygga en ångmaskin i Sverige år 1725", *Daedalus* 1936, 79–94. Björkbom's original paper has been translated into English, though abridged. See: idem, "A Proposal to Erect an Atmospheric Engine in Sweden in 1725", *Transactions of the Newcomen Society* 18 (1937–38), 75–85. Cf. Svante Lindqvist, "Nya bidrag till ångmaskinens historia", *Daedalus* 1976, 74–78 (this arti-

cle contains the results of some of my early studies of O'Kelly and his project).

48. Graham J. Hollister-Short, "The Introduction of the Newcomen Engine into Europe", *Transactions of the Newcomen Society* 48 (1976–77), 11–24; idem, "A New Technology and Its Diffusion: Steam Engine Construction in Europe 1720–c.1780 (Part 2)", *Industrial Archaeology* 13 (1978), 103–128.

49. It seems that Björkbom only studied the correspondence with the Swedish Government.

50. Björkbom 1936, 81 f.

51. Georges Hansotte, "L'introduction de la machine à vapeur au pays de Liège (1720)", *La Vie Wallone* 24 (1950), 47–55.

52. Jean Prosper Désiré O'Kelly d'Aghrim, *Annales de la Maison d'HyMancy, issue des anciens rois d'Irlande et connu depuis le XIᵉ siècle sous le nom de O'Kelly* ... (La Haye, 1830), 43; Isidore de Stein d'Altenstein, *Annuaire de la Noblesse de Belgique* (Bruxelles, 1859), 278–279.—I am indebted to the following institutions in Belgium for answering my questions on the Belgian branch of the O'Kelly family, and for providing me with these references: Archives Générales du Royaume, Bibliothèque Royale Albert Iᵉʳ, and Service de la Noblesse.—See also: *Burke's Genealogical and Heraldic History of the Landed Gentry of Ireland* (London, 1958, 4th ed.), 407.

53. Loc. cit.

54. Archives de l'Etat à Liège, Wanzoulle correspondence, Letter from John O'Kelly to Baron Berthold de Wanzoulle dated March 17, 1721.

55. Hollister-Short 1978, 105.

56. Björkbom 1936, 90–92.

57. Ibid., 90.

58. Lionel T. C. Rolt & John S. Allen, *The Steam Engine of Thomas Newcomen* (Hartington, 1977), 144; John Kanefsky & John Robey, "Steam Engines in 18th-Century Britain: A Quantitative Assessment", *Technology and Culture* 21 (1980), 166–172.

59. Hansotte op. cit.

60. I am indebted to Georges Hansotte for showing me O'Kelly's letters in Archives de l'Etat à Liège in 1976. John S. Allen kindly provided me with English translations of the letters. Cf. Rolt & Allen, 9; Hollister-Short 1976–77, 22 n. 5.

61. Archives de l'Etat à Liège, Wanzoulle correspondence, Letter from John O'Kelly to Baron Berthold de Wanzoulle dated February 18, 1721. Original in French: "et j'ai non seulement honte de leur demander encore, mais crainte d'un Refus. et jai dans ma jeunesse appri une mauvaise habitude de manger donc je ne sai pas men defaire et je ne vois pas icy du jour la continuer. je vien denvoyer ma montre pour la vendre, et quand je me serois defait de ce que j'ai d'util, me associes icy ont assez d'Egard pour moi, et assez de generosité pour me laisser crever."

62. A watch represented as much as ten years' income to a servant girl in the 1720s. This is why the gold watch was of particular symbolic value to Moll Flanders, the heroine of Daniel Defoe's novel in 1722. Moll always felt safer when she was wearing a gold watch—it was her badge as a gentlewoman. See: Samuel C. Macey, *Clocks and the Cosmos: Time in Western Life and Thought* (Hamden, 1980), 49–54.

63. Hansotte, 54.

64. Ibid., 55.

65. See n. 52.

66. See n. 45.

67. Björkbom 1936, 82–85.

68. Hollister-Short 1978, 107.

69. See n. 33.

70. Björkbom 1936, 81.

71. See n. 45. Preiss wrote of the supposed engine at the silver mine: "the engine built in Sweden by his workman, whom he called Saunders". Original in Swedish: "den af hans Werkdräng, som han nämnde Saunders i Swerige upbyggda Machinen".

72. Loc. cit.

73. Erik Holmkvist, *Bergslagens gruvspråk* (Uppsala, 1941), 90; idem, *Bergslagens hyttspråk* (Uppsala, 1945), 105.

74. *Svenska Akademiens Ordbok*, Vol. 7 (Lund, 1925), D 2257–2259.

75. See n. 45. Original in Swedish: "At hwad bemelte Wärkdräng widkommo, han woro en ärlig, arbetsam och förståndig karl, som wäl öfwat sig at utarbeta ock laga hwad honom förelägges, som han några år å rad honom lärt hafwer, ock om året 100 pund Sterling lön förtient, Men för det öfriga des ingen grund hade om sielfwa wetenskapens fundament ock icke det ringaste förstodo i proportions uträkningar hwarefter den machinen till alla sina delar bör inrättas."

76. Ibid., Letter from John O'Kelly as an enclosure.

77. Loc. cit.

78. Loc. cit. Original in Swedish: "At alt ankommo på en proportions Regel hwilcken i Engelland med långa uträkningar plägades sökas, men han igenom förfarenheten brackt til

den fullkomligheten, at den i några linier kunde utfinnas. At dens fundament woro talet siu och nio, hwarmedelst några Radices *quadrates* borde utsökias, och där efter proportionen inrättas."—For early English methods of calculating the power of Newcomen engines, see: Henry W. Dickinson, *A Short History of the Steam Engine* (Cambridge, 1939), 43–45.

79. See n. 45.

80. Loc. cit. Original in Swedish: "Det är, Allernådigste Konung, hwad bemelte OKelly föredragit hafwer, med mycken beskedlighet, och utan at med skrytande rosa sin Wetenskap."

81. Loc. cit. Original in Swedish: "af allerunderdånigaste wördnad för Eders Kongeliga Majestät och Kjerlek för Swenska nationen".

82. Loc. cit.

83. If D = The diameter of the cylinder
    w = The weight of the pumps and the water in the pumps
    we can express O'Kelly's rule of proportions as:

$$D_{50} = \sqrt{\frac{w_{50}}{9}} \text{ for 50 fathoms depth, and}$$

$$D_{100} = \sqrt{\frac{w_{100}}{7}} \text{ for 100 fathoms depth}$$

with A = The cross sectional area of the cylinder, i.e. $A = \pi\left(\frac{D}{2}\right)^2$, and assuming that $w_{100} = 2\,w_{50}$, i.e. that the weight of the pump rods and the water in the pumps is twice as great for 100 fathoms as for 50, we get:

$$\frac{A_{100}}{A_{50}} = 2.57$$

Any engineer who wanted to use O'Kelly's rule would probably visualize it as a diagram, either on a piece of paper or in his mind, which would look like this:

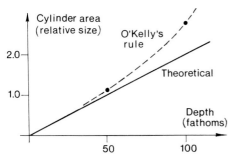

In theory, the relationship between the depth and the area of the cylinder should be a linear function, but experience had shown that it

was non-linear. The value of O'Kelly's rule of proportion was that it expressed this non-linear function as a simple rule of thumb.

84. RA, BKA, Huvudarkivet, EIV:169, Brev och suppliker 1725:I, 753–757, Letter from Joakim Fredrik Preiss to the Government dated March 9, 1725. Also in: RA, Hollandica, Vol. 364.

85. Loc. cit. Original in Swedish: "hwilken äfwenledes uti Physiken hade studerat och såwäl Theorien som Praxin af denna Physiska Machinens construction förstår".

86. See n. 52.

87. See n. 84. O'Kelly's description is published in Björkbom 1936, 90–94.

88. See n. 84. Original in Swedish: "effter den planen som derom i Engeland tryckt är, och sällias, men många stycken är upprinckad mindre för nyttan skuld, efter de ej äro nödiga, än till at förwilla de fremmandes begärelse des Construction at utleta".

89. See Chapter 7 and Figs. 6.3 & 7.2.

90. Cf. Björkbom 1936, 86.

91. RA, Bergskollegium till Kungl. Maj:t, Vol. 20, 445. Original in Swedish: "som innehafwer rätta wettenskapen af samma Machins construction".

92. See n. 84.

93. Loc. cit. Original in Swedish: "sex hästar kunna draga".

94. Loc. cit. Original in Swedish: "elden således i een stadigare grad kan bibehållas".

95. RA, Hollandica, Vol. 364, Letter from Joakim Fredrik Preiss to the Government dated April 6, 1725.

96. Preiss' first letter, dated March 2, reached the Government in Stockholm on March 12, and the second, dated March 9, arrived on March 19. Three days after each letter had arrived, they had been read and forwarded to the Board of Mines. The letters were considered by the Board for the first time at its meeting on April 7, when it was presumably decided to summon de Valair to the meeting on April 11 (RA, BKA, Huvudarkivet, AI:71, Protokoll 1725, 1435). Preiss' third letter, with the drawing and dated April 6, reached the Government on April 16, was referred to the Board on April 19, and was considered by the Board on April 28. All the dates of the procedure are carefully noted on top of each letter.

97. RA, BKA, Huvudarkivet, AI:71, Protokoll 1725, 435.

98. The Belgian biographical sources note only

that he was "capitaine au service d'Angleterre en 1711". See n. 52.

99. See n. 97. Original in Swedish: "af dylika wärkan".

100. RA, BKA, Huvudarkivet, EIV: 169, Brev och suppliker 1725: I, 758. De Valair's handwriting is difficult to read since his German is poor, and the letter seems to have been written while in a state of emotion.

101. Loc. cit.

102. For a description of the area in the eighteenth century, see: Hülphers 1762, 78–98.

103. Ibid., 92.

104. RA, Bergskollegium till Kungl. Maj:t, Vol. 20, 442–450.

105. Björkbom 1936, 86–89.

106. RA, BKA, Huvudarkivet, AI: 71, Protokoll 1725, 1641–1643.

107. Loc. cit. Original in Swedish: "Assessoren Swedenborgs tanckar öfwer denna machinens både nyttighet och beswärligheter."

108. Loc. cit. Original in Swedish: "Häruppå discourerades åtskilligt."

109. Loc. cit. Original in Swedish: "så wida Valaire sielf ej posiderar den wetenskapen utan tänckt betiäna sig af en OKellys lärodräng, som ej eller äger full komna kunskap om berörda machines construction".

110. RA, Bergskollegium till Kungl. Maj:t, Vol. 20, 444. Original in Swedish: "woro den samma som med eld förrättas, och långt för detta i England uppfunnen och practicerad är, och redan på många orter bekant".

111. Loc. cit. Original in Swedish: "om han finner god lägenhet och har sin utkomst".

112. Loc. cit.

113. RA, BKA, Huvudarkivet, AI: 71, Protokoll 1725, 1813–1816.

114. Loc. cit. Original in Swedish: "till någon wiss tid kunde med sin machine wisa något prof, efter han derå erhållit privilegium, på det, om han ingenting nyttigt dermed kunde åstadkomma, andra, som torde hafwa lust at accordera med öfwersten OKelly, ej måge i genom hans privilegium blifwa hindrade".

115. There are numerous references to de Valair's steelworks in the minutes of the Board in the mid-1720s. See n. 17.

116. Sten Lindroth, *Gruvbrytning och kopparhantering vid Stora Kopparberget intill 1800-talets början*, Vol. 2 (Uppsala, 1955), 326.

117. Ibid., 327–328.

118. See n. 104.

119. RA, Hollandica, Vol. 706, Letter from the Government to Joakim Fredrik Preiss dated May 31, 1725. Original in Swedish: "huruwida han förmehnar att denne Machinen här om wintertiden kan brukas och hållas igång".

120. The letter was signed by the King.

121. RA, Hollandica, vol. 690 (Preiss' registry of his correspondence with the Swedish Government, under the heading "Manufacturer i Sverige").

122. Loc. cit. and RA, BKA, Huvudarkivet, AI: 71, Protokoll 1725, 765.

123. Björkbom 1936, 89; Conrad Matschoss, *Die Entwicklung der Dampfmaschine*, Vol. 1 (Berlin, 1908), 242. Matschoss wrote: "Alte englische Geschichtswerke wissen zu berichten, dass bereits 1726 für Toledo eine Newcomen-Maschine in England gebaut worden sei." There was obviously an early interest in Spain in taking advantage of the new technology, since Triewald wrote in his book on the Dannemora engine that the Spanish Ambassador to the Court of St. James' came from London with a large entourage to see Thomas Newcomen's first engine of 1712. See: Mårten Triewald, *Kort beskrifning, om eld- och luft-machin wid Danmora grufwor* (Stockholm, 1734), 5.—The Swedish mining engineer Reinhold Rüttger Angerstein (1718–1760) visited Spain in 1752, and saw the remains of several unsuccessful Newcomen engines. See: Gustaf A. Granström, "Svensk bergsmannagärning i Spanien under 1700-talet", *Med Hammare och Fackla* 1 (1928), 161.

124. See n. 52.

9. TECHNOLOGY ON TRIAL, 1725

The discussion on agents of technological change in this chapter has been inspired by the concept "social carriers" introduced by Charles Edquist and Olle Edqvist. See: Charles Edquist & Olle Edqvist, *Social Carriers of Techniques for Development* (Swedish Agency for Research Cooperation with Developing Countries, SAREC Report No. R.3: Stockholm, 1979). For an abridged version with the same title and by the same authors, see: *Journal of Peace Research* 16 (1979), 313–331.

1. Emanuel Swedenborg, *Opera quaedam aut inedita aut obsoleta de rebus naturalibus*, Vol. 1 (Stockholm, 1907), 262 f.

2. RA, BKA, Huvudarkivet, AI:62, Protokoll 1716, 11.

3. Ibid., 439.

4. Swedenborg attended the meetings on April 8, 9, 10, 12, 13, 15, 16 and 17.

5. RA op. cit., 555.

6. Sten Lindroth, *Svensk lärdomshistoria*, Vol. 2 (Stockholm, 1975), 558.

7. Swedenborg 1907, 282.

8. Lindroth 1975, 559.

9. Swedenborg 1907, 291 f., 295 f.

10. Ibid., 292. Swedenborg had sent his description of blast furnaces and their operation to the Board of Mines the day before, on November 2, 1719. For his report, see: Emanuel Swedenborg, "Beskrifning öfver swenska masvgnar och theras blåsningar", *Noraskogs arkiv*, Vol. 4 (Stockholm, 1901–1903), 200–232.

11. Hjalmar Sjögren, "Några ord om Swedenborgs manuskript: 'Nya anledningar til grufvors igenfinnande' etc.", *Geologiska Föreningens Förhandlingar* 29 (1907), 436–443.

12. RA, BKA, Huvudarkivet, AI:65, Protokoll 1719. Swedenborg attended the meetings on November 5, 6, 14, 17 and 18.

13. Ibid., 1501. Original in Swedish: "Inkom Extraordinarie Assessoren Hr. Emanuel Swedenborg och aflade des trohets ock ämbetsed."

14. RA, BKA, Huvudarkivet, EIV:160, Brev och suppliker 1720:III, 2.

15. Lindroth 1975, 561.

16. Swedenborg 1907, 308.

17. Ibid., 306–308.

18. Lindroth 1975, 561.

19. RA, BKA, Huvudarkivet, AI:69, Protokoll 1723, 465–470, 550–554, esp. 554.

20. Ibid., 677–698.

21. RA, BKA, Huvudarkivet, AI:70, Protokoll 1724, 904, 1039, 1387, and esp. 1869.

22. RA, BKA, Huvudarkivet, EIV:169, Brev och suppliker 1725:I, 270–272.

23. Swedenborg 1907, 209, 214, 220, 228.

24. RA op. cit. Original in Swedish: "som igenom en sådan *antlia pneumatica* monga sorters experimenter kunna wisas, som angår wäder, eld, och watn".

25. RA, BKA, Huvudarkivet, AI:71, Protokoll 1725, 138, 144–146.

26. Loc. cit. Original in Swedish: "nyttige machin".

27. Ibid., 1689.

28. RA, BKA, Huvudarkivet, EIV:169, Brev och suppliker 1725:I, 766. Original in Swedish: "til landets allgemena nytta och bästa".—A similar air pump was bought from London in the 1740s by the Royal Swedish Academy of Sciences, see: Gunnar Pipping, *The Chamber of Physics: Instruments in the History of Sciences Collections of the Royal Swedish Academy of Sciences, Stockholm* (Stockholm, 1977), 177, Plate 1. Cf. Sten Lindroth, *Kungl. Svenska Vetenskapsakademiens historia 1739–1818*, Vol. 1 (Stockholm, 1967), 457, 470 f.

29. For O'Kelly's drawing, see: RA, Hollandica, Vol. 364, Preiss t. K.M:t. 6/17 april 1725 (kartavd. m. format). The drawing arrived together with Preiss' letter of April 6, 1725. Cf. n. 96, Chapter 8.

30. RA, BKA, Huvudarkivet, AI:71, Protokoll 1725, 1642.

31. RA, Hollandica, Vol. 364, Letter from Joakim Fredrik Preiss to the Government dated May 18, 1725.

32. Loc. cit. Original in Swedish: "ehuruwäl han hafft stor möda, aldenstund folcket där warit så öma ock afwundsfulle at de ej welat tillåta honom att bese Machinen så noga som det wäl elliest hade kunnat skie".

33. Loc. cit.

34. For Bergenstierna's copy of O'Kelly's drawing, see: RA, Kommerskollegii gruvkartor, maskiner, nr. 6. This drawing is reproduced in: Carl Björkbom, "Ett projekt att bygga en ångmaskin i Sverige år 1725", *Daedalus* 1936, 83.

35. RA, Hollandica, Vol. 364, Letter from Joakim Fredrik Preiss to the Government dated March 9, 1725. Also in: RA, BKA, Huvudarkivet, EIV:169, Brev och suppliker 1725:I, 754–756. O'Kelly's description is published in Björkbom 1936, 90–94. Original in French: "les principes de son mouvement sont la rarefaction et la condensation de l'air, contenu dans le Cylindre; car un vacuum (ou une espece de vacuum, pour ne pas facher Mess. les Philosophes, qui soutiennent, quod non datur vacuum) y étant fait, l'Atmosphere ne trouvant aucune resistance, fait sentir tout son poid sur le piston; Et si le poid de l'Atmosphere soit plus grand, que n'est celui de la colonne d'eau, qu'on se propose par ce moyen de lever, il est constant, qu'on rëuissera."

36. See n. 39.

37. Op. cit. Original in French: "On peut donc facilement par la simple regle de trois decouvrir la pressure de l'atmosphere sur tout corps cylindrique. Car comme le diametre du cylindre d'un pouce (:) a 30 (disons) le cylindre dont vous vous servez, (::) est 14.livr: etc (:) poid que vous cherchez."

38. RA, Hollandica, Vol. 364, Letter from Joakim Fredrik Preiss to the Government dated March

2, 1725. Also in: RA, BKA, Huvudarkivet, EIV: 169, Brev och suppliker 1725: I, 752.

39. O'Kelly may have referred to French, Liégeois, or British pounds and inches. In modern British units atmospheric pressure is 14.7 pounds per square inch.

40. See n. 35. Original in French: "Mais, Monsieur, la regle, que je vous ai deja envoyée, est non seulement courte, mais aussie le frottement y est compris, et elle est calculée sure une longue experience."

41. RA, BKA, Huvudarkivet, EIV: 169, Brev och suppliker 1725: I, 759–765. It is neither dated nor signed, but it is quite clear that it is by Swedenborg from the minutes of the Board; see: RA, BKA, Huvudarkivet, AI: 71, Protokoll 1725, 1642.

42. Ibid., 759. Original in Swedish: "en jemn och *penetrant* eld efter som watnet skal jemt och starkt delas uti *vapores*".

43. Loc. cit. Original in Swedish: "goda kohl".

44. Ibid., 760. Original in Swedish: "sådana [proportioner] som intet allenast hafwa i *mechaniquen* sina *reglor*, utan jämwäl i *Physiquen*".

45. Loc. cit. Original in Swedish: "någon gemen konstmästare".

46. Sten Lindroth, *Christopher Polhem och Stora Kopparberget: Ett bidrag till bergsmekanikens historia* (Uppsala, 1951), 60–81.

47. RA op. cit., 761. Original in Swedish: "men i fall sådane gifwes, så wore denna *machinen* nyttig och nödig".

48. Ibid., 762. Original in Swedish: "som ock på de orter där ingen ström eller lägenhet för konsthjul är wid handen, utan god ock ymnig skog".

49. Loc. cit. Original in Swedish: "3° Dessutom är den af en särdeles *curieusitet*, i det at et nyt *principium*, neml. eld, *evaporatio aqua* och *vacuum operera* hela rörelsen, som elljest skjer igenom andra *moventia mechanica*".

50. Loc. cit. Original in Swedish: "Men denne Regel kommer at ändras i så måtto, at i stället för *Cylinderns diameter* bör sättas *Cylinderns area* eller *qvadratum diametri*, ty wädrets tryckning på *Cylindriske* kroppar [...] har ingen *proportion* till *diametern* på *Cylindern* utan till dess *area*."

51. Ibid., 764. Original in Swedish: "men för skubbningen eller det som kallas *frottement*, blifwer den större, hwilket man endast kan hafwa af *Praxi*".

52. Loc. cit. Original in Swedish: "än att bruka desse tal 9 och 7, allenast man gjörer sig några *Cirklar* på följjande sätt neml:".

53. Ibid., 765. Original in Swedish: "Man kan *com-*

ponera en *figure* af sådana *Cirklar*, at man strax där af kan finna *diametern* till alla djup."

54. Loc. cit. Original in Swedish: "*diametern* tagas altid något större til, än som wisat är, för skubbningen, eller det så kallade *frottements* skull, efter det intet skadar at *Cylindern* är större men wäl att den är mindre".

55. RA, Bergskollegium till Kungl. Maj:t, Vol. 20, 442–450.

56. Björkbom 1936, 86–89.

57. RA, BKA, Huvudarkivet AI: 71, Protokoll 1725, 1641–1643.

58. Ibid., 551, 1664.

59. Ibid., 555.

60. RA, Bergskollegium till Kungl. Maj:t, Vol. 20, 442–450. Original in Swedish: "som alt igenom linier lärer kunna demonstreras".

61. Loc. cit. Original in Swedish: "Utaf alt detta, allernådigste Konung, täktes Eders Kongl. Majestät allernådigast intaga, hwad nytta denna *Machin* å ena sidan hafwer med sig, därest den kan ställas uti wärket, och göra den utlofwade efecten, samt å den andra, hwad swårigheter synes där wid möta."

62. RA, BKA, Huvudarkivet, EIII: 10, Utländska bergverksrelationer 1718–1727.

63. Ibid., 25.

64. Ibid., 26. Original in Swedish: "Hwad berörda Machine angår, så war han för något mer än 20 år först påtänkt af en wid namn *Savery*, ok sedermera efter många försök bragt til fullkomlighet ok på åtskilliga ställen i England upsatt, hwar brist på fall ej tillåter Wattukonsters inrättande. Det är ej utan at hon säges wara mångfaldig reparation underkastat, hälst på diupare diup."

65. Kalmeter used the Swedish unit "*oxhufwud*", which was approximately equal to a British hogshead. See Chapter 7.

66. RA op. cit. Original in Swedish: "warande storleken af *Cylindern* inrättad efter diupet ok tyngden af Watnet".

67. The equivalent of one pound in Swedish currency was 38.81 *copperdaler*. Cf. *Sveriges riksbank 1668–1924: Bankens tillkomst och verksamhet*, Vol. 5 (Stockholm, 1931), 140–143.

68. The equivalent of one *gulden* in Swedish currency was approx. 3.4 *copperdaler*. I am indebted to Lars Lagerqvist, Royal Coin Cabinet National Museum of Monetary History, Stockholm, for the information in nn. 67–68.

69. Sven Rydberg, *Svenska studieresor till England under frihetstiden* (Lychnos-Bibliotek, No. 12: Uppsala, 1951), 155.

70. UUB, MS G 130, Alströmerska brevsamlingen,

Letter from Henrik Kalmeter to Jonas Alströmer dated Liège September 10, 1725.

71. Ibid., Letter from Henrik Kalmeter to Jonas Alströmer dated Aix la Chapelle October 5, 1725.

72. Dietrich Hoffmann, "Die frühesten Berichte über die erste [*sic*] Dampfmaschine auf dem europäischen Kontinent", *Technikgeschichte* 41 (1974), 118–131. Cf. Graham J. Hollister-Short, "The Introduction of the Newcomen Engine into Europe", *Transactions of the Newcomen Society* 48 (1976–77), 11–24; idem, "A New Technology and its Diffusion: Steam Engine Construction in Europe 1720–c.1780", *Industrial Archaeology* 13 (1978), 9–41, 103–128 (the paper is in two parts).—Schönström's drawings are reproduced by Hoffmann (p. 124 f.) and Hollister-Short 1976–77 (p. 16).

73. Hoffmann, 127.

74. Bertil Boëthius & Åke Kromnow, *Jernkontorets historia*, Vol. 2:2 (Stockholm, 1968), 980; Gustaf Elgenstierna, *Den introducerade svenska adelns ättartavlor*, Vol. 7 (Stockholm, 1932), 118 f.

75. RA, BKA, Huvudarkivet, AI:68, Protokoll 1722, 1893.

76. I have not been able to find any reference to von Schönström in the minutes of the Board of Mines for 1725.

77. Swedenborg 1907, 318 f.

78. Hollister-Short 1976–77, 11.

79. See n. 96, Chapter 8.

80. *Biographiskt lexicon öfver namnkunnige svenska män*, Vol. 11 (Upsala, 1845), 362–364, *Svenska Män och Kvinnor*, Vol. 6 (Stockholm, 1949), 169. Cf. Stig Jägerskiöld, *Sverige och Europa 1716–1718: Studier i Karl XII:s och Görtz' utrikespolitik* (Ekenäs, 1937), passim.

81. Sten Lindroth, *Gruvbrytning och kopparhantering vid Stora Kopparberget*, Vol. 1 (Uppsala, 1955), 685 f.

82. Rydberg, 151–154.

83. Henrik Schück, ed., *Bokwetts gillets protokoll* (Uppsala, 1918), passim.

84. For the history of the Swedish Ironmasters' Association, see: Bertil Boëthius & Åke Kromnow, *Jernkontorets historia*, 3 Vols. (Stockholm, 1947–1968).

85. Ibid., Vol. 1 (Stockholm, 1947), 487–490.

86. See n. 76, Chapter 8.

87. See n. 75, Chapter 8.

88. See n. 85, Chapter 8.

89. See n. 44.

90. RA, BKA, Huvudarkivet, EIV:169, Brev och suppliker 1725:I, 760 f. Original in Swedish:

"som *mechaniska* så wäl som de *Physikaliska* Reglorna förstår".

91. Henrik Sandblad, ed., *Christopher Polhems efterlämnade skrifter: Teknologiska skrifter* (Lychnos-Bibliotek, No. 10.1: Uppsala, 1947), 277–307 ("Samtahl emällan Fröken Theoria och Byggmästar Practicus om sitt förehafvande").

92. Ibid., 277. Original in Swedish: "Theoria: Huem komer där med sitt svarta förskinn? Visit af sådant fålk ähr intet iag vahn vijd."

93. Ibid., 278. Original in Swedish: "En skiön fröken som I mig kallen, och Edert svarta förskin, tiäna väl tillsamans."

94. Loc. cit. Original in Swedish: "Th: Jag har dåk hört honom äga många barn som han har stort beröm uti. Pr: Jag kan väll det icke neka, men beklagar att de icke blifva så särdeles hederliga ansedde så länge man vet dem vara oächta. Men om I min skiöna fröken blefve moder för andra mina barn, så skulle de komma i långt högre anseende."

95. Loc. cit. Original in Swedish: "för publici tiänst skull", "Imedlertijd får iag betänka mig om det öfriga."

96. Sten Carlsson, *Fröknar, mamseller, jungfrur och pigor: Ogifta kvinnor i det svenska ståndssamhället* (Studia Historica Upsaliensia, No. 90: Uppsala, 1977), 16.

10. ENTER MÅRTEN TRIEWALD

An earlier version of this chapter was read at the "Newcomen Society Seminar to Commemorate the 250th Anniversary of the Death of Thomas Newcomen in 1729" at Imperial College, London, on June 23, 1979. It was later published under the title "The Work of Martin Triewald in England", *Transactions of the Newcomen Society* 50 (1978–79), 165–172. I am indebted to John S. Allen for comments on that version.

1. Sten Lindroth, *Svensk lärdomshistoria*, Vol. 3 (Uppsala, 1978), 338.

2. Bengt Hildebrand, *Kungl. Svenska Vetenskapsakademien: Förhistoria, grundläggning och första organisation* (Stockholm, 1939), 139.

3. NMA, Skråarkivalier, Hovslagareämbetet i Stockholm, Protokoll 1676–1698, April 6, 1687.—There were 20 masters and 38 apprentices in the Guild of Blacksmiths in 1689, and 80 % of the masters were of foreign descent;

see: Ernst Söderlund, *Stockholms hantverkarklass 1720–1772: Sociala och ekonomiska förhållanden* (Monografier utgivna av Stockholms kommunalförvaltning: Stockholm, 1943), Tab. 2.

4. Mårten Triewald, "Fortsättning om stenkolswettenskapen", *KVAH* 1740, 235.

5. Op. cit., October 8, 1688.

6. NMA, Skråarkivalier, Hovslagareämbetet i Stockholm, In- och utskrivningsbok 1623–1700, December 31, 1688.

7. NMA, Skråarkivalier, Hovslagareämbetet i Stockholm, In- och utskrivningsbok 1685–1732, January 3, 1689.

8. NMA, Skråarkivalier, Hovslagareämbetet i Stockholm, Protokoll 1698–1725 & 1726–1761.

9. Nils Staf, ed., *Borgarståndets riksdagsprotokoll från frihetstidens början*, Vol. 4 (Stockholm, 1958), passim.

10. NMA, Skråarkivalier, Hovslagareämbetet i Stockholm, Protokoll 1676–1698, October 21, 1689.

11. Ibid., November 26, 1689.

12. Ibid., July 1, 1690.

13. Bengt Hildebrand, "Till släkten Triewalds historia", *Personhistorisk Tidskrift* 39 (1938), 153.

14. Loc. cit.

15. SSA, Tyska församlingens kyrkoarkiv, CI1:B, Dopbok 1689–1734.

16. NMA, Skråarkivalier, Hovslagareämbetet i Stockholm, Protokoll 1676–1698.

17. NMA, Skråarkivalier, Hovslagareämbetet i Stockholm, Protokoll 1698–1725, 1726–1761.

18. NMA, Skråarkivalier, Hovslagareämbetet i Stockholm, In- och utskrivningsbok 1685–1732.

19. SSA, Tyska församlingens kyrkoarkiv, LIh 1a, Längder över uppburna bänkavgifter.—For the German parish, see: Emil Schieche, *400 Jahre Deutsche St. Gertruds Gemeinde im Stockholm 1571–1971* (Stockholm, 1971).

20. Op. cit.

21. Hildebrand 1938, 152.

22. NMA, Skråarkivalier, Hovslagareämbetet i Stockholm, In- och utskrivningsbok 1685–1732.

23. NMA, Passansökningar och pass, Hovslagareämbetet i Stockholm, 1695–1792, No. 158, April 17, 1727.

24. Ibid., No. 648, 1700.

25. NMA, Skråarkivalier, Hovslagareämbetet i Stockholm, Protokoll 1698–1725.

26. Lars Laurel, *Åminnelsetal öfwer [...] Mårten Triewald* (Stockholm, 1748), 6.—Lars Laurel (1705–1793), professor of philosophy at the University of Lund, became a member of the Royal Swedish Academy of Sciences in 1747, the year in which Triewald died. In the introduction to his memorial address (p. 2), Laurel writes: "it is not unknown how much I am lacking in needful information about so deserving a man. But I have, as best I have been able, gathered some facts from his friends, and thereof made a summary." The address was given in 1748, when Linnaeus, Alströmer and Kalmeter were alive. They were probably among those who supplied Laurel with information.

27. Johann Georg Rüdling, *Supplement till thet i flor stående Stockholm* (Stockholm, 1740), 109.

28. There are no records of the German School for these years in the archives of the German Parish in SSA.

29. Hildebrand 1939, 156.

30. Martin Lamm, "Samuel Triewalds lif och diktning", *Samlaren* 28 (1907), 113 ff.

31. Hildebrand 1939, 138 f.

32. Loc. cit.

33. Loc. cit.

34. Laurel, 7.

35. SSA, Bouppteckningar, 1745/2:544.

36. NMA, Skråarkivalier, Hovslagareämbetet i Stockholm, In- och utskrivningsbok 1685–1732 & 1726–1794.—The average was 1.5 apprentices per master in the period 1689–1740, see: Söderlund, Tab. 1, Tab. 15.

37. Lamm, 123 f.

38. Hildebrand 1939, 159.

39. Ibid., 161.

40. Mårten Triewald, *Kort beskrifning, om eld- och luft-machin wid Dannemora grufwor* (Stockholm, 1734), 3.

41. Mårten Triewald, *Föreläsningar öfwer nya naturkunnigheten*, Vol. 1 (Stockholm, 1735), preface.

42. Laurel, 7.

43. Anne-Marie Brötje, "Samuel Worster: En frihetstida Stockholmsköpman", *Personhistorisk Tidskrift* 41 (1942–1945), 95.

44. Ibid., 94 ff.

45. Ibid., 99.

46. Triewald 1735, 166 n.

47. Brötje, 104.

48. Loc. cit.

49. SSA, Bouppteckningar, 1747/2:271. Cf. Brötje, 100 ff.

50. Brötje also states that Triewald had the opportunity to "use instruments" in Worster's household. But this cannot have been of importance, because according to the estate inventory of Samuel Worster, made in 1747, his

collection of instruments did not amount to much more than a dozen dividers and rulers, a compass, a sundial and a broken prism.

51. Brötje, 104.
52. Lamm, 113.
53. Triewald 1734, 6.
54. Brötje, 95.
55. Loc. cit.
56. Hildebrand 1939, 534 f.
57. Hildebrand 1938, 158.
58. Ibid., 152.
59. Laurel, 11.
60. Hildebrand 1938, 151.
61. Based on a count of ships arriving in 1727 and 1728 according to: SSA, Stockholms stads verifikationsbok 1727 & 1728.
62. Laurel, 8.
63. Lamm, 114 f.
64. Laurel, 8.
65. Loc. cit.
66. Ibid., 9.
67. Ibid., 11.
68. Loc. cit.
69. Loc. cit.
70. Sven Rydberg, *Svenska studieresor till England under frihetstiden* (Lychnos-Bibliotek, No. 12: Uppsala, 1951), 405.
71. Laurel, 11 f.
72. Loc. cit.
73. Rydberg, 117.
74. Mårten Triewald, *Konsten at lefwa under watn* (Stockholm, 1734), 17. For Triewald and the diving-bell, see: Anna Beckman, "Två svenska experimentalfysiker på 1700-talet: Mårten Triewald och Nils Wallerius", *Lychnos* 1967–1968, 198 f.; Bo Cassel, *Havet, dykaren, fynden under 2000 år* (Stockholm, 1967), 29 f., 42–45; idem, "Dykarkonstens utveckling i Sverige fram till 1850-talet", *Sjöhistorisk årsbok* 1975–1976, 9–36.—For Edmund Halley and the diving-bell, see: Colin A. Ronan, *Edmond Halley: Genius in Eclipse* (London, 1970), 102–105. Cf. F. W. Heinke & W. G. Davis, *A History of Diving from the Earliest Times to the Present Date* (London, 1873, 4th ed.), 11–13 (Triewald on p. 13).
75. Mårten Triewald, "An Improvement of the Diving Bell", *Philosophical Transactions* 39 (1735–36), 377–383.
76. Carl Forsstrand, "Mårten Triewald och hans Stockholmsträdgård", *Svenska Linné-Sällskapets Årsskrift* 10 (1927), 59.
77. Lindroth 1978, 342.
78. Laurel, 13.
79. Loc. cit.

80. Stig Jägerskiöld, *Sverige och Europa 1716–1718: Studier i Karl XII:s och Görtz' utrikespolitik* (Ekenäs, 1937), 196.
81. Lamm, 123.
82. Rydberg, 202.
83. It may be mentioned in passing that Mårten's youngest brother, Daniel, was to be more directly affected by Sweden's relations with England. Daniel Triewald, who had enlisted in the Navy, was assigned to the frigate *Illerim* in 1717. On June 16 the vessel was engaged by four English men-of-war off Gotland, and Daniel was wounded in the thigh. The *Illerim* was captured and handed over to the Danes, but the crew were landed at Helsingborg. See: Hjalmar Börjeson, *Biografiska anteckningar om Örlogsflottans officerare 1700–1799* (Stockholm, 1942).
84. Mårten Triewald, "Walk- eller klädemakarlers grufwornes beskaffenhet uti Bedfordshire i England", *KVAH* 1742, 10.
85. Mårten Triewald, "En ny ler-älta särdeles tiänlig för taktegel bruk", *KVAH* 1742, 211–214.
86. On a certificate dated Wedevåg, January 25, 1732, Triewald states that in 1730 he "organized the whole Brickworks" at Wedevåg Ironworks, so that it could now turn out 50 000 bricks a year, instead of the former maximum of 1 000–1 500. See: KTHB, Intyg utfärdat av Mårten Triewald den 25/1 1732.
87. Laurel, 14.
88. Loc. cit.
89. For Desaguliers' biography, see: A. Rupert Hall, "John Theophilus Desaguliers", in: Charles C. Gillispie, ed., *Dictionary of Scientific Biography*, Vol. 3 & 4 (New York, 1981), 43–46. Cf. *Dictionary of National Biography*, Vol. 14 (London, 1888), 400 f.—For Desaguliers as a Freemason, see: Albert F. Calvert, *The Grand Lodge of England 1717–1917: Being an Account of 200 Years of English Freemasonry* (London, 1917), 10–30.—For Desaguliers as a lecturer in experimental physics, see: John L. Heilbron, *Electricity in the 17th and 18th Centuries: A Study of Early Modern Physics* (Berkeley, 1979), esp. 161; A. E. Musson & Eric Robinson, *Science and Technology in the Industrial Revolution* (Manchester, 1969), esp. 37–40.
90. Calvert, 10.
91. Ibid., 20.
92. Ibid., 12 f.
93. I am indebted to Mr. Harald Qvistgaard, himself a high-ranking Freemason, for drawing my attention to the possibility that Triewald may

have become a Freemason during his time in London. Mr. Qvistgaard asked the librarian of the United Grand Lodge of England whether Triewald was to be found in the lists of members, and the librarian in London wrote on February 26, 1976: "I regret to report that I have not been able to find the name of Martin Triewald in the membership lists of our Grand Lodge in the 1720s. These, however, are by no means complete for proper membership registers were not instituted by the premier Grand Lodge until 1770. The early lists (1723, 1725 and 1730) were entered in the first Minute Book of the Grand Lodge and are very much 'hit and miss' affairs [...] In A.Q.C. Vol. 80 (1967) there is a paper by Bro. J. R. Clarke entitled 'The Royal Society and Early Grand Lodge Freemasonry' in which he prints lists of Freemasons who were also Fellows of the Royal Society but, again, Triewald's name does not appear among them." Mr. Qvistgaard has however pointed out that Triewald was in the habit of finishing his signature with a little cross stroke, the stylized cross with which Freemasons used to indicate their solidarity with the order. See: Harald Qvistgaard, "Frimureriska namnteckningar från Gustaf III:s tid till våra dagar", *Meddelanden från Svenska Frimurare Orden* 34 (1962), No. 5, 2–12; idem, "Frimureriska namnteckningar och deras ursprung", ibid., 46 (1974), No. 1, 10–17.

94. J. C. Poggendorff, *Biographisch-Literarisches Handwörterbuch*, Vol. 1 (Leipzig, 1863).

95. Mårten Triewald, "Queries Concerning the Cause of Cohesion of the Parts of Matter, Proposed in a Letter to Dr. Desaguliers, F.R.S. By Fr. [sic] Triewald, Director of Mechaniks in the Kingdom of Sweden", *Philosophical Transactions* 36 (1729–30), 39–43.

96. Mårten Triewald, *Nyttan och bruket af wäderwäxlings-machin på Kongl. maj:ts och riksens örlogs flotta* (Stockholm, n.d.). Cf. idem, "Väderväxlings machin påfunnen och ingifven", *KVAH* 1744, 251–260.—For the history of ventilation in Sweden, see: Sten Lindroth, *Kungl. Svenska Vetenskapsakademiens historia 1739–1818*, Vol. 1 (Stockholm, 1967), 352–355. Cf. David G. E. Allan & Robert E. Schofield, *Stephen Hales: Scientist and Philanthropist* (London, 1980), 81–91; A. E. Clark-Kennedy, *Stephen Hales, D. D., F.R.S.: An Eighteenth Century Biography* (Cambridge, 1929), esp. 152f.—The economic incentive for the interest of Triewald and a "philanthropist" like Hales in the ventilation of ships was the slave trade. The slaves were often packed very tightly under deck and this caused high mortality, approx. 15 percent, on the journey from Africa to America. Those of the slavers who believed in the principle of "tight-packing" calculated that although the loss of life might be considerable on each voyage, so too were the net receipts from a larger cargo. See: Daniel P. Mannix, *Black Cargoes: A History of the Atlantic Slave Trade 1518–1865* (London, 1963), 104–130, esp. 116f. Triewald said so himself in a letter to an unknown correspondent in London, dated Stockholm April 22, 1743 (RA, Anglica, Vol. 317): "If my affairs would allow me to take a tripp to England, I know what reward I could obtain for this my Invention, particularly of those concern'd in the Slave Trade from Africa, which Trade by this Engine in the Ships that carry blacks to the West Indies, would be many Thousand pounds gaines every Year. Notwithstanding all this, if Doctor Hales will deposite in the hands of Mr. Spalding a [Swedish] Merchant in London 100 Guineas, I will send immediately over a compleat Engine for a Man of Warr of 90 Gunns, with full directions & draughts of all its uses by Sea and Land."

97. C. Stewart Gillmor, *Coulomb and the Evolution of Physics and Engineering in Eighteenth-Century France* (Princeton, 1971), 124: "In 1725 Desaguliers became acquainted with 'some experiments with cohesion in two balls of lead made by Mr. Trievall at Newcastle and Edinburgh'." Cf. John L. Heilbron, *Physics at the Royal Society during Newton's Presidency* (Los Angeles, 1983), 97.

98. Op. cit.

99. Laurel, 13.

100. Triewald 1734, 5f.

101. Jägerskiöld, 204, 274.

102. Viscountess Ridley, ed., *Cecilia: The Life and Letters of Cecilia Ridley 1819–1845* (London, 1968), 41ff.; Northumberland County Record Office, ZRI, Ridley (Blagdon) MSS, Introduction ("Blagdon and the Ridleys by Ursula, Viscountess Ridley"). Cf. Edward Hughes, *North Country Life in the Eighteenth Century, the North East 1700–1750* (London, 1952).

103. Triewald 1734, 5f. Cf. Lionel T. C. Rolt & John S. Allen, *The Steam Engine of Thomas Newcomen* (Hartington, 1977), 64. Allen writes, in his revised edition of Rolt's book, that the engine "was built at Byker about 1717/18 for Richard and Nicholas Ridley of Newcastle-on-Tyne. Although the agreement for the engine was dated 4 November 1718, royalties were to

be paid from 24 June 1718 so that it appears that the engine building work started well before the legal documents were completed." Cf. John S. Allen, "The Introduction of the Newcomen Engine from 1710 to 1733: Second Addendum", *Transactions of the Newcomen Society* 45 (1972–73), 223–226.—For the history of early Newcomen engines in the Newcastle area, see: Edward Hughes, "The first Steam Engines in the Durham Coalfield", *Archaeologica Aeliana*, 4th s., 27 (1949), 29–40; Henry Louis, "Early Steam-Engines in the North of England", *Transactions of the Institution of Mining Engineers* 82 (1932), 526–539; Arthur Raistrick, "The Steam Engine on Tyneside, 1715–1778", *Transactions of the Newcomen Society* 17 (1936–37), 131–164; Marie B. Rowlands, "Stonier Parrott and the Newcomen Engine", ibid., 41 (1968–69), 49–67.—The findings of these articles are included in Rolt & Allen, 61–70.

104. Triewald 1734, 6.
105. Loc. cit. Cf. Rolt & Allen, 64.
106. Triewald 1734, 7.
107. KB, MS I.t.18., (Triewald, M.), Handlingar i rättegången … 1726–1745. Cf. Rolt & Allen, 65, 9 (acknowledgements).
108. Triewald 1734, 8.
109. Northumberland County Record Office, ZRI. 23/2, Agreement between Triewald, Ridley, Calley and Prior, 1722.—The agreement was located in 1970 by Torsten Althin.
110. City Libraries, Newcastle-upon-Tyne, Minute Book of Goldsmiths Company, May 1721, and Church Register, 1726–1759; John C. Hodgson, "An Alphabetical Catalogue of the Goldsmiths of Newcastle", *Archaeologica Aeliana*, 3rd s., 11 (1914), 66–67.
111. *Newcastle Courant* 1724, February 1.—When Prior died in 1759, the *Newcastle Courant* wrote on April 14: "Last week died the ingenious and mathematical Mr. William Prior, Assay Master of the Plate Office here, and for the counties of Durham and Northumberland, eminent for musical instruments and toys."—I am indebted to the City Libraries, Newcastle-upon-Tyne, and the Central Library, Gateshead, for the information in nn. 110–111.
112. See n. 109.
113. (Mårten Triewald), *Triewald's patent A.D. 1722 No. 449: Engine for Drawing Coals and Waters from Mines, Supplying Water to Towns, &c.* (London, 1857); Bennett Woodcroft, *Alphabetical Index of Patentees of Inventions* (1854), facsimile ed.

(London, 1969), 575.—Triewald's original application for a patent is kept in the Public Record Office, London, SP 35/31, but it contains no additional information. I am indebted to Stephen Buckland for this information.
114. Rolt & Allen, 65.
115. KB, I.t.18., (Triewald, M.), Handlingar i rättegången … 1726–1746.
116. Rydberg, 203 n. 6.
117. Mårten Triewald, "Ytterligare fortsättning om stenkåls wetenskapen", *KVAH* 1740, 384.
118. Mårten Triewald, "Om alt hwad som länder till kundskapen om stenkol efter många års rön framgifwit", *KVAH* 1739, 100.
119. Triewald 1740 (b), 379.
120. Frank Atkinson, "Some Northumberland Collieries in 1724", *Transactions of the Architectural and Archaeological Society of Durham and Northumberland* 11 (1965), Parts 5 & 6, 425–434.
121. Ibid., 429.
122. Ibid., 430.
123. Ibid., 431 f.
124. Ibid., 432 n. 14.—Clerk illustrated his diary with a sketch of Triewald's design (ibid.), which is similar to Triewald's illustration of the same in the *Proceedings of the Royal Swedish Academy of Sciences* in 1741; see: Mårten Triewald, "Och sidsta fortsättningen af stenkåls wetenskapen: Beskrifning af et påfund, hwarigenom den dödeliga luften på en kort tid, utur et shakt blef dragen", *KVAH* 1741, 96–100, Tab. II.
125. Triewald's activities in Newcastle are also mentioned in another source, see: John Brand, *History and Antiquities of Newcastle-upon-Tyne*, Vol. 2 (London, 1789), 686. Brand wrote: "About the year 1713 or 1714, the first fire engine on the north side of the river Tyne is said to have been erected at Biker Colliery, the property of Richard Ridley, Esq. The engineer was the reputed son of a Swedish Nobleman, who taught mathematics at Newcastle." Brand's account appeared more than sixty years after Triewald left Newcastle, and was based on second-hand information.
126. Triewald 1734, 7.
127. Hildebrand 1939, 141.
128. KVA, Letter from Mårten Triewald to Henrik Kalmeter dated Newcastle December 6, 1723.
129. Triewald 1735, 145 n.
130. Ibid., 199 n.
131. Triewald 1735, preface.
132. Lindroth 1978, 343.
133. Brötje, 102.
134. Laurel, 15.

135. The library left by Triewald has been discussed by Rydberg 1951 (p. 205) and is the subject of a study by Torsten Althin. The library was sold at a book auction in Stockholm in 1752, and the printed catalogue contains 427 lots. See: Torsten Althin, *Capitaine-Mechanici wid Fortificationen Herr Mårten Triewalds wäl conditionerade Bibliotheque år 1747 jämte hans bibliografi* (Stockholm Papers in History and Philosophy of Technology, TRITA-HOT-3001: Stockholm, 1976).

136. Robert Boyle, *Experimentorum novorum physicomechanicorum* (London, 1680), Royal Institute of Technology Library copy, call No. Cc-1. Cf. Althin, 28.

137. See n. 128.

138. René Antoine Ferchault de Réaumur, *L'art de convertir le fer forgé en acier, et l'art d'adoucir le fer fondu* (Paris, 1722). Cf. *Réaumur's Memoirs on Steel and Iron* (1722), English translation by Annelise Grünhaldt Sisco, with an introduction and notes by Cyril Stanley Smith (Chicago, 1956).

139. See, for example, *Newcastle Courant* 1724, No. 213 (July 18), which listed 24 books, and among them: Robert Boulton, *Some Thoughts Concerning the Unusual Qualities of the Air: Containing Some Cautions Necessary to Prevent Malignant and Pestilential, or Contagious Distempers*; Philip Miller, *The Gardener's and Florist's Dictionary, or a Complete System of Horticulture*. These were both topics on which Triewald was to write numerous articles after his return to Sweden.

140. Triewald 1735, preface.

141. Heilbron 1979, 158. See also: Robin E. Rider, *The Show of Science* (Keepsakes issued by the Friends of the Bancroft Library, No. 31: Berkeley, 1983), 9–27.

142. Laurel, 17.—Triewald even called himself "Professor Honorarius in Edinburg" in a letter to the Society of Science in Uppsala in 1729 (Hildebrand 1939, 165).

143. *Newcastle Courant* 1724, No. 213 (July 18), 11.—I am indebted to Stephen Buckland for this reference.

144. I have been unable to establish the identity of John Thorold, or how he and Triewald came to be partners. I would, however, surmise that he may have been the John Thorold (1703–1775), who obtained his B.A. in Oxford in 1724. This was the eldest son of Sir John Thorold, 7th baronet, of Gainsborough, Lincolnshire. He succeeded his father as the 8th baronet in 1748. It seems probable that Thorold was in the Newcastle area in the late 1720s, and perhaps became acquainted with Triewald then, because in 1730 he married a young lady from West Herrington, Co. Durham, only ten miles from Newcastle. Thorold published a book on Bernard de Mandeville's *A Fable of the Bees* in 1726 and a number of books on religious subjects in the 1760s and 1770s. See: *Burke's Peerage and Baronetage* (London, 1978); *Burke's Genealogical and Heraldic History of the Landed Gentry* (London, 1937); *Alumni Oxoniensis 1715–1886*, Vol. 4 (Oxford, 1888), 1415; *British Museum General Catalogue*, Vol. 238 (London, 1964), 567 f.

145. *Caledonian Mercury* 1725, March 25, March 29, March 30, April 1, April 5, April 6, April 8. Cf. John R. R. Christie, "The Origins and Development of the Scottish Scientific Community, 1680–1760", *History of Science* 12 (1974), 122–141; Steven Shapin, "The Audience for Science in Eighteenth Century Edinburgh", ibid., 95–121.

146. This was the common practice and the established fee in the 1720s, see: Heilbron 1979, 163.

147. Op. cit.

148. Triewald 1735, 253 n.

149. Ibid., 228 n.

150. Lorentz Leopold von Horn, *Biografiska anteckningar*, Vol. 2 (Örebro, 1937), 485 f.; Adam Lewenhaupt, *Karl XII:s officerare: Biografiska anteckningar*, Vol. 2 (Stockholm, 1921), 712.

151. SSA, Stockholms Stads Verifikationsbok, 1726.

152. Loc. cit.

153. Triewald 1735, preface: "I hurried home in 1726 with my collected apparatus." On June 3, 1726, Triewald applied to the Board of Mines for a patent on the Newcomen engine, see: RA, BKA, Huvudarkivet, EIV:172, Brev och suppliker 1726:II, 332.

154. See n. 8.

155. See n. 9.

156. Hildebrand 1938, 153 f.; Lamm, 122.

157. UUB, MS G. 130, Alströmerska brevsamlingen, Letter from Henrik Kalmeter to Jonas Alströmer dated London September 24, 1723.

158. UUB, MS G. 130, Alströmerska brevsamlingen, Letter from Henrik Kalmeter to Jonas Alströmer dated London October 23, 1723.

159. KB, MS I.t.18., (Triewald, M.), Handlingar i rättegången ... 1726–1746. Original in Swedish: "iag är den förste och endaste af alla utlänningar, som fått i ängeland fritt tillträde till denna härliga inventionen".

160. *Newcastle Courant* 1724, No. 227 (October 24), 4: "They write from Stockholm, that since the Declaration which the King has given in Behalf of such Foreigners as shall come and settle in that Kingdom, His Majesty has publish'd another, particularly in favour of the Calvinists, wherein 'tis exprest, that all such of that Religion as shall have any Talents for Arts, Manufactures and Trade, and who shall come and settle in Sweden, shall, besides the Advantages already promised of the free Exercise of their Religion, and the Protection granted to Foreigners settled in that Kingdom, enjoy, during a certain Number of Years, the Exemption from all Offices and Taxes, and shall receive other Encouragements, every one in Proportion to his Capacity in his Profession."

11. THE SOCIAL CONTEXT: THE DANNEMORA MINES

1. Antoine Gabriel Jars, *Voyages métallurgiques, ou recherches et observations sur les mines & forges de fer*, Vol. 1 (Lyon, 1774), 120.
2. The only exhaustive history of the Dannemora Mines is: Johan Wahlund, *Dannemora grufvor: Historisk skildring* (Stockholm, 1879). It is, however, not a historical work in the modern sense, and it consists mainly of extracts from various documents. It is structured in a way that makes it difficult to use. See also: Johan O. Carlberg, *Historiskt sammandrag om svenska bergverkens uppkomst och utveckling samt grufvelagstiftningen* (Stockholm, 1879), 401–423. For an older geological description of the mines, including a short history and maps, see: Axel Erdman, "Dannemora jernmalmsfält i Upsala län, till dess geognostiska beskaffenhet skildradt", *KVAH* 1850, 1–138, Tab. I–XVI. For a bibliography of literature on the Dannemora Mines up to 1918, see: Carl Sahlin, "Dannemora-litteratur", *Teknisk Tidskrift: Kemi och bergsvetenskap* 48 (1918), 166 f.—For a short history in English, see: Gösta A. Eriksson, *The Iron Mines of Dannemora and Dannemora Iron* (Meddelanden från Uppsala universitets geografiska institution, No. 161: Uppsala, 1961).—The most recent works are: Per Olov Holmstrand, "Dannemora gruvor 500 år", *Jernkontorets Annaler* 165 (1981), No. 4,

28–30; Sven Rydberg, *Dannemora genom 500 år* (Fagersta, 1981).
3. It was mentioned in Agricola's *De veteribus et novis metallis*, see: Carl Sahlin, "Det svenska bergsbruket i Georg Agricolas skrifter", *Med Hammare och Fackla* 1 (1928), 103–109.
4. See, for example: Carl Sahlin, *Svenskt stål före de stora götstålsprocessernas införande* (Stockholm, 1931), 133–143; idem, "Det svenska järnets världsrykte, I: Från äldsta tider till omkring år 1850", *Daedalus* 1932, 47–57.—See also Le Play's discussion on Dannemora iron in: Frédéric Le Play, "Mémoire sur la fabrication de l'acier en Yorkshire, et comparasion des principaux groupes d'aciéries européennes, *Annales des mines*, quatrième série, Tome III, 1843 (Paris, 1843), 583–714, esp. 605 f. Le Play's article was translated into Swedish and published in: idem, "Om ståltillverkningen i Yorkshire, samt jemförelse mellan de förnämsta ståltillverkningssorter i Europa", *Tidskrift för Svenska Bergshandteringen* 3 (1845), 1–127.
5. There is a large literature on the ironworks, in contrast to the meagre literature on the Dannemora Mines. See, for example: Eli F. Heckscher, *Sveriges ekonomiska historia från Gustav Vasa*, Vol. 1:2 (Stockholm, 1936), 462–506.—For references to the history of the various ironworks, see under their respective names in the following bibliography of Swedish industrial history: Inga-Britta Sandqvist, *Litteratur om svenska industriföretag* (IVA-Meddelanden, No. 227: Stockholm, 1979).
6. SKCA, DGA, Vol. 1002, Protokoll 1726–1736. This volume will be referred to here as *Partners*. Cf. Wahlund, 4–14.
7. Bo Molander, "Ryska härjningar under Nordiska kriget knäckte svensk järnindustri", *Daedalus* 1978–1979, 51–62; Harry Molin, *Karlholms bruks bok: En krönika kring ett upplandsbruk* (Karlholm, 1950), 53–81; Heribert Seitz, "Den stora förödelsen år 1719 och striderna vid Södra Stäket", *Karolinska Förbundets Årsbok* 1960, 149–175; Sven Sjöberg, *Rysshärjningarna i Roslagen* (Stockholm, 1981); Wahlund, appendix 304–314.
8. Wahlund, appendix 304–314.
9. Marie Nisser, ed., *Industriminnen: En bok om industri- och teknikhistoriska bebyggelsemiljöer* (Stockholm, 1979), 94–98.
10. Quoted from Rydberg 1981, 20.
11. Wahlund, 57–144.
12. Ibid., 50. Cf. Erdman, 64–66.
13. Johan Axel Almquist, *Bergskollegium och bergslagsstaterna 1637–1857: Administrativa och biografiska anteckningar* (Meddelanden från Svenska

Riksarkivet, n. s. No. 2.3: Stockholm, 1909), 69, 149 f.

14. Ibid., 149, 168; Alfred Bernard Carlsson, ed., *Uppsala universitets matrikel*, Vol. 2 (Uppsala, 1919–1923), 40, 78; Bo V:son Lundqvist, ed., *Västgöta nation i Uppsala från år 1595*, Vol. 1 (Uppsala, 1928–1946) 145 f., 277, 367.

15. SKCA, DGA, Vol. 2350, Letter from Mine Bailiff Thomas Kröger to Mine Inspector Thor Bellander, dated December 13, 1729.

16. Sten Lindroth, *Svensk lärdomshistoria*, Vol. 2 (Stockholm, 1975), 340–342, 362–366.

17. Gustaf Elgenstierna, *Den introducerade svenska adelns ättartavlor*, Vol. 3 (Stockholm, 1927), 667 f.

18. Georg Wittrock, "Nils Bielke", *Svenskt Biografiskt Lexikon*, Vol. 4 (Stockholm, 1924), 241–257. See also: Oscar Malmström, *Nils Bielke och kriget mot turkarna, 1684–1687* (Stockholm, 1895); idem, *Nils Bielke såsom generalguvenör i Pommern, 1687–1697* (Stockholm, 1896); idem, *Högmålsprocessen mot Nils Bielke* (Stockholm, 1899); Per Sondén, *Nils Bielke och det svenska kavalleriet 1674–1679* (Stockholm, 1883); Georg Wittrock, "Förräderipunkten i Nils Bielkes process 1704–1705", *Karolinska Förbundets Årsbok* 1917, 40–80; idem, "Nils Bielkes underhandling i Brandenburg 1696", *Karolinska Förbundets Årsbok* 1918, 46–105.

19. Alf Åberg, "Gustaf Horn", *Svenskt Biografiskt Lexikon*, Vol. 19 (Stockholm, 1971–1973), 361–364.

20. Wittrock 1924, 243.

21. For a description of Salsta Manor, see: *Slott och herresäten i Sverige: Uppland*, Vol. 2 (Malmö, 1967), 84–100; *Svenska slott och herresäten vid 1900-talets början: Uppland* (Stockholm, 1909), 21–26.

22. *Partners*, passim.—Her letters, in the original or in copy, appear in most of the archives I have studied. Salsta archives, which are still kept at Salsta Manor (1984), contain much of her correspondence in connection with Wattholma Ironworks and the Dannemora Mines. I am indebted to Baron Gustaf von Essen for allowing me to examine the archives at Salsta Manor in 1975.

23. Sten Carlsson, *Ståndssamhälle och ståndspersoner 1700–1865: Studier rörande det svenska ståndssamhällets upplösning*, (Lund, 1973, rev. ed.); Jalmar Furuskog, *De värmländska järnbruken: Kulturgeografiska studier över den värmländska järnhanteringen under dess olika utvecklingsskeden* (Filipstad, 1924), 318–324; Eli F. Heckscher, *Svenskt arbete och liv* (Stockholm, 1942), 127 f., 169, 239 f.; Karl-Gustaf Hildebrand, *Fagerstabrukens historia: Sexton- och sjuttonhundratalen* (Fagerstabrukens historia,

Vol. 1: Uppsala, 1957), 165–182; idem, "Brukshistoria och brukskultur", *Blad för Bergshandteringens Vänner* 33 (1958), 90–105; Gösta Selling, 'Bruken som kulturmiljö", ibid., 106–128; Tom Söderberg, *Bergsmän och brukspatroner i svenskt samhällsliv* (Det levande förflutna, Svenska Historiska Föreningens Folkskrifter, No. 12: Stockholm, 1948).

24. Cf. the discussion of different interpretations of the meaning of "utility" in section 5, Chapter 15.

25. *Partners*, passim.

26. Almquist, 150, 229.

27. Wahlund, 50.

28. Ibid., 52.

29. Rydberg 1981, 10.

30. Wahlund, 52.

31. Ibid., 53 f.—This refers to conditions in 1656, but it is likely that they were similar in the 1720s. Cf. Bertil Boëthius, *Gruvornas, hyttornas och hamrarnas folk: Bergshanteringens arbetare från medeltiden till gustavianska tiden* (Den svenska arbetarklassens historia, Vol. 9: Stockholm, 1951), esp. 155–161, 177–181. Carl-Herman Tillhagen, *Järnet och människorna: Verklighet och vidskepelse* (Stockholm, 1981), 59–102; Emil Sommarin, *Bidrag till kännedom om arbetareförhållanden vid svenska bergverk och bruk i äldre tid till omkring år 1720* (Lund, 1908), 106.

32. Carl Heijkenskjöld, "Krigsfångar som arbetare i Sala gruva och annorstädes", *Med Hammare och Fackla* 8 (1937), 121–130.

33. Ibid., 124.

34. Ibid., 127; Wahlund 54 f.

35. Loc. cit. (quoted from Wahlund).

36. Per Hallman, "Tillståndet i Uppland under det stora nordiska kriget", *Karolinska Förbundets Årsbok* 1919, 264 f.

37. *Partners*, August 17, 1726.

38. Loc. cit. Original in Swedish: "Wist en stuzighet och motspeenighet".

39. Ibid., August 18, 1726.

40. Loc. cit. Original in Swedish: "hafwa de doch icke längre derwid *continuerat* än till klåckan half åtta, då de sig samsatte, gjorde sig studsige, och med owettige *expressioner*, gingo utur lafwarne och öfwergåfwo arbetet".

41. Loc. cit. Original in Swedish: "at å ämbetes wägnar tillhålla dem at fortfara med arbetet".

42. Loc. cit. Original in Swedish: "hwarpå de swarade, att de ej mächtade med så swårt arbete, utan wille wähl arbeta, men ingaledes annorlunda än på förra sättet, och med de föra Bälljorna, och som de hwarken med goda ord eller hårda tillsägelser stodo at bringas till arbete

med de nya Bälljorna, utan gjorde sig lika mot-spännige med otidige ord som tillförende, så nödgades Krono Gruffogden falla på det förslag at tillåta dem bruka de gambla Bäljorne, men doch med det wilkor, at [...] skulle de ej blifwa fler än 8 personer i hwar wind".

43. Loc. cit. Original in Swedish: "fåfängt förlopp".

44. Loc. cit. Original in Swedish: "fant man nödigt at låta förwara desse anklagade öfwer natten i kistan, efttersom de elliest torde sig undanhålla".

45. Boëthius 1951, 361 f. Lars Levander, *Brottsling och bödel* (1933), facsimile edition, (Stockholm, 1975), 154–156; Tillhagen, 80, 100.—The use of the coffin (*"kistan"*) in Swedish mines is documented as early as in 1354, in the second of the mining statutes issued by the state. See: *Bergsordningar* 1736, 9 and index (*"kista"*). Cf. Sigfrid Wieselgren, *Sveriges fängelser och fångvård från äldre tider till våra dagar* (Stockholm, 1895), 6.

46. Op. cit. Original in Swedish: "det owäsende och den studzighet".

47. Loc. cit. Original in Swedish: "at han intet wet af något annat owäsende, än at han med des medarbetare måste lemna arbetet med nya Bülljorna efter som de warit för swåra, men elliest nekade han enständigt, at han skall upstudzat någon eller sielf wist någon motspän-nighet, och ehuru man sökte at få honom at tillstå sitt brott så war dock alt förgiäfwes".

48. Loc. cit. Original in Swedish: "men de påstodo enständigt at alla skohla warit ense och enhäl-lige, ock alla med en min begynt att säga, det de intet stodo ut med arbetet".

49. Loc. cit. Original in Swedish: "hwilket Pirgo icke neka kunde, men wille icke säga om någon bedit honom det giöra, eller till hwad ända han det gjorde utan förmelte at det skall skiedt för ro skull, och skall han icke heller sedan sagt något der om för någon annan".

50. Loc. cit. Original in Swedish: "detta för hela Bergslagen ganska angelägne arbetet icke med det efttertryck kunnat drifwas som weder-bordt".

51. Loc. cit. Original in Swedish: "sig sielfwa till straff och androm till warnagel".—For the Royal Mining Statute of July 6, 1649, see: *Bergsordningar*. 1736, 154–172. Cf. Boëthius 1951, 372 f.

52. Loc. cit. Original in Swedish: "första orsaken".

53. Loc. cit. Original in Swedish: "och med det straff blifwa ansedd, som det brått förtienar".

54. Loc. cit. Original in Swedish: "allenast bedit at arbetet skulle få blifwa efter förra wanlighe-ten".

55. Ibid., Appendix to the Minutes of the Partners August 17–19, 1726. Original in Swedish: "bedit gråtande".

56. Ibid., Minutes of the Partners August 19, 1726. Original in Swedish: "blef straxt *exequerat*".

57. Cf. Michel Foucault, *Discipline and Punish: The Birth of the Prison* (Harmondsworth, 1979), 24 f.; David S. Landes, *The Unbound Prometheus: Technological Change and Industrial Development in Western Europe from 1750 to the Present* (Cambridge, 1969), 96 f.

58. SKCA, DGA, Vol. 2350, Letter from Mine Bailiff Thomas Kröger to Mine Inspector Thor Bellander dated July 9, 1726. Original in Swedish: "Ehuru gierna iag söckt, att willja effter-lefwa högachtade Hr. Bergmästaren ..."

59. Wahlund, 22–43, appendix passim. See also: Jonas Beronius, "Utdrag af de vid Dannemora grufvor befintliga handlingar, relationer och rä-kenskaper, rörande arbeten, som tid efter annan blifvit verkstäldte, till vattnets afhållande från grufvorna, och om deras tömmande efter den år 1795 timade öfversvämningen", *Jern-Kontorets Annaler* 2 (1818), 77–98; Gustaf af Uhr, *Betän-kande till herrar fullmäktige i Jern-Contoret, innehål-lande en öfversigt af bergs-mechanikens tillstånd* (Stockholm, 1817), 78–89. Cf. Lennart Liljen-dahl, "Upplands bruksdammar", *Uppland* 1979, 41–57; Carl Sahlin, "Länshållningen i Dannemora grufvor i äldre tider", *Teknisk Tid-skrift: Kemi och bergsvetenskap* 48 (1918), 156–158.

60. Beronius, 77 f.; af Uhr 1817, 78 f.

61. Sten Lindroth, *Christopher Polhem och Stora Kop-parberget: Ett bidrag till bergsmekanikens historia* (Uppsala, 1951), esp. 13, 22–24, 30–34; idem, *Gruvbrytning och kopparhantering vid Stora Koppar-berget intill 1800-talets början*, Vol. 1 (Uppsala, 1955), esp. 323 f.

62. Holmstrand, 29.

63. Wahlund, 43–46, appendix 267 f.

64. Ibid., 85.

12. THE TECHNOLOGY: THE TRIEWALD ENGINE

An earlier version of this chapter was read at the Annual Meeting of the Midlands Branch of the Newcomen Society in Birmingham on October 7, 1981. I am indebted to John S. Allen, Stephen Buckland and Jennifer Tann for their comments on that version. A revised version was read at the meeting of the Newcomen Society in London on

January 11, 1984, and is to be published in the *Transactions of the Newcomen Society* 55 (1983–84).

1. Patrick M. Malone, "Teaching the History of Technology in Museums", *Humanities Perspectives on Technology* (Curriculum Newsletter of the Lehigh HPT Program), 1978, No. 9, 3–6.
2. The drawings are kept in the Central Archives of Stora Kopparbergs Bergslags AB, Falun. I am indebted to Rune Ferling, Tommy Forss and Sven Rydberg for the loan of these drawings.
3. For the second project in Dannemora, see: Torsten Althin, "Sveriges andra ångmaskin", *Daedalus* 1939, 53; Bertil Boëthius & Åke Kromnow, *Jernkontorets historia*, Vol. 1 (Stockholm, 1947), 487–490; Johan Wahlund, *Dannemora grufvor: Historisk skildring* (Stockholm, 1879), 36f., appendix 137–149.
4. Torsten Althin, "Eric Geisler och hans utländska resa 1772–1773", *Med Hammare och Fackla* 26 (1971), 58. For Geisler as a minesurveyor, see: Sten Lindroth, *Gruvbrytning och kopparhantering vid Stora Kopparberget intill 1800-talets början*, Vol. 1 (Uppsala, 1955), 690ff.
5. I did this study in 1975 as a paper in Torsten Althin's course in the history of technology at the Royal Institute of Technology, Stockholm.
6. Mårten Triewald, *Kort beskrifning, om eld- och luft-machin wid Dannemora grufwor* (Stockholm, 1734).
7. Cf. James H. Andrew, "The Copying of Engineering Drawings and Documents", *Transactions of the Newcomen Society* 53 (1981–82), 1–15.
8. This was a common method at the time, see: Peter J. Booker, *A History of Engineering Drawings* (London, 1963), 37–47. Cf. Alois Nedoluha, "Kulturgeschichte des technischen Zeichnens", *Blätter für Technikgeschichte* 19 (1957), part 1 (pp. 1–51) as appendix; 20 (1958), part 2 (pp. 53–108) as appendix; 21 (1959), part 3 (pp. 109–183) as appendix.
9. Lindroth 1955, 659–699, esp. 664. Cf. Ulla Ehrensvärd, "Gruvor på kartor", in: Nicolai Herlofson et al., eds., *Vilja och kunnande: Teknikhistoriska uppsatser tillägnade Torsten Althin på hans åttioårsdag den 11 juli 1977 av vänner* (Uppsala, 1977), 171–188.
10. Cf. Eugene S. Ferguson, "The Mind's Eye: Nonverbal Thought in Engineering", *Science* 197 (1977), 827–836.
11. Sam Owen Jansson, *Måttordbok: Svenska måttstermer före metersystemet* (Stockholm, 1950), 9.
12. This odd figure is probably due to the fact that the paper has shrunk. The shrinkage of eigh-

teenth-century paper varies between 2 and 4%. I am indebted to Edo G. Loeber and Elisabet Penn for this information.

13. The terms for the principal parts of the engine are, on the first series of drawings, the following: "*Maschin*" for the engine, "*Zelynder*" for the cylinder, "*Piston*" for the piston, "*Panna*" for the boiler, and "*Regulator*" for the automatic valve mechanism.
14. For Swedish literature on filigranology, see: Gösta Liljedahl, "Om vattenmärken och filigranologi", *Historisk Tidskrift* 76 (1956), 241–274; idem, "Om vattenmärken i papper och vattenmärkesforskning (filigranologi)", *Biblis* 1970, 91–129. See also: Gunilla Dahlberg, "Datering av skrifter med hjälp av vattenmärken", *Samlaren* 95 (1974), 81–111.
15. By fitting together the pieces of paper with the help of fragmentary watermarks and the uneven edges, it is possible to reconstruct the original sheets, the size of which was approx. 30×40 cm. The watermarks show that it is Dutch paper (see n. 16).
16. There are two watermarks in the paper of the drawings, the "*Dutch Lion*" and "*VRYHEYT + Propatria Eiusque Libertate*". These were two of the most common watermarks in Holland during the eighteenth century. The first is recorded as having been used during the period ca. 1650–1735, and the second ca. 1718–1833. Since both watermarks appear in the paper in the first series of drawings, it is reasonable to assume that the drawings were made in the period ca. 1718–1735. There are also several smaller watermarks with initials: "*IV*" and "*TVH*" in the first series, and "$H_L^B M$" in the second series.

The first two provide no information since they were commonly imitated. The third, however, is recorded in the collection of Adam Lewenhaupt (p. 341) in the Swedish National Record Office, and dated ca. 1725. (The pattern is similar, but the size is different.) I am indebted to the Dutch paper historian Edo G. Loeber for his generous help in examining copies of the watermarks, and the information above has kindly been provided by Mr. Loeber. Cf. Henk Voorn, *De papiermolens in de provincie Noord-Holland* (De geschiedenis der Nederlandse papierindustrie, No. 1: Haarlem, 1960).—It was not possible to use the ^{14}C method to date the drawings. It would have demanded far too much paper (in fact, burning the entire set of drawings to ashes), and the concentration of ^{14}C in the atmosphere has varied considerably over the last 400 years, which makes datings in this period inconclusive. It might, however, be possible to date them more accurately once accelerator datings can be obtained. I am indebted to Ingrid U. Olsson for this information.

17. SKCA, DGA, Vol. 1002, Protokoll 1726–1736, Minutes of the Partners on August 17, 1726.
18. Loc. cit. Original in Swedish: "wid wästra grufwebädden af norra Silfberg Grufwan".
19. Loc. cit. Original in Swedish: "icke allenast der updraga alt wattnet utan och med samma *Machine* på ett lätt och behändigt sätt all malmen upfordra".
20. Ibid., Minutes of the Partners, October 24, 1726.
21. Ibid., Minutes of the Partners, August 17, 1726. The passage reads in Swedish: "och hwad Malm upfordringen angår, så will han derwid bruka den *invention* at och samma *Machine* skall berörde upfordring i alla desse Silfberg Grufwor tillika förrätta, hwartill allenast twänne Malm bälljor i hwart schacht skola giöra till fyllest ock dertill antingen kedior eller garw linor".
22. Loc. cit.
23. Cf. Figs. 6.3 (Beighton), 7.3 (Barney), 15.1 (Triewald) and 15.2 (Nicholls).
24. See Chapter 14.
25. Wahlund, appendix 148f. Cf. Boëthius & Kromnow 1947, Vol. 1, 490.
26. TMA, Vol. 39:b, Dannemora gruvor, "Iakttagelser i eld- och luftmaskinshuset under arbetets gång" (Notes by Torsten Althin dated November 15, 1930, during the investigation and excavation in the engine house in Dannemora).
27. Torsten Althin, "Memorial to Martin Triewald", *Transactions of the Newcomen Society* 11 (1930–31), 163, Plate 22; Carl-Th. Thäberg, "Föreningen Tekniska Museet under år 1932", *Daedalus* 1933, 9, 49. Cf. Marie Nisser, ed., *Industriminnen: En bok om industri- och teknikhistoriska bebyggelsemiljöer* (Stockholm, 1979, 24f., 60–63).
28. Torsten Althin, "Tekniska Museet under år 1931", *Daedalus* 1932, 20, 27.
29. See n. 26.
30. SKCA, DGA, Vol. 2388, "Pro Memoria" (giving the chronology of the whole project. It is neither signed nor dated, but obviously by the Mine Bailiff and part of the case against Triewald).—It is confirmed by several other sources that the house was built in May 1727.
31. Althin 1939, 49–66; Boëthius & Kromnow 1947, Vol. 1, 487–489.
32. John Kanefsky & John Robey, "Steam Engines in 18th-Century Britain: A Quantitative Assess-

ment", *Technology and Culture* 21 (1980), 169 (Table 2).
33. Cf. Lionel T. C. Rolt & John S. Allen, *The Steam Engine of Thomas Newcomen* (Hartington, 1977), 89–106.
34. Samuel Lilley, "Technological Progress and the Industrial Revolution 1700–1914", in: Carlo M. Cipolla, ed., *The Fontana Economic History of Europe: The Industrial Revolution* (The Fontana Economic History of Europe, Vol. 3: Huntington, 1973), 204.
35. Carlo M. Cipolla, "Introduction", ibid., 11.
36. Lilley, 204.
37. John Farey, *A Treatise on the Steam Engine: Historical, Practical, and Descriptive* (London, 1827), 296–306. Cf. Henry W. Dickinson, *A Short History of the Steam Engine* (Cambridge, 1939), 62–65; Richard L. Hills, *Power in the Industrial Revolution* (Manchester, 1970), 134–164; Terry S. Reynolds, *Stronger Than a Hundred Men: A History of the Vertical Water Wheel* (Baltimore, 1983), 321–325.
38. Farey, 296.
39. Ibid., 406–422; Robert Stuart, *Historical and Descriptive Anecdotes of the Steam Engine* (London, 1829), Vol. 1, 279f., Vol. 2, 329–340, 624–626. Cf. A. E. Musson & Eric Robinson, *Science and Technology in the Industrial Revolution* (Manchester, 1969), 398f.; George M. Watkins, "The Development of the Steam Winding Engine", *Transactions of the Newcomen Society* 50 (1978–1979), 11–24.
40. Cf. Hills 1970, 141f.
41. Farey, 407f. See, for example, the description and drawing of Oxley's patent: (Joseph Oxley), *Oxley's patent A.D. 1763 No. 795: Machinery for Drawing Coals out of Pits, &c.* (London, 1855). I am indebted to George M. Watkins for this reference.

13. AGREEMENT, CONSTRUCTION AND OPERATION, 1726–1730

1. See "The Return to Sweden: Reasons and Timing", Chapter 10.
2. RA, BKA, Huvudarkivet, EIV:172, Brev och suppliker 1726:II, 332.
3. RA, Bergskollegium till Kungl. Maj:t, June 21, 1726.
4. RA, BKA, Huvudarkivet, EIV:172, Brev och suppliker 1726:II, 429.

5. Mårten Triewald, *Föreläsningar öfwer nya natur-kunnigheten*, Vol. 1 (Stockholm, 1735), 166 n.

6. SKCA, DGA, Vol. 1002, Protokoll 1726–1736, Minutes of the Partners, August 17, 1726. Original in Swedish: "en *Mechanicus Mårten Triewald* wid namn i från ängeland hit i landet ankommit". This volume will be referred to here as *Partners*.—Triewald had visited Dannemora and met Mine Bailiff Thomas Kröger for the first time at the end of July, 1726. He had then continued to Gävle to call on Mine Inspector Thor Bellander. See: SKCA, DGA, Vol. 2350, Letters from Mine Bailiff Thomas Kröger to Mine Inspector Thor Bellander 1724–1734, July 30, 1726. This volume will be referred to here as *Kröger*.

7. *Partners*, August 17, 1726. Original in Swedish: "redan 33 slika *Machiner* uti Ängeland äro i deras fulla gång, och wisa sin stora wärkan".

8. The equivalent of 24 000 *copperdaler* in English currency was £618 (see n. 67, Chapter 9). This was below the average range for early Newcomen engines in England (see Chapter 6).

9. One *stafrum* was traditionally a cubic stack of wood fuel with sides of 3 Swedish ells (*alnar*), i.e. approx. 6 m³. But there were numerous local variations and changes over time. See: Sam Owen Jansson, *Måttordbok: Svenska måttstermer före metersystemet* (Stockholm, 1950), 82 f.; Sten Lindroth, *Gruvbrytning och kopparhantering vid Stora Kopparberget intill 1800-talets början*, Vol. 1 (Uppsala, 1955), 266 f., 521; Sven Rinman, *Bergwerks lexicon*, Vol. 2 (Stockholm, 1789), 794 f.

10. *Partners*, August 17, 1726. Original in Swedish: "Som om *Machinen* och elliest *Mechaniske* wettenskaper kan äga någon erfarenhet".

11. Loc. cit. Original in Swedish: "kunde giöra arbetet uti de beswärlige Silfbergs Grufworne lättare, och förorsaka någon god besparing uti den anseenlige omkostnad, som åhrligen derpå har måst anwändas, och härtills lärer bestigit till inemot 9000 dahler om åhret, och det allenast till wattnets afhållande, oberäcknadt den omkostnad som till malmens upkörande åtgått".

12. Johan Wahlund, *Dannemora grufvor: Historisk skildring* (Stockholm, 1879), appendix 107.

13. There are copies of the agreement in several of the archives; for example in: *Partners*, October 24, 1726. See also: KB, MS I.t. 18, (Mårten Triewald), Handlingar i rättegången ... 1726–1746. This is a collection in the Royal Library, Stockholm, of the documents in the lawsuit between Triewald and the Partners. It will be referred to here as *Documents*. The agreement is *Documents*, No. 2.

14. *Partners*, October 24, 1726. Original in Swedish: "till nästkommande Junii månad kunna hafwa *Machinen* så färdig, att iag då både med wattu och malmupfordringen kan giöra begynnelse".

15. SKCA, DGA, Vol. 2388, Om Triewalds eld- och luftmaskin 1726–1731, undated "Pro Memoria" (3 pp.). This is a chronology, presumably drawn up for the Partners when they were preparing their case against Triewald in 1731. The dates are given in various other sources as well, but since they coincide this one alone will be used and will be referred to here as *Chronology*.

16. *Chronology*.

17. Alfred Bernard Carlsson, ed., *Uppsala universitets matrikel*, Vol. 2 (Uppsala, 1919–1923), 158, 208; Julius Lagerholm, ed., *Södermanland-Närkes nation: Biografiska och genealogiska anteckningar om i Uppsala studerande södermanlänningar och närkingar 1595–1900* (Sunnansjö, 1933), 168.

18. Per Hebbe, "Anders Gabriel Duhres 'Laboratorium mathematico-oeconomicum'", *Kungl. Lantbruksakademiens Handlingar och Tidskrift* 72 (1933), 585 f., 590.

19. *Documents*, No. 4, appendix K, Letter from Mårten Triewald to Olof Hultberg, dated December 20, 1729.

20. *Chronology*. (The details were given by Triewald in his pleading at the trial in 1731, see Chapter 14.)

21. Loc. cit.

22. Loc. cit.

23. Triewald probably travelled by the regular postillion service. The horses of the post chaise were changed every other Swedish mile (10.7 km.), and a traveller could cover a distance of 75–100 kilometres in a day. It would have been possible to make the journey Stockholm–Dannemora in one day, but it is more likely that it took two days. The journey Stockholm–Wedevåg or Wedevåg–Dannemora took at least two, and probably three days. See: Göran Andolf, "Resandets demokratisering: Skjutsväsendet" (mimeographed paper at the Department of History, University of Gothenburg, 1977); idem, "Resandets revolutioner", *Fataburen* 1978, 49–72; Carl Fredrik Ström, *Karta öfver landsvägarna uti Sverige och Norrige* (Stockholm, 1846).—I am indebted to Göran Andolf for this information and the references.

24. Bengt Hildebrand, *Kungl. Svenska Vetenskapsakademien: Förhistoria, grundläggning och första or-*

ganisation (Stockholm, 1939), 148–151. Cf. idem, "Till släkten Triewalds historia", *Personhistorisk Tidskrift* 39 (1938), 154–156 n. 7.

25. *Chronology*; LAU, BMA, FI:5, Dannemora bergstingshandlingar och protokoll 1731–1733, Minutes of the Court of Mines on June 28, 1727, § 37.

26. *Chronology*.

27. Loc. cit. Original in Swedish: "till sin utwärtes skapnad war helt ferdigt, då och kittelen war äfwenähls inmurad".

28. Hildebrand 1939, 149–151.—Rumours of Triewald's appointment reached Thomas Kröger in the middle of July, but Triewald denied it as idle talk when questioned by him (*Kröger*, July 22, 1727). Earlier, Kröger had sometimes referred to Triewald as "the Professor", a title which Triewald probably had bestowed upon himself on the strength of his series of public lectures in the university city of Edinburgh. After his appointment had been confirmed he could, however, justly claim the right to a high-sounding official Swedish title, and thereafter Kröger always referred to him as "the *Directeur*" (*Kröger*, passim).

29. Quoted from Hildebrand 1939, 151. Original in Swedish: "i behörigt stånd, så att publicum deraf kan hafua den tillförmodande nytta".

30. *Chronology*.

31. Loc. cit.

32. Hildebrand 1939, 151; Sten Lindroth, *Kungl. Svenska Vetenskapsakademiens historia 1739–1818*, Vol. 1 (Stockholm, 1967), 78 f.

33. SSA, Stockholms stads verifikationsbok 1727:1, 507.—The bill of lading in: SKCA, DGA, Vol. 2388, Om Triewalds eld- och luftmaskin 1726–1731.

34. *Stockholmske Post Tidningar* 1727, No. 40, October 2.

35. *Chronology*.

36. *Partners*, December 12–13, 1727, appendix, Letter from Mårten Triewald to Thor Bellander, dated November 20, 1727.

37. For Meijer's biography, see: Hildebrand 1939, 585–587; Edvard F. Runeberg, *Åminnelse-tal öfver ... Gerhard Meijer* (Stockholm, 1798). Cf. *Biographiskt lexicon öfver namnkunnige svenska män*, Vol. 9 (Upsala, 1843), 92–98.—For the Royal Gun Foundry, see: Nils Lundequist, *Stockholms stads historia, från stadens anläggning till närwarande tid*, Vol. 3 (Stockholm, 1829), 304–307; Ludvig Hammarskiöld, "Kopparkanoner i Sverige och deras tillverkning", *Med Hammare och Fackla* 18 (1949–1950), 25–64.—Meijer is well known for his horizontal boring machine,

constructed in 1763, see: Sten Carlqvist, "Kanonborrningsmaskin från 1700-talet", *Daedalus* 1934, 116–118; Gerhard Meijer, "Beskrifning och ritning på en mindre bårrmachine til massift gutne canoner", *KVAH* 1782, 276–284, Tab. IX; Sven Rinman, *Afhandling rörande mechaniquen, med tillämpning i synnerhet til bruk och bergwerk*, Vol. 2 (Stockholm, 1794), 505–574, Tab. XLVIII–LIII.—A model of Meijer's boring machine is kept in the Army Museum, Stockholm (No. F 681).—For the boring of the Dannemora cylinder, see n. 48.

38. See n. 36. Original in Swedish: "P.S. Jag förglömde andraga det *Metall* blir *Metall* och efter 100:e Åhr bruk af samma wärde". Cf. n. 54, Chapter 7.

39. *Partners*, December 12–13, 1727.

40. *Stockholmske Post Tidningar* 1727, No. 51, December 18.

41. Ibid. 1728, No. 4, January 22. Cf. Hildebrand 1939, 151–158; Sten Lindroth, *Svensk lärdomshistoria*, Vol. 3 (Stockholm, 1978), 338–346. See also: Anna Beckman, "Två svenska experimentalfysiker på 1700-talet: Mårten Triewald och Nils Wallerius", *Lychnos* 1967–1968, 186–201, 214 (English summary); Stig Nilsson, "Materieuppfattningens termer i svenskan fram till år 1800", *Lychnos* 1975–1976, 129–136, 115 f. (English summary).—Beckman passes a verdict on Triewald's originality and scientific achievements from the standpoint of present-day criteria and knowledge.

42. Nils Viktor Emanuel Nordenmark, *Anders Celsius: Professor i Uppsala, 1701–1744* (Lychnos-Bibliotek, No. 1: Uppsala, 1936), 11 f.

43. Loc. cit. Cf. Daniel Menlös, *Kort beskrifning af den hydrostatiske wåg-balken* (Stockholm, 1728), preface. Menlös writes that *Studiosus Mechanices* Olof Hultberg had made the hydrostatic balances.

44. Hildebrand 1939, 157 f.

45. Loc. cit.

46. Mårten Triewald, *Föreläsningar öfwer nya naturkunnigheten*, 2 Vols. (Stockholm, 1735–36), list of subscribers. Cf. Kurt Samuelsson, *De stora köpmanshusen i Stockholm 1730–1815: En studie i den svenska handelskapitalismens historia* (Skrifter utgivna av Ekonomisk-historiska institutet i Stockholm: Stockholm, 1951).

47. *Chronology*.

48. Loc. cit. for the dates.—That the cylinder had been bored is reported in a letter from Mine Bailiff Thomas Kröger to Mine Inspector

Thor Bellander, see: *Kröger*, March 23, 1728. Kröger writes that Triewald has told him in a letter that Meijer has been boring the cylinder "for fourteen days with four horses and men from morning until night, but has not passed more than one borer through. The other two or three borers will then pass quickly. It is very well cast". Original in Swedish: "på 14 dagars tid, med 4 hästar och folck, ifrån morgonen till qwällen, ej fåt mera än en Borr igenom Cylindren, dhe öfrige 2:ne eller 3:ne Borrar går sedan fordt, men träffelig wähl ähr den guten".—For the manufacturing of cylinders for the early Newcomen engines, see: Arthur Raistrick, *Dynasty of Ironfounders: The Darbys and Coalbrookdale* (Newton Abbot, 1970), 128–138.

49. Hildebrand 1939, 151.—For Menlös' background, see: *Biographiskt lexicon öfver namnkunnige svenska män*, Vol. 9 (Uppsala, 1843), 78–87.
50. *Documents*, No. 4.
51. Hildebrand 1939, 167; Johan G. Tandberg, "Die Triewaldsche Sammlung am Physikal. Institut der Universität zu Lund und die Original-Luftpumpe Guerickes", *Lunds Universitets Årsskrift*, n. s. avd. 2 bd. 16. No. 9 (Lund, 1920), 30 pp. + 1 p. bibliography; idem, "Historiska instrument i Lund" *Kosmos* 1922 (Fysiska uppsatser utgivna av Svenska fysikersamfundet: Stockholm, 1922), 194–211. Cf. Arvid Leide, *Fysiska institutionen vid Lunds universitet* (Acta Universitatis Lundensis, Sectio I, Theologica Juridica Humaniora 8: Lund, 1968), 26–60.—Tandberg 1920 contains a list of Triewald's collection of instruments. No. 68 was a model of the Dannemora engine, but it is not among the items that have been preserved.
52. *Kröger*, June 17, 1728. Original in Swedish: "Och hwad Herr *Direkteur Triewalds Machins* wärk angår, så förfares dermed dageligen, och förmenas det blifwa färdigt förutom uppfordring wärcket, till nästkommande Bärgting, han har hafft något beswär denne wecka med sistern eller wattukistan, effter dhen intet waret så täter giord som han åstundat, men änteligen dhen swårigheten öfwerwunnet, så att sistern och kittelen, nu står fulla med watten, har och under giordt begynnelse medh stora pumpars ihopafogande, men måste först göra en hand pump, hwar med han will utpumpa till en dhell watnet uhr grufwan, innan han får tilfälla at neder sättia dhe stoora pumparne".
53. *Chronology*.—The date is confirmed by several independent sources. See, for example: KB, MS D 850, Brev till fru Eva Juliana Insenstierna på Harg, innehållande politiska ny-

heter från Stockholm och utlandet, emellan d. 9 juni 1727 och d. 7 nov. 1729, p. 102. Cf. Otto Sylwan, *Svenska pressens historia till statshvälfningen 1772* (Lund, 1896), 117.—I am indebted to Gunnar Broberg for these references.
54. Anders Celsius, ed., *Almanach på skått-året ifrån Jesu Christi födelse 1728. Efter den gamla och nya stylen uträknad til Stockholms längd* (Stockholm, n.d.). Copy in KB.
55. *Chronology*. Original in Swedish (the complete passage): "d. 4 Juli, släptes Machin första gången till gång af Hr. Directeuren. 6 dagar före Bergtinget".
56. SKCA, DGA, Vol. 2388, Om Triewalds eld- och luftmaskin 1726–1731. Original in Swedish: "i afräkning på den utlofwade discretion".
57. Kröger does not seem to have accepted Hultberg's title, since he usually referred to him by his surname or as *Monsieur*; a title sometimes used ironically for someone claiming to be a gentleman (*Kröger*, passim).
58. *Kröger*, August 10, 17 & 24, 1728.
59. Ibid., August 17, 1728. Original in Swedish: "ingen anstalt är ännu giordt till upfordringwärckets byggande, och som synees lärer der medh i åhr hafwas anståndh".
60. Ibid., August 24, 1728.
61. Loc. cit. Original in Swedish: "hwad *Machinen* angår, så är den ännu icke fulkomblig, och Gudh wet när den kommer i rätt tilstånd, ty altid är något som brister och fehlar, och så förwetter det äfwen pumparne, dhen har redan upödt 57 stafrum Wedh".
62. Ibid., August 17, 1728. Original in Swedish: "et treffeligit minage för Bärgslagen".
63. Kröger might, however, have referred to some local variation of *stafrum* as a unit of measure for wood fuel; smaller than the general *stafrum*. It is likely that Triewald and Kröger chose the unit of measure that supported their respective point of view. Cf. n. 9.
64. See Chapter 14.
65. Johan Hinric Lidén, ed., *Brefwäxling imellan ärke-biskop Eric Benzelius den yngre och dess broder, censor librorum Gustaf Benzelstierna* (Linköping, 1791), 23 f.
66. Loc. cit. Cf. Hildebrand 1939, 145 f.—Mine Bailiff Kröger reported the visit of the Austrian Ambassador Count Freytag to Mine Inspector Bellander, and added that strangers arrived daily at Dannemora to see the engine. See: *Kröger*, August 10 & 17, 1728.
67. *Acta literaria Sveciae* 2 (1725–1729), 3rd. quarter 1728, 453. Original in Latin: "machina tum conformationis elegantia, tum successu admi-

rabilis".—For the reaction of the Society to Kalmeter's report in 1720, see "The Society of Science in Uppsala" in Chapter 7.

68. KB, MS M 25:2, Daniel Tilas, "Sokne-skrifvare, eller svenska resesamlingar", 165. Original in Swedish: "Cylinder till den första Eld-Machin i Swerige bygd af Mart. Triewald: guten af ger. Meyer i Stockholm A° MDCCXXVIII".

69. *Kröger*, September 28, 1728. Original in Swedish: "någorlunda hållet".

70. Loc. cit. Original in Swedish: "mycket ostadig i sin gång, för altid går något sönder, som straxt måste lagas, och har den ännu aldrig gådt 3 hela dygn i sänder".

71. Loc. cit. Original in Swedish: "iag fruktar, när wintern kommer, fryser hela wärket tilsammans".

72. *Chronology*.

73. *Stockholmske Post Tidningar* 1728, No. 37, September 9; Mårten Triewald, *Notification, om trettijo publique föreläsningar, öfwer nya naturkunnigheten, illustrerad genom mechaniske, hydrostatiske, aërometriske och optiske experimenter och försök, som på Riddare-huset i Stockholm komma at taga sin början den 15:de Octobris innewarande åhr 1728* (Stockholm, 1728). Copy in KB. The preface is dated Dannemora Mines, September 2, 1728.

74. *Documents*, No. 4, appendix F, Letter from Mårten Triewald to the Board of Mines, dated October 2, 1728; ibid., appendix G, Letter from Thor Bellander to Mårten Triewald, dated September 9, 1728; *Kröger*, November 9, December 7, 1728.

75. UUB, MS Okat. Royal Society (418g:2), Letter from Mårten Triewald to John T. Desaguliers, dated November 20, 1728 (photocopy of letter in RS).

76. Mårten Triewald, "Queries, concerning the cause of cohesion of the parts of matter, proposed in a letter to Dr. Desaguliers, F.R.S. By Fr. [sic] Triewald, Director of Mechanicks in the Kingdom of Sweden", *Philosophical Transactions* 36 (1729–30), 39–43. Cf. n. 97, Chapter 10.

77. Mårten Triewald, *Nödig tractat om bij, deras natur, egenskaper, skiötzel och nytta, utur åtskillige nyares wittra anmärkningar samt egne anstälte rön och försök sammanfattad, jemte beskrifning på en ny inrättning av bijstockar* (Stockholm, 1728). Cf. Albert Sandklef, "Äldre biskötsel i Sverige och Danmark: Bidrag till kännedomen om sydskandinavisk biskötsel före mitten av 1800-talet", *Göteborgs Kungl. Vetenskaps- och Vitterhets-*Samhälles Handlingar, 5:e följd., ser. A, bd. 6, No. 3 (Göteborg, 1937), 86–94, 118.—The British Library's copy (call no. 958. a 19) has once belonged to Joseph Banks.

78. *Stockholmske Post Tidningar* 1728, No. 36, September 2.

79. Sixten Rönnow, *Wedevågs bruks historia* (Stockholm, 1944), 116–154, esp. 141–144.

80. Ibid., 141 f.

81. *Chronology*; *Kröger*, January 25, 1729.

82. *Chronology*; *Kröger*, March 15, 1729.

83. UUB, MS G 130, Alströmerska brevsamlingen, Letter from Henrik Kalmeter to Jonas Alströmer, dated March 18, 1729.—They corresponded in English.

84. *Stockholmske Post Tidningar* 1729, No. 11, March 17. Cf. Hildebrand 1939, 152 (reproduction of the advertisement).

85. Loc. cit. Original in Swedish: "i brist af klara dagar".

86. *Chronology*.

87. Loc. cit. Original in Swedish: "reste Hr. Directeuren bort til sit Brölop".

88. Hildebrand 1938, 158.

89. Hildebrand 1939, 534.

90. RA, Biografica, Vol. T 19, Mårten Triewald. Cf. Hildebrand 1938, 157; Mårten Triewald, *Konsten at lefwa under watn eller en kort beskrifning om de påfunder, machiner och redskap hwarpå dykeri- och bärgnings-societetens privilegier äro grundade, hwarmed de anstält profwen under twenne riks-dagar för Sweriges rikes högl. ständers herrar deputerade* (Stockholm, 1734 b), 52–62.—For Triewald and the history of diving, see n. 74, Chapter 10.

91. Hildebrand 1938, 157; Triewald 1734 b, 52.

92. Triewald 1734 b, 62.

93. *Kröger*, June 13, 1729.

94. LUB, MS Brevsamling Kilian Stobæus, Letter from Linnaeus to Kilian Stobæus, dated June 23, 1729. Original in Swedish: "Uti Dannemora såg jag continuerligen på Trivalds eld Machin; hwars krafft är så stor att hela grufwan, huset och all des behör ristas; bielken på hwilken Axy sitter som är af 3 timmerstockar swichtar. Jag skulle mehra tala om denna konst, där jag icke förmodade att hon woro så wähl Hr Professoren tillforne bekant, ty jag menar att jag aldeles förstår huru hon drifwes". Cf. Ewald Ährling, ed., *Carl von Linnés svenska arbeten i urval*, Vol. 2 (Stockholm, 1880), 18, 20 n. 21.—Linnaeus also mentioned Triewald's engine in one of his autobiographies, see: Elis Malmeström & Arvid Hj. Uggla, eds., *Vita Caroli Linnæi: Carl von Linnés*

självbiografier (Stockholm, 1957), 52. This autobiography was written in August 1734, before Triewald's book on the engine had been published (it was advertised in *Stockholmske Post Tidningar* in December). Linnaeus wrote, quite correctly, that the engine was powered by atmospheric pressure. The autobiography also gives the dates of his visit to Dannemora as from May 24 until June 10, 1729.—Both passages above are referred to in: Theodor M. Fries, *Linné: Lefnadsteckning*, Vol. 1 (Stockholm, 1903), 56 and n. 4.

95. Cf. Carolyn Merchant, "Mining the Earth's Womb", in: Joan Rothschild, ed., *Machina Ex Dea: Feminist Perspectives on Technology* (New York, 1983), 99–117.

96. LAU, BMA, BIII:7, Bergmästarens konceptböcker, 1727–1729, Letter from Mine Inspector Thor Bellander to Mårten Triewald, dated August 28, 1729. Cf. *Kröger*, August 9, 1728.

97. *Partners*, September 15–17, 1729.

98. See n. 52.

99. SKCA, DGA, Vol. 2388, Om Triewalds eld- och luftmaskin 1726–1731, "No. 1, Förslag på de *materialer* och Omkåstningar, som kunna erfordras till den wid Dannemora Grufwa warande *Eld Machinens* fulbordan och behörige i gångbringande hwad wattudriften angår". Original in Swedish: "icke något på sielfwa *Machinen* är befunnit, som fordrar att blifwa förbättrat eller ändrat, undantagandes *Injections* tappen, som i första början ej skall blifwit wähl giorder och förfärdigad, utan altid warit mycket otäter".

100. See n. 16, Chapter 6.

101. See n. 75.

102. *Partners*, September 16, 1729, § 1.

103. Loc. cit. Original in Swedish: "*Hultberg* blef inkallad, hwilken på efterfrågan, wille beskylla Crono-grufwefogden *Kröger*, det skall han fördt om honom hårda ord ock utlåtelser, bestående i synnerhet deruti, at *Kröger* skall sagt, at när *Hultberg* kommer under Bergslagens *disposition*, skulle *Kröger* wisa honom annat eller tuchta honom bättre, med mehra; hwartill dock *Kröger* närwarande, enständigt nekade, påståendes detta ej annat wara, än ett ogrundat sqwaller, som aldrig skall kunna bewisas".

104. Loc. cit. Original in Swedish: "på sin sahlighet".

105. Loc. cit. Original in Swedish: "ifall han finner sig dertill skiähl hafwa".

106. Loc. cit. Original in Swedish: "att de hwarken sielfwa giordt honom det ringaste förnär, eller hördt någon annan det giöra, utan bemött honom med all höflighet ock skaffat honom tillhanda, alt hwad han till byggnaden påkallat".

107. Ibid., September 17, 1729, § 4. Original in Swedish: "Konstmästaren *Olof Hultberg* inkom ock gaf Herrar Bergslags Interessenterne wid handen, det wore han sinnad at begiära afskied ifrån sin beställning wid *Machinen*; Mend förwistes till Herr *Directeuren Triewald*, at först sig derom hoos honom anmäla, under hwilkens *disposition* han står".

108. *Documents*, No. 4, appendix H, Agreement between Mårten Triewald and Olof Hultberg, dated September 18, 1729.

109. See n. 56.

110. *Chronology*.

111. Hildebrand 1939, 167.

112. Quoted from Hildebrand loc. cit.

113. *Kröger*, October 11, 1729. Original in Swedish: "har *Machinen* warit sedan sammankomsten ändades 3:e gånger till gång, första gången d 1 October gick allenast 6 Minuter ungefär, dagen effter ifrån 2 effter Middagen till 6 om afttonen, 3:e gången d 8 October ifrån 2 till 7 dito om afttonen, dock hwar gång gåt sönder, eller watnet blifwet slut i sistern".

114. *Documents*, No. 4, appendix J, Extract of letter from Olof Hultberg to Mårten Triewald, dated October 26, 1729.

115. Loc. cit. Original in Swedish: "Opfordrings wärcket har redan proberats ock wist en bättre förhoppning än jag kunde föreställa mig, det Hr Directeuren lärer finna [...] var med skiäl begiäres 100 dukater efter utfästelse, dåck lärer jag ännu intet drista mig at sätta baljor uti linorne för än bättre försökt det samma. Owännen måste nu om Machin och Opfordringen yttra sig: 'Nå så tar mig tusend etc: går icke det nu' [...] fick dåck mot slutet tid at röka en pip tobak, medan jag satt och förnöjde mig af dess gång [...] Stora linkorgen gick omkring så lätt för kuggarne som en spinstols rulla".—It is confirmed by a letter from Mine Bailiff Thomas Kröger to Mine Inspector Thor Bellander that the hoisting machinery functioned at a test run on October 24. See: *Kröger*, October 25, 1729.

116. *Kröger*, November 15, December 6 & 13, 1729.

117. Ibid., December 6, 1729.

118. BM, Sloane MSS, 4025, p. 293 f., Letter from Mårten Triewald to Hans Sloane. Copy in: RS, Register Book Copy, Vol. 16.—Triewald's English is corrected in the copy.

119. Mårten Triewald, "An Extraordinary Instance of the Almost Instantaneous Freezing of

Water; and giving an Account of Tulips, and such Bulbous Plants, Flowering much Sooner when their Bulbs are Placed upon Bottles Filled with Warm Water, than when Planted in the Ground", *Philosophical Transactions* 37 (1731–32), 79–81.

120. *Documents*, No. 4, appendix K, Letter from Mårten Triewald to Olof Hultberg, dated December 20, 1729.

121. Loc. cit. Original in Swedish: "någorlunda heder".

122. Hildebrand 1939, 159 f.

123. Loc cit.

124. Ibid., 160–166.

125. BM, Sloane MSS, 4053, p. 293–296, Letter from Mårten Triewald to Hans Sloane, dated July 25, 1734. Copy in: RS, Letter Book Copy, Vol. 21, pp. 269–275.

126. Lindroth 1967, Vol. 1, 496–500.

127. Op. cit.—This was one of his few letters to the Royal Society that were not published.

128. BM, Sloane MSS, 4053, p. 294.

129. *Kröger*, November 15, 1729. Original in Swedish: "ibland går den sönder, än drager den intet watten, än fastnar Piston i Cylindern så att hela wärket stanar [...] så at man intet vet hwad man om detta wärket widare skall tänka".

130. LAU, FBA, FIII:7, Handlingar rörande egendomar 1640–1805, Letter from Mine Inspector Thor Bellander to the Partners, dated March 2, 1730.

131. LAU, BMA, BIII:8, Bergmästarens konceptböcker 1730–1732, Letter from Mine Inspector Thor Bellander to Mårten Triewald, dated March 2, 1730.—There is an earlier letter from Bellander, dated February 16. It is still polite in tone (he begins by wishing Triewald a Happy New Year), but implicitly ominous. The draft is filled with alterations and additions. It is clear that Bellander did his best to persuade Triewald to come to Dannemora to put the engine in order.

132. In his pleading at the Court of Mines in 1731, see Chapter 14.

133. *Documents*, No. 3, Letter from the Partners in the Dannemora Mines to the Board of Mines.

134. Loc. cit.—It was read at the meeting of the Board on April 7, see: RA, BKA, Huvudarkivet, AI:76, Protokoll 1730, 673.

135. See n. 118.—The letter was dated Stockholm, April 4, 1730.

136. *Documents*, No. 4, Letter from Mårten Triewald to the Board of Mines.—The Board received

Triewald's letter on June 1, 1730; see: RA, BKA, Huvudarkivet, AI:76, Protokoll 1730, 1300.

14. TECHNOLOGY ON TRIAL, 1730–1736

1. RA, BKA, Huvudarkivet, AI:76, Protokoll 1730, 2546, 2591–2596, 2602, 2669, 2824–2832, 2839–2842.

2. Ibid., 2852–2857.

3. LAU, FBA, FIII:7, Handlingar rörande egendomar 1640–1805, Letter from Mårten Triewald to the Partners in the Dannemora Mines (undated, but clearly written in the spring of 1731).

4. Loc. cit. Original in Swedish: "det är ju intet annat än omogna tankar, som flutit ifrån en främmande rå hiärna uppå det papperet jag framburit, och förtiena således snarare att öfwersees än beifras".

5. Loc. cit. original in Swedish: "tiena mitt fädernes land samt alla redeliga *Patrioter*".

6. Loc. cit. Original in Swedish: "och för *Publice* nyttigt wärk".

7. Loc. cit. Original in Swedish: "den Edla Wettenskapen som redan tillbracht, och än dageligen tillskyndar främmande *Nationer* en obeskrifelig nytta, skulle tillika med mig fruchtlöst förqwäfjas och nedergräfwas, utan att lämbna annat eftter sig, än den med respectiwe herrar Interessenternes stora bekostnad redan updragne *Machinen* så som ett *Epitaphicem*, hwarpå förmoderligen efterkommande torde sättia denna öfwerskriften *Laudanda Voluntas*".—Triewald was referring to a quotation from Ovid's *Epistulae ex Ponto*: "Though the strength is lacking, yet the willingness is to be praised" (*Ut desint vires, tamen est laudanda voluntas*).

8. It is not known where in Dannemora the hearing was held, but it was presumably in the building in which the Partners held their meetings.

9. RA, BKA, Huvudarkivet, AI:76, Protokoll 1730, 2904–2906.

10. Ibid., 689, 706 f.

11. Johan Axel Almquist, *Bergskollegium och bergslagsstaterna 1637–1857: Administrativa och biografiska anteckningar* (Meddelanden från Svenska Riksarkivet, n. s. No. 2.3: Stockholm, 1909), 96, 196. Cf. nn. 81–83, Chapter 9.

12. The following is based on a copy of the judg-

ment in the archives of Forsmark Ironworks. It runs to fifty-six pages and is signed by Göran Wallerius and dated Dannemora, September 28, 1731. See: LAU, FBA, FIII:7, Handlingar rörande egendomar 1640–1805.—For courts of mines in general, see: Andreas Nicolaus Tunborg, *Om bergs-domstolar* (Upsala, 1799).

13. Triewald was elected a Fellow of the Royal Society on July 1, 1731; see: *The Record of the Royal Society of London* (London, 1912), 335.—He had written to Sir Hans Sloane, the President of the Royal Society, in 1730 and asked to be elected since he had heard that the Swedish physicist Samuel Klingenstierna was about to be elected. This practice does not seem to have been uncommon, for later that year another Swede, Jacob Serenius, wrote and asked Sloane for the same favour since he had heard that Triewald was about to be elected. (Klingenstierna, Triewald and Serenius were all elected.) See: BM, Sloane MSS, 4051, Letter from Mårten Triewald to Hans Sloane, dated May 16, 1730, & Letter from Jacob Serenius to the same, dated December 8, 1730. Cf. Sven Rydberg, *Svenska studieresor till England under frihetstiden* (Lychnos-Bibliotek, No. 12: Uppsala, 1951), 85.

14. LAU, FBA, FIII:7, Handlingar rörande egendomar 1640–1805, The judgment of the Board of Mines in the lawsuit between Mårten Triewald and the Partners in the Dannemora Mines, dated September 28, 1731, 1 (my page numbering). Original in Swedish: "efterlåtenhet och felachtigheter".

15. Ibid., 2. Original in Swedish: "*directe* och *indirecte* brutit *Contractet*, och medelst åtskillige i wägen lagda hinder til egen skada och des förklening eluderat dess goda uppsåt, samt anwände kostnad, möda och beswär".

16. Ibid., 5. Original in Swedish: "den *promitterade* och i *Contractet* utfästa wärkan".

17. Ibid., 7. Original in Swedish: "en ansenlig besparing".

18. Ibid., 8. Original in Swedish: "drägliga summa".

19. Ibid., 10. Original in Swedish: "et klart *Contract*, hwilcket skall wara sielfwa grunden och *fundamentet*, hwar efter *Contrahenterne* skola böra å begge sidor sig endast och allenast rätta".

20. Ibid., 12. Original in Swedish: "sedermera".

21. Loc. cit. Original in Swedish: "ifrån den tid Machinen först kom igång".

22. Loc. cit. Original in Swedish: "en hederlig och emot Machins wärdighet swarande recompance".

23. Ibid., 12 f. Original in Swedish: "som alle *Contractus onerosi* grunda sig *in aqualitale*, så att obligationerne måtte ömse stå i jemnwigt emot hwarannan så skall herrar Intressenter ålegat, at utur wägen rödia alla de hinder, som kunde infalla och honom befordra uti alt det som til Contractets fullgiörande kunde erfordras".

24. Ibid., 16. Original in Swedish: "ej så hastigt kunnat komma till fullbordan och sättias i det stånd, at det, kunde hållas i jemn och stadig gång".

25. Ibid., 18. Original in Swedish: "Men som han här i landet ej skall funnit så fast wircke".

26. Ibid., 19. Original in Swedish: "förmodat, som orden lyda, at Machinen til följande Juni Månad skulle kunna wara färdig".

27. Ibid., 20. Original in Swedish: "hwarcken med *Contract* eller annat bewis intygas kan".

28. Ibid., 23. Original in Swedish: "Eld- och Luft *Machinen* med alt för store och *ecessive* omkostningars påförande welat widhänga; hwarigenom denne *invention* till *Publici*' otienst torde här i Riket discrediteras".

29. Ibid., 25 f. Original in Swedish: "Men sådant oachtat skall *Hultberg* här i orten lidit hwarjehanda widrighet, och blifwit för försummelig beskylt, så at han wid sådan sysla trött upp".

30. Ibid., 26. Original in Swedish: "efter han ingen tilwand i dess stelle få kunnat".

31. Ibid., 26 f. Original in Swedish: "försee wercket med en ny konstmästare i den förres ställe, som blifwit så afspänstig giord, at han ingenting med nöje kunde taga sig före".

32. Ibid., 27 f. Original in Swedish: "hwar med Herrar Interessenter dess resa förekommit, och med ett sådant inkastat hinder des hitkomst förebydt. Ty sedan han således för sin möda, beswär och arbete blifwit bemött [...] så skall han hållit betänckeligt at widare lägga handen wid et werck som skada och förtretlighet af sig kastat och ingen belöning, i anseende hwartil han ock sedan ej mera welat taga sig någon del af detta werckets skiötsel eller widare inrättning".

33. Ibid., 28. Original in Swedish: "om den någorlunda skall proportioneras efter inventionens wärdighet".

34. Ibid., 29. Original in Swedish: "i detta målet ej böra ansees som en *Commissionair*, utan såsom en *Contrahent*, den der påtagit sig ej allenast omsorger, utan ock alt answar för *Machinens* fasthet och skall således ensam böra wid kännas en sådan skada".

35. Ibid., 30. Original in Swedish: "efter han som en förståndig *Mechanicus* bort se förut alt sådant

innan han engagerat sig med Bergslagen uti så stora *promesser*".

36. Ibid., 31. Original in Swedish: "sedan han wercket kommit igång, så skall han åtskillige förfall och undan flyckter haft at förebära".

37. Loc. cit. Original in Swedish: "förrän det blifwit stält i beständig gång".

38. Ibid., 32. Original in Swedish: "tydelige ord".

39. Ibid., 35. Original in Swedish: "at de på Herr Triewalds *persuassion* lika som i siön utkastat öfwer 52 000 daler reda penningar, och skola til *datum* icke haft en styfwers nytta deraf".

40. Loc. cit. Original in Swedish: "utan hålla de sig til Hr. Triwald som skall vara Berglagens man, ensam och allena".

41. Ibid., 36. Original in Swedish: "med all modestie".

42. Loc. cit. Original in Swedish: "til at giöra Herrar Interessenter nöje".

43. Ibid., 37. Original in Swedish: "för än *Machinen* giort den promitterade nyttan och besparingen som ej ännu skall skiedt".

44. Ibid., 40. Original in Swedish: "at han under arbetet warit flitig, och med arbets folcket haft god upsickt".

45. Loc. cit. Original in Swedish: "nyss i Riket inkommen och om *Materialers* och dagwerckens pris med mera här å orten okunnig".

46. Ibid., 41. Original in Swedish: "Warandes ock ej owanligt at uti bygnader, kostnaderne öfwerstiga förslagen hwilcket uti rörlige wercks omställande så mycket mer kan *pardonneras*, som man der alla mötande swårigheter ej förutse kan".

47. Loc. cit. Original in Swedish: "hwad effect och wärkan Eld och Luft Machinen här intil har kunnat giöra".

48. Ibid., 42. Original in Swedish: "en skiälig och god effect".

49. Ibid., 43 f. Original in Swedish: "at *Machinen* haft en ostadig och owiss gång i det han under tiden gådt 1 och 1½ dygn, under tiden 1 och ½ tima, ofta och ej mer än några slag allenast, hwar på han genast afstanat stundom och, at ehuru man undertänt elden, och sökt släppa honom åstad, han har dock ej kunnat bringas til någon gång; Ett lika skick uti dess gång har ock rätten under påstundande Bergting jemwäl hos *Machinen* funnit".

50. Ibid., 45. Original in Swedish: "en beständig och säker gång".

51. Loc. cit. Original in Swedish: "then *promitterade* nyttan och besparingen uti omkostnaderne wid Silfberg grufworne".

52. Ibid., 48 f. Original in Swedish: "Emot hwilcket alt Herr Directeuren til denne dag och på 4:e åhret efter *Machinens* första igångstellande ej mer *prästerat* än at grufwan en gång warit uttömd af watn, men står nu til en god del under watn, hwilcket sedan den tiden för *Machins* ostadiga gång skul och Herr Directeurens eftersättiande uti des förbättring ej kunnat utdragas, så at med den nytta och besparing som Herr Directeuren *promitterat* Herrar Interessenter ej *soulageras* eller hugnas kunnat".

53. Ibid., 50. Original in Swedish: "ampla löften".

54. Loc. cit. Original in Swedish: "på långt när".

55. Ibid., 51. Original in Swedish: "goda *intention*, at en så ädel och i naturens konst förborgad *inwention*, här i riket införa".

56. KB, MS I.t. 18, (Mårten Triewald), Handlingar i rättegången mellan Dannemora grufvas intressenter och direktören Mårten Triewald ... 1726–1746, Document No. 15, p. 28, Surety for Mårten Triewald by Martin Triewald, dated October 28, 1731.

57. Ibid., Document No. 15, passim.

58. RA, BKA, Huvudarkivet, AI:81, Protokoll 1734:I, 79–82, 252–257.—The Board of Mines had inspected the engine in December 1733. The nine-page report on the condition of the engine was signed by the Mine Inspector and Triewald. See: KB, MS I.t. 18, (Mårten Triewald), Handlingar i rättegången ... 1726–1746, Documents Nos. 11 & 12.

59. Loc. cit. Original in Swedish: "för *Publicii* räkning".

60. Ibid., 330–335.

61. Ibid., 352–357.

62. LAU, BMA, BIII:9, Bergmästarens konceptböcker 1733–1736, Letter from Mine Inspector Thor Bellander to Mine Bailiff Thomas Kröger, dated February 28, 1734.

63. Ibid., Letter from Mine Inspector Thor Bellander to the Board of Mines, dated March 14, 1734.

64. Ibid., Letter from Mine Inspector Thor Bellander to Mine Bailiff Thomas Kröger, dated May 8, 1734.

65. RA, BKA, Huvudarkivet, AI:81, protokoll 1734:I, 1280–1284.

66. Österby bruks arkiv, Room B, Shelf II:4, Envelope marked "Div. äldre handlingar 1734–1864", "Utdrag af *Journalen* öfwer *Dannemora* Eld och Luft *machin* i May Månad 1734". Copy in TM, Vol. 1899:a, Biografica, Mårten Triewald.

67. Loc. cit. Original in Swedish: "Att föregående *Journal* blifwit för oss uppläst, som ifrån början till slutet warit brukade wid *machin*, samt at alt

således är *passerat*, som denne *Journalen* förmäler, kunna wi med wår lifliga ed bekräfta der så fordras skulle. Dannemora d. 30 Maj 1734. (*Signed with marks:*) Andreas Widberg, konstsmed; Per Bång, konstknekt; Eric Anderson Gräse, konstknekt; Per Bom, Grufwedräng; Eric And: Söderberg, Grufwedräng".

68. Loc. cit. Original in Swedish: "om morgonen half tre".

69. RA, BKA, Huvudarkivet, AI:81, Protokoll 1734:I, 1367–1369.

70. Ibid., 1387–1389. Original in Swedish: "at den senares berättelse *differerade* i från den förres".

71. Olof von Dalin, ed., *Then swänska Argus* 2 (1734), No. 24, last page.

72. RA, BKA, Huvudarkivet, AI:81, Protokoll 1734:I, 1393–1400.

73. RA, BKA, Huvudarkivet, AI:83, Protokoll 1735:I, 3131–3135.

74. Ibid., 3131–3135, 3137–3139, 3140–3155.

75. KB, MS I.t. 18, (Mårten Triewald), Handlingar angående rättegången ... 1726–1746, Document No. 14, Letter from the Board of Mines to Mine Inspector Thor Bellander, dated December 22, 1735. Original in Swedish: "*Oeconomiske disposition*".

76. LAU, FMA, FIII:9, Handlingar rörande egendomar 1694–1752, Letter from Mine Inspector Thor Bellander to Mine Bailiff Thomas Kröger, dated March 12, 1736. Cf. SKCA, DGA, Vol. 1002, Protokoll 1726–1736, Minutes of the Partners in the Dannemora Mines, March 12, 1736, § 2.

77. Mårten Triewald, "Rön och försök angående möjligheten at Svea Rike kunda äga egit rådt silke", *KVAH* 1745, 22–29.

78. BM, Sloane MSS, 4054, p. 244f., Letter from Mårten Triewald to Hans Sloane, dated May 22, 1736 (original); RS, Letter Book Copy, Vol. 23, pp. 26f. (copy).

79. Loc. cit.

80. BM, Sloane MSS, 4054, p. 254, Letter from Mårten Triewald to Hans Sloane, dated June 17, 1736 (letter); RS, Classified Papers, Vol. III (2), No. 37 (description); RS, Letter Book Copy, Vol. 23, pp. 28–38 (copies of letter and description). Cf. Mårten Triewald, *Föreläsningar öfwer nya naturkunnigheten*, Vol. 2 (Stockholm, 1736), 208–217, Tab. XXI; idem, "Nova inventio Follium Hydravlicorum", *Acta literaria et scientiarum Sveciae* 4 (1735–1739), 67–72, Tab.; idem, "A Description of a New Invention of Bellows, Called Water-Bellows", *Philosophical Transactions* 40 (1737–1738), 231–238, Tab. Cf. Joseph Needham, *Science and Civilization in China*,

Vol. 4, Part 2 (Cambridge, 1965), Plate CCXXXV.

81. Johan Wahlund, *Dannemora grufvor: Historisk skildring* (Stockholm, 1879), appendix 115. Cf. LAU, BMA, BIII:9, Bergmästarens konceptböcker 1733–1736, Letter from Mine Inspector Thor Bellander to the Board of Mines, dated July 29, 1736.

82. LAU, BMA, BIII:9, Bergmästarens konceptböcker 1733–1736, Letter from Mine Inspector Thor Bellander to Mine Bailiff Thomas Kröger, dated September 16, 1736.

83. Ibid., November 18, 1736.

84. Loc. cit. Original in Swedish: "Elliest fägnar det mig at wid grufwan alt wähl tillstår".

85. *Stockholmske Post Tidningar* 1734, No. 48, December 2.—Mårten Triewald, *Kort beskrifning, om eld- och luft-machin wid Dannemora grufwor* (Stockholm 1734). Also published in an English translation by the Newcomen Society, London, in 1928 as Extra Publication No. 1, see: *Mårten Triewald's Short Description of the Atmospheric Engine. Published at Stockholm 1734. Translated from the Swedish with Foreword, Introduction and Notes* (London, 1928).

86. Patrick M. Malone, "Teaching the History of Technology in Museums", *Humanities Perspectives on Technology* (Curriculum Newsletter of the Lehigh HPT Program), 1978, No. 9, 3–6. Cf. Carroll W. Pursell, Jr., "The History of Technology and the Study of Material Culture", *American Quarterly* 35 (1983), 304–315.

87. Sune Ambrosiani, "Bergsmansyxor och bergsmanskäppar i Norden", *Med Hammare och Fackla* 2 (1930), 21–49; Bertil Waldén, "Några ämbetsstavar från svenska bergslager", *Med Hammare och Fackla* 10 (1939), 95–101.

88. Carl Björkbom, "Ett projekt att bygga en ångmaskin i Sverige år 1725", *Daedalus* 1936, 85 n. 3.

89. Nicholls' engraving was published a few weeks before Triewald departed for Sweden in 1726, see: Lionel T. C. Rolt & John S. Allen, *The Steam Engine of Thomas Newcomen* (Hartington, 1977), 81.

15. CRITICAL FACTORS IN TECHNOLOGY TRANSFER

This chapter has grown out of a paper first read in 1977 at the symposium "Technology and its Impact on Society", arranged by the National Museum of Science and Technology, Stockholm. I would like to acknowledge the comments I received on that paper in 1978 at seminars in the history of technology at the University of Delaware and Case Western Reserve University. I am indebted to Olle Edqvist for valuable comments and suggestions on an earlier version of this chapter.

1. Nathan Rosenberg, "Selection and Adaptation in the Transfer of Technology: Steam and Iron in America—1800–1870", in: *L'acquisition des techniques par les pays non-initiateurs* (Colloques Internationaux du Centre National de la Recherche Scientifique, No. 538: Paris, 1973), 68. Also published in: Nathan Rosenberg, *Perspectives on Technology* (Cambridge, 1976), 173–188.

2. Johan Wahlund, *Dannemora grufvor: Historisk skildring* (Stockholm, 1879), 50 f., appendix 265–268.—Gunpowder blasting was introduced slowly in Sweden during the eighteenth century, and often used in combination with the traditional fire-setting until the end of the nineteenth century; see: Chapter 3. Even a large, and comparatively highly centralized, mine like the Great Coppermine in Falun only consumed 50 kg. of gunpowder per year in 1730, compared with 15 000 kg. per year in 1780; see: Sten Lindroth, *Gruvbrytning och kopparhantering vid Stora Kopparberget intill 1800-talets början*, Vol. 1 (Uppsala, 1955), 503.

3. Wahlund, appendix 265–268.

4. Bertil Boëthius & Åke Kromnow, *Jernkontorets historia*, Vol. 1 (Stockholm, 1947), 487–490.

5. SKCA, DGA, Vol. 2388, Om Triewalds eld- och luftmaskin 1726–1731, "Några oförgripeliga Anmärckningar öfwer *Eld Machinen* uti Bergsstaden *Königsberg* i *Ungern* och det i anledning af höglofl. styckgjutaren Hr. Leopolds Bref daterad Wien d. 25 Juny 1733 till Hr. Gerhard Meyer och densamma jämfört med *Dannemora Eld Machine*".—It is quite clear that this estimate is by Triewald, since it says "my royalty".—Triewald quoted Leopold's letter in his book on the Dannemora engine in 1734, but used it only to compare the output of the two engines. He left out everything concerning the economics of his engine.

6. Loc. cit. According to the quoted letter, the Königsberg engine consumed 6 *Klafter* wood fuel a day. It was managed by four men; two in the day and two at night. "Mr Potter, who built the fire engine, receives for supervision, and in accordance with his Imperial patent, 164 *fl.* or 736 *copperdaler* a month".

7. Graham J. Hollister-Short, "A New Technology and its Diffusion: Steam Engine Construction in Europe 1720–c. 1780", *Industrial Archaeology* 13 (1978), 122.

8. Brooke Hindle, "The Transfer of Power and Metallurgical Technologies to the United States, 1800–1880: Processes of Transfer, with Special Reference to the Role of the Mechanics", in: *L'acquisition des techniques par les pays non-initiateurs* 1973, 407–428. Cf. David J. Jeremy, *Transatlantic Industrial Revolution: The Diffusion of Textile Technologies between Britain and America, 1790–1830* (Cambridge, Mass., 1981), esp. 254–257.

9. Rosenberg 1973, 55 f.

10. Lynn White, Jr., "Technology Assessment from the Stance of a Medieval Historian", in: idem, *Medieval Religion and Technology* (Berkeley, 1978), 261–277. Originally published in *American Historical Review* 79 (1974), 1–13.

11. Ibid., 276.

12. Karin Johannisson, "Naturvetenskap på reträtt: En diskussion om naturvetenskapens status under svenskt 1700-tal", *Lychnos* 1979–1980, 109–154 (English summary, 153 f.: "Natural Science in Retreat: A Discussion on the Status of Science in Sweden during the 18th Century"). Cf. Andrew Jamison, *National Components of Scientific Knowledge: A Contribution to the Social Theory of Science* (Research Policy Institute, University of Lund: Lund, 1982), 231–258.

13. The letter: UUB, MS Okat. Royal Society (418g:2), Letter from Mårten Triewald to John T. Desaguliers, dated November 20, 1728 (photocopy of letter in RS).—The book: The British Library's copy (call no. 536k. 12.(1.)) has once belonged to Hans Sloane.

14. Ernst Ericsson & Gustaf Rabe, *Kungl. Fortifikationens historia*, Vol. 4:1 (Stockholm, 1930), esp. 86 (appointment), 89–93 (rank and salary), 285, 524–536 (teaching).—For Triewald's commission, see: KRA, Biographica, Mårten Triewald.—For Triewald's teaching, see also: Wilhelm Sjöstrand, *Den militära undervisningen i Sverige intill år 1792* (Uppsala, 1941), 450–458. Cf. Marie Nisser, "Byggnadsteknisk debatt och utbildning i Sverige under 1600- och 1700-talen" (mimeographed fil. lic. dissertation at the Department of History of Art, University of Uppsala, 1966), 159.

15. Carl Forsstrand, "Mårten Triewald och hans Stockholmsträdgård", *Svenska Linné-Sällskapets Årsskrift* 10 (1927), 57–67; Nils Lundequist, *Stockholms stads historia, från stadens anläggning till närvarande tid*, Vol. 3 (Stockholm, 1829), 195.—For the drawing of Triewald's house

(Fig. 15.1), see: SSA, Stockholms byggnads-nämnds byggnadslovsritningar 1739–50, 212.

16. Sven Almqvist, "Mårten Triewalds malmgård Marieberg", *Fälttelegrafisten* 25 (1958), No. 3, 7–11; Forsstrand, 63 f.—For a map of Triewald's estate, see: SSA, Stockholms stads-ingenjörs arkiv, AId:2, Figurbok, fria, 1740–1744, 320.

17. Bengt Hildebrand, *Kungl. Svenska Vetenskapsaka-demien: Förhistoria, grundläggning och första organi-sation* (Stockholm, 1939), 279–290.

18. Letter No. 1: UUB, MS Okat. Royal Society (418g:2), dated November 20, 1728; No. 2: BM, Sloane MSS, 4025, dated April 4, 1730; No. 3: BM, Sloane MSS 4051, dated May 16, 1730; No. 4: RS, Register Book Copy, Vol. 16, dated May 29, 1731; No. 5: UUB, MS Okat. Royal Society (418g:2), dated November 1, 1732; No. 6: BM, Sloane MSS, 4053, dated July 25, 1734; No. 7: RS, Letter Book Copy, Vol. 22, dated April 24, 1735; No. 8: ibid., dated July 12, 1735; No. 9: BM, Sloane MSS, 4054, dated May 22, 1736; No. 10: ibid., dated June 17, 1736 (letter), RS, Classified Papers, Vol. III (2), No. 37, dated May 12, 1736 (description).—Anders Celsius, ed., *Almanach* ... 1728, 1730, 1731, 1732, 1734, 1735, 1736. Copies in KB.

19. E. W. Dahlgren, ed., *Svenska Vetenskapsakademiens protokoll för åren 1739, 1740 och 1741*, Vol. 1 (Stockholm, 1918), 44 f. Cf. Hildebrand 1939, 320.

20. Linnaeus later changed his mind after an exper-ience which convinced him that the divining rod could detect metals, see: Carl von Linné, *Skånska resa, på höga öfwerhetens befallning förrättad år 1749* ... (Stockholm, 1751), 160 f.

21. Dahlgren, 45. Original in Swedish: "Ja, sade Hr Triwaldt, de fleste ting i verlden hafva mer af en händelse än a priori blifvit uptäckte".

22. Hildebrand 1939, 208, 304 f., 322 f. Hildebrand remarks that Triewald, Linnaeus and Anders Johan von Höpken—the three foremost of the founders of the Royal Swedish Academy of Sciences—shared exactly the same physicotheo-logical belief. The Academy may in this per-spective be seen as an institutionalization of their religious belief.—For the relationship between religion and economy in eighteenth century Sweden, see: Tore Frängsmyr, "Den gudomliga ekonomin: Religion och hushållning i 1700-talets Sverige", *Lychnos* 1971–1972, 217–244 (English summary, 243 f.).—For phy-sicotheology in general, see: A. D. Atkinson, "William Derham, F.R.S. (1675–1735)", *Annals of Science* 8 (1952), 368–392; Clarence J.

Glacken, *Traces on the Rhodian Shore: Nature and Culture in Western Thought from Ancient Times to the End of the Eighteenth Century* (Berkeley, 1976), 375–428, 504–550; Wolfgang Philipp, "Physi-cotheology in the Age of Enlightenment: Ap-pearance and History", *Studies on Voltaire and the Eighteenth Century* 57 (1967), 1233–1267.

23. Torsten Althin, *Capitain-Mechanici wid Fortifica-tion Herr Mårten Triewalds wäl conditionerade bib-liotheque år 1747 jämte hans bibliografi* (Stockholm Papers in History and Philosophy of Techno-logy. TRITA-HOT-3001: Stockholm, 1976), 23–25.

24. Mårten Triewald, *Föreläsningar öfwer nya natur-kunnigheten*, Vol. 1 (Stockholm, 1735), 38, 61, 63, 64, 106, 196, 198; Vol. 2 (Stockholm, 1736), "Oration" n. x, 85.

25. Althin 1976, 22, 27. The books by Wolff in his library were: *Anfangsgründe aller matematischen Wissenschaften*, Vol. 3 (Halle, 1717); *Metaphysik, oder vernünfftige Gedancken von Gott, der Welt und der Seele des Menschen* (Halle, 1720); *Vernünfftige Ge-dancken von den Würkungen der Natur* (Halle, 1723); *Anfangsgründe aller matematischen Wissen-schaften* (Halle, 1710).

26. Tore Frängsmyr, *Wolffianismens genombrott i Upp-sala: Frihetstidens universitetsfilosofi till 1700-talets mitt* (Acta Universitatis Upsaliensis, No. C.26: Uppsala, 1972), with an English summary: The Emergence of Wolffianism at Uppsala.

27. Tore Frängsmyr, "Christian Wolff's Mathema-tical Method and Its Impact on the Eighteenth Century", *Journal of the History of Ideas* 36 (1975), 666.

28. This is most evident from his proposal in 1729 to the Society of Science in Uppsala. See Chapter 13 and Hildebrand 1939, 159–166.

29. See n. 25.

30. Mårten Triewald, *Konsten at lefwa under watn* ... (Stockholm, 1734b), 1. Original in Swedish: "Den stora Guden är den endaste och sanna Uphofsmannen af alt; hwar emot wi usla Men-niskior allenast då och då, kunna bli warse något af Guds oendeliga Wishet i de skapade ting: och lempa oss de samme til nytta, då när wi wåre tankar et wist syftemål föresättia, samt förut wel giort oss alla naturens Lager bekante, hwilka Lagers naturens HErre så stadfäst, at de aldrig här i tiden af oss kunna rubbas eller ändras; wi kunna således mycket mindre till-skrifwa oss något mera, när vi vad nyttigt är påfinna, än at wi igenom eftersinnande kommit på den wägen, hwarpå wi Guds Werk blifwit warse." Cf. Fig. 15.4.

31. Philipp, 1266.—I am indebted to Allan Ellenius

for his comments on the illustration in Triewald's book (Fig. 15.4).—The publisher, Benjamin Gottlieb Schneider in Stockholm, had used this illustration before; see: Olof von Dalin, ed., *Then swänska Argus* 1 (1732), No. 2 and passim 1733–1734.

32. Triewald 1734 b, 2. Original in Swedish: "lenda til Guds Ära och Menniskians nytta".

33. Loc. cit. Original in Swedish: "igenom sina påfund frambrakt Guds ära, och så märkeligen tient ther med det Menneskeliga släcktet, at kunna lindra de beswärligheter i detta usla lefwerne finnas".—A similar passage appears in the first volume of his lectures, published the following year. See: Triewald 1735, 285 n. God has, Triewald writes, rightfully imposed the punishment for man's sins saying "in the sweat of thy face shalt thou eat bread". But He has also given man several means to facilitate this work, as well as rationality to "invent" them and ability to apply them through human muscle power or other forces of nature.

34. Mårten Triewald, *Kort beskrifning, om eld- och luft-machin wid Dannemora grufwor* (Stockholm, 1734), 2 f. The English translation is quoted from *Mårten Triewald's Short Description of the Atmospheric Engine. Published at Stockholm, 1734* (Newcomen Society Extra Publication No. 1: London, 1928), 2 f.

35. See, for example: Lionel T. C. Rolt & John S. Allen, *The Steam Engine of Thomas Newcomen* (Hartington, 1977), 42 f.

36. Quoted from the English translation in: Sten Lindroth, "The Two Faces of Linnaeus", in: Tore Frängsmyr, ed., *Linnaeus: The Man and His Work* (University of California Press: Berkeley, 1983), 12.

16. EPILOGUE

I am indebted to Göran Blomberg for his demonstration of the Cahman organ at Leufsta Ironworks, and for his comments on the last part of this chapter.

1. KB, MS I.t. 18, (Mårten Triewald), "Handlingar i rättegången mellan Dannemora grufvas intressenter och direktören Mårten Triewald angående den senares 'Eld- och luftmaskin' för vattenpumpningen ur 'Silfbergsgrufvan' vid Dannemora. Afskrifter från Salsta slottsarkiv, skänkta t. K.B. 1874. Originaler, från Svea Hof-

rätts archiv aflemnade till K.B. 1864".—This is the complete title of this volume, which has previously been referred to in an abbreviated form as "Handlingar i rättegången ... 1726–1746". It contains copies of some of the more important documents in the case between Triewald and the Partners, as well as a collection of original documents in the lawsuit from the Svea Court of Appeal. The last document is dated February 22, 1746.

2. Bengt Hildebrand, "Till släkten Triewalds historia", *Personhistorisk Tidskrift* 39 (1938), 158.

3. Ibid., 158 f.

4. SSA, Ulrika Eleonora församlings kyrkoarkiv, FI:1, Begravningsbok 1688–1778, 425. Original in Swedish: "på et hemligt rum wardt plötzligen död sittiande funnen".

5. UUB, Linnaean collections, Linnaeus' copy of: Mårten Triewald, *Föreläsningar öfwer nya naturkunnigheten*, Vol. 2 (Stockholm, 1736).—I am indebted to Gunnar Broberg for this reference.

6. Alfred Bernhard Carlsson, ed., *Uppsala universitets matrikel*, Vol. 2 (Uppsala, 1919–1923), 208.

7. SKCA, DGA, Vol. 2388, Om Triewalds eld- och luftmaskin 1726–1731, Letter from Olof Hultberg to Mine Inspector Thor Bellander, dated April 13, 1730.—The dates of Hultberg's stay in Stockholm appear in the chronology in the same volume. Cf. n. 15, Chapter 13.

8. Julius Lagerholm, ed., *Södermanland-Närkes nation: Biografiska och genealogiska anteckningar om i Uppsala studerande södermanlänningar och närkingar 1595–1900* (Sunnansjö, 1933), 168.

9. LAU, Uppsala rådhusrätt och magistrat, FI:17, Bouppteckningar 1751–1755, Inventory of Olof Hultberg's estate, dated October 29, 1752.

10. Nils Viktor Emanuel Nordenmark, *Anders Celsius, professor i Uppsala 1701–1744* (Lychnos-Bibliotek, No. 1: Uppsala, 1936), 124. Cf. Gunnar Pipping, *The Chamber of Physics: Instruments in the History of Sciences Collections of the Royal Swedish Academy of Sciences* (Stockholm, 1977), 67.

11. Lagerholm, 168.

12. Daniel Tilas, "Dag bok öfwer Hans Kongl. Höghet Cronprintsen Prints Gustavs resa igenom bergslagerne in sept: 1768", *Noraskogs Arkiv*, Vol. 6: 1 (Stockholm, 1928), 9 f.

13. Ibid., 9.

14. Loc. cit.

15. See n. 3, Chapter 12.

16. Johan Wahlund, *Dannemora grufvor: Historisk skildring* (Stockholm, 1879), appendix 148 f.

17. RA, Bertil Boëthius lappregister till Jernkontorets arkiv, Rön och försök, Eld- och luftmaskiner.

18. LAU, FBA, FIII:9, Handlingar rörande egen-domar 1694–1752, Letter from Johan J. Ur-lander to Mine Bailiff Anders Kalmeter, dated May 15, 1751.

19. Johan Eric Fant & August Theodor Låstbom, *Upsala ärkestifts herdaminne*, Vol. 1 (Upsala, 1842), 276 f.

20. LAU loc. cit. Original in Swedish: "sinnad, at samma Eld och Luft *Machine* låta *reparera*". Cf. Wahlund, appendix 120.

21. Einar Erici, *Inventarium över bevarade äldre kyrkorg-lar i Sverige* (Stockholm, 1965), 111 f.; R. Axel Unnerbäck, "Leufsta bruks-orgeln restaur-erad", *Kyrkomusikernas Tidning* 30 (1964), 128–131; Bertil Wester, "Orgelverket i Leufsta bruks kyrka", *Fornvännen* 28 (1933), 224–245.

22. Bengt Kyhlberg, "Orgelbyggarefamiljen Cah-man, Hülphers och orgeln i Trefaldighetskyr-kan i Kristianstad", *Svensk Tidskrift för Musik-forskning* 27 (1945), 61–75; Torild Lindgren, "Från högbarock till senbarock. Stilstudier i Cahmanskolan", *Svensk Tidskrift för Musikforsk-ning* 38 (1956), 111–134; Bertil Wester, "Studier i svensk orgelkonst under 1600- och 1700-ta-len", *Svensk Tidskrift för Musikforskning* 13 (1931), 45–72.

23. For eighteenth-century organ-building techno-logy, see: François Bédos de Celles, *L'art du facteur d'orgues* (1766–1770).—This book is in three parts, and consists in total of 536 folio pages and 79 engravings. It was published by Académie Royal des Sciences as a part of its major work *Descriptions des arts et métiers*. This was an attempt to describe all existing technolo-gies, and the description of organ building ac-counts for approx. 5 % of the entire encyclopae-dia. Copy in KTHB. Cf. n. 22, Chapter 2.

24. The Cahman organ at Leufsta Ironworks may be heard on a record made in 1981 and issued by Bluebell of Sweden. On this record (Bell 133) the musicologist and organist Göran Blomberg plays compositions by Dietrich Bux-tehude and Johann Sebastian Bach.

25. Johann Philip Bendeler, *Organopoeia, oder: Unter-weisung wie eine Orgel nach ihren Hauptstücken ... aus mathematischen Gründen zuerbauen* (Franckfurt und Leipzig, n.d., ca. 1690), copy in KTHB; Abraham Abrahamsson Hülphers, *Historisk af-handling om musik och instrumenter särdeles om org-werks inrättningen i allmänhet, jemte kort beskrifning öfwer orgwerken i Swerige* (Westerås, 1773). —That organ building was an empirical craft is illustrated by the outcome of the attempt in 1757 to introduce State supervision of the com-petence of organ builders. The Government ap-pointed the Royal Swedish Academy of Sciences as an examining body, but the Academy only reluctantly performed its duties since few of the fellows were competent to act as examiners. See: Sten Lindroth, *Kungl. Svenska Vetenskapsaka-demiens historia 1739–1818*, Vol. 1 (Stockholm, 1967), 159–161. The empirical nature of organ-building technology is also illustrated by a letter from Christopher Polhem to the organ builder Daniel Stråle (1700–1746), who had been a pu-pil of Johan Niclas Cahman. See: Henrik Sandblad, ed., *Christopher Polhems efterlämnade skrifter*, Vol. 1 (Lychnos-Bibliotek, No. 10.1: Uppsala, 1947), 160–163.—In the early nine-teenth century, the teaching of organ building was considered a part of the responsibilities of *Technologiska Institutet* (the present Royal Insti-tute of Technology). See: Lars-Erik Sanner, "Musiklitteratur i Kungl. Tekniska Högskolans Bibliotek, Stockholm", *Nordisk Tidskrift för Bok- och Biblioteksväsen* 54 (1967), 94–97.

26. Michael Praetorius, *Syntagma musicum*, Vol. 2 (Wolfenbüttel, 1619), 85. Copy in the library of the Royal Swedish Academy of Music, Stock-holm.

27. Heinrich Philip Johnsen, "Kort orgwerks-beskrifning", in: Hülphers 1773, 320.

Sources and Literature

ARCHIVAL SOURCES

Archives de l'Etat à Liège, Liège:
Wanzoulle correspondence

British Museum Library and Archives, London (BM):
Sloane MSS 4025, 4051, 4053, 4054

City Libraries, Newcastle upon Tyne:
Church Register 1726–1759
Minute Book of Goldsmiths Company 1721

Landsarkivet i Uppsala, Uppsala (LAU):
Bergmästarämbetets i Gävleborgs, Uppsala och
    Stockholms län arkiv (BMA):
BIII:7–9, Bergmästarens konceptböcker, 1727–
    1736
FI:5, Dannemora bergstingshandlingar och pro-
    tokoll 1731–1733
Forsmarks bruks arkiv (FBA):
FIII:7, Handlingar rörande egendomar 1640–
    1805
FIII:9, Handlingar rörande egendomar 1694–
    1752
Uppsala rådhusrätt och magistrat, FI:17, Boupp-
    teckningar 1751–1755

Military Archives, Stockholm (KRA):
Biographica, Mårten Triewald
Krigskollegii brevböcker 1743, Vol. 4

National Museum of Science and Technology, Stockholm
(TM):
Vols. 39:a–d, Dannemora gruvor
Vol. 1899:a, Biografica, Mårten Triewald
MS 7404, Cronstedtska planschsamlingen
MS 7405, Carl Johan Cronstedt's skissbok 1729
Carl Sahlins bergshistoriska samlingar:
    Vols. 64–65, Dannemora gruvor
    Vols. 200–209, Reseskildringar

National Record Office, Stockholm (RA):
Bergskollegiums arkiv (BKA), Huvudarkivet:
    AI:62–83, Protokoll, 1716–1735:I
    BI:86, Registratur 1719
    EI:10, Kungl. brev 1717–1721
    EIII:1–12, Utländska bergverksrelationer 1718–
        1727
    EIV:157–172, Brev och suppliker, 1718–1726:II
Bergskollegium till Kungl. Maj:t, 1718, 1719, 1726
Biografica, Vol. T:19
Bertil Boëthius lappregister till Jernkontorets arkiv
Diplomatica:
    Anglica, Vol. 317
    Hollandica, Vols. 364, 690, 706
Registratur Inrikes Civilexpeditionen, 1723
Sammansatta kollegier till Kungl. Maj:t, Vol. 246 b,
    1723

Nordiska Museets arkiv, Stockholm (NMA):
Passansökningar och pass, Hovslagareämbetet i
    Stockholm 1695–1792
Skråarkivalier, Hovslagareämbetet i Stockholm:
    In- och utskrivningsböcker, 1623–1700, 1685–
        1732, 1726–1794
    Protokoll, 1676–1698, 1698–1725, 1726–1761

*Northumberland County Record Office, Newcastle upon
Tyne:*
ZRI, Ridely (Blagdon) MSS
ZRI, 23/2, Agreement between Triewald, Ridley,
    Calley, and Prior, 1722

Royal Institute of Technology Library, Stockholm
(KTHB):
MS 7, Intyg utfärdat av Mårten Triewald på Wede-
    våg, 25 januari 1732
MS Pf-38, Giöran Wallerius Haraldson, "Kårtt och
    ungefärlig relation med des derhoos tillhörige
    rijtningar, angående de fyra af Hr. Directeuren
    Påhlhammar Inventerade och af Hr. Markschei-

dern Buschenfelt förfärdigade mekaniske machiner med des experimenter och bijfogade tabeller. Hwilcka af bemelte Hr. Markscheider jemte mig undertecknad och flere äro genomgångne, och sedan efter Inventoris egen disposition och underrättelse på följande sätt deducerade af Giöran Vallerius Haraldson A:o 1705"

Royal Library, Stockholm (KB):
D 850, Brev till fru Eva Juliana Insenstierna på Harg, innehållande politiska nyheter från Stockholm och utlandet, emellan 9 juni 1727 och 7 november 1729
Depos. 69, Carl Daniel Burén, Dagbok, Vol. 1, 1790–1792
Ep. H2:3b, Brev till Lorenzo Hammarsköld från Pehr Lagerhjelm, 7 juni 1813
I.l. 1b., Pehr Lagerhjelm, Självbiografiska anteckningar
I.t. 18, (Mårten Triewald), Handlingar i rättegången mellan Dannemora grufvas intressenter och direktören Mårten Triewald ... 1726–1746
M 25:2, Daniel Tilas, "Sokne-skrifvare, eller svenska resesamlingar"
M 218, Jonas Alströmer, Resedagbok från resa i England 1719–1720
M 249:1–5, Henrik Kalmeter, Resedagböcker från resor i Tyskland, Holland, Frankrike och England 1718–1726, samt Tyskland 1729–1730

Royal Society Library, London (RS):
Classified Papers, Vol. III (2)
Letter Book Copy, Vols. 21–23
Register Copy Book, Vol. 16

Stockholm City Archives, Stockholm (SSA):
Bouppteckningar, 1745/2:544, 1747/2:271
Stockholms byggnadsnämnds byggnadslovsritningar 1739–1750
Stockholms stadsingenjörs arkiv, AId:2, Figurbok, fria, 1740–1744
Stockholms stads verifikationsböcker, 1726–1728
Tyska församlingens kyrkoarkiv:
LIh1a, Längder över uppburna bänkavgifter
CI1:B, Dopbok 1689–1734
Ulrika Eleonoras församlings kyrkoarkiv:
FI:1, Begravningsbok 1688–1778

Stockholm University Library with the Library of the Royal Academy of Sciences, Stockholm (KVA):
Brev till Jacob Henrik af Forselles från Pehr Lagerhjelm
Brev från Mårten Triewald till Henrik Kalmeter, 6 december 1723

Stora Kopparbergs Bergslags AB Centralarkiv, Falun (SKCA):
Dannemora gruvors arkiv (DGA):
Vol. 1002, Intressenternas sammanträdesprotokoll 1726–1736
Vol. 2350, Brev från Kronogruvfogde Thomas Kröger till Bergmästare Thor Bellander
Vol. 2388, Om Triewalds eld- och luftmaskin 1726–1731

University of Lund Library, Lund (LUB):
Brevsamling Kilian Stobæus, Brev till Stobæus från Linné, 23 juni 1729

University of Uppsala Library, Uppsala (UUB):
A 147, Göran Wallerius, "Kårtt och ungefärlig relation ... af Jöran Wallerius Haraldsson"
G 130, Alströmerska brevsamlingen, Brev från Henrik Kalmeter till Jonas Alströmer
Linnésamlingarna, Linnés exemplar av Triewald's föreläsningar
Okat. Royal Society (418g:2), Brev från Mårten Triewald till J. T. Desaguliers (photocopies)
X 376, Jonas Alströmer, Resedagbok från resa i England 1719–1720

Österby bruks arkiv, Österbybruk:
Div. äldre handlingar 1734–1864 (Kuvert i rum B på hylla II:4)

PRIMARY AND SECONDARY LITERATURE

Acerbi, Joseph, *Travels through Sweden, Finland, and Lapland, to the North Cape in the years 1798 and 1799,* Vol. 1 (London, 1802).
L'acquisition des techniques par les pays non-initiateurs (Colloques Internationaux du Centre National de la Recherche Scientifique, No. 538: Paris, 1973).
Acta literaria (et scientiarum) Sveciae, 1 (1720–1724), 2 (1725–1729), 3 (1730–1734), 4 (1735–1739).
Allan, David G. E. & Schofield, Robert E., *Stephen Hales: Scientist and Philanthropist* (London, 1980).
Allen, John S., "A Chronological List of Newcomen Society Papers on Thomas Newcomen and the Newcomen Engine", *Transactions of the Newcomen Society* 50 (1978–79).
— "The 1712 and Other Newcomen Engines of the Earls of Dudley", *Transactions of the Newcomen Society* 37 (1964–65).
— "The Introduction of the Newcomen Engine from 1710 to 1733: Second Addendum", *Trans-*

actions of the Newcomen Society 45 (1972–73), 223–226.

Almquist, Johan Axel, *Bergskollegium och bergslagssta-terna 1637–1857: Administrativa och biografiska an-teckningar* (Meddelanden från Svenska Riksarki-vet, n. s. No. 2.3: Stockholm, 1909).

Almqvist, Sven, "Mårten Triewalds malmgård Marieberg", *Fälttelegrafisten* 25 (1958), No. 3.

Althin, Torsten, "Memorial to Martin Triewald", *Transactions of the Newcomen Society* 11 (1930–31).

— "Tekniska Museet under år 1931", *Daedalus* 1932.

— "Sveriges andra ångmaskin", *Daedalus* 1939.

— "Omdömen om det tekniska museet i Stockholm för 150 år sedan", *Daedalus* 1940.

— "Stationary Steam Engines in Sweden 1725–1806", *Daedalus* 1961.

— "Rinman's Treatise on Mechanics", *Technology and Culture* 10 (1969).

— "Eric Geisler och hans utländska resa 1772–1773," *Med Hammare och Fackla* 26 (1971).

— *Capitaine-Mechanici wid Fortificationen Herr Mårten Triewalds wäl conditionerade Bibliotheque år 1747 jämte hans bibliografi* (Stockholm Papers in His-tory and Philosophy of Technology, TRITA-HOT-3001: Stockholm, 1976).

(Althin, Torsten), "The Leonardo da Vinci Me-dal", *Technology and Culture* 20 (1979).

Alumni Oxonienses 1715–1886, Vol. 4 (Oxford, 1888).

Ambrosiani, Sune, "Bergsmansyxor och bergs-manskäppar i Norden", *Med Hammare och Fackla* 2 (1930).

Anderberg, Rudolf, *Grunddragen av det svenska tekniska undervisningsväsendets historia* (Skrifter utgivna av Ingeniörsvetenskapsakademien, Meddelanden No. 5: Stockholm, 1921).

Andolf, Göran, "Resandets demokratisering: Skjutsväsendet" (mimeographed paper present-ed at the Department of History, University of Gothenburg, 1977).

— "Resandets revolutioner", *Fataburen* 1978.

Andrew, James H., "Some Observations on the Thomas Barney Engraving of the 1712 Newco-men Engine", *Transactions of the Newcomen Society* 50 (1978–79).

— "The Copying of Engineering Drawings and Do-cuments", *Transactions of the Newcomen Society* 53 (1981–82).

Annerstedt, Claes, *Upsala universitets historia*, Vol. 3:1 (Upsala, 1913).

Areen, Ernst E., "Mårten Triewald och luft- och eldmaskinen vid Dannemora gruvor. Något om en märkesman inom svenskt näringsliv under förra hälften av 1700-talet", *Upsala Nya Tidning* 1927, December 17.

Arpi, Gunnar, *Den svenska järnhanteringens träkolsför-sörjning 1830–1950* (Jernkontorets Bergshistoriska Skriftserie, No. 14: Stockholm, 1951).

Artz, Frederick B., *The Development of Technical Edu-cation in France* (Cambridge, Mass., 1966).

Atkinson, A. D., "William Derham, F.R.S., (1675–1735)", *Annals of Science* 8 (1952).

Atkinson, Frank, "Some Northumberland Collieries in 1724", *Transactions of the Architectural and Ar-chaeological Society of Durham and Northumberland* 11 (1965).

Attelid, Tore, "Världspremiär i Dannemora", *Up-sala Nya Tidning* 1977, December 4.

Beck, Ludwig, *Die Geschichte des Eisens in technischer und kulturgeschichtlicher Beziehung*, Vol. 3 (Braunschweig, 1897).

Beckman, Anna, "Två svenska experimentalfysiker på 1700-talet: Mårten Triewald och Nils Walle-rius", *Lychnos* 1967–1968.

Bedini, Silvio A., "The Evolution of Science Mu-seums", *Technology and Culture* 6 (1965).

Bédos de Celles, François, *L'art du facteur d'orgues* (Descriptions des arts et métiers: Paris, 1766–1770).

de Bélidor, Bernard Forest, *Architecture hydraulique*, 2 Vols. (Paris, 1737–1753).

Bendeler, Johann Philip, *Organopoeia, oder: Unterwei-sung wie eine Orgel nach ihren Hauptstücken ... aus mathematischen Gründen zuerbauen* (Franckfurt und Leipzig, n.d., ca 1690).

Benzelstierna, Lars, "Berättelse om åtskillige nyare malm- ock mineral upfinningar i riket", *KVAH* 1741.

Berch, Anders, *Åminnelsetal öfver Henric Kalmeter* (Stockholm, 1752).

Bergman, Torbern, *Inträdestal, om möjeligheten at före-komma åskans skadeliga verkningar* (Stockholm, 1764).

— "Anledningar at tilverka varaktigt tegel", *KVAH* 1771.

Berndes, Per Bernhard, "Försök att använda brän-bar alunskiffer såsom bränsle i ställe för ved, till åtskillige hushållsbehof", *KVAH* 1802.

Berner, Boel, *Teknikens värld: Teknisk förändring och ingenjörsarbete i svensk industri* (Arkiv avhandlings-serie, No. 11: Lund, 1981).

— "Experiment, teknikhistoria och ingenjörens fö-delse", *Daedalus* 51 (1982).

Beronius, Jonas, "Utdrag af de vid Dannemora grufvor befintliga handlingar, relationer och rä-kenskaper, rörande arbeten som tid efter annan blifvit verkställdte, till vattnets afhållande från grufvorna, och om deras tömmande efter den år

1795 timade öfversvämningen", *Jern-Kontorets Annaler* 2 (1818).

Berzelius, Jöns Jacob, *Tabell som utvisar vigten af större delen vid den oorganiska kemiens studium märkvärdiga enkla och sammansatta kroppars atomer, jemte deras sammansättning, räknad i procent. Bihang till tredje delen av läroboken i kemien* (Stockholm, 1818).

Biographiskt lexicon öfver namnkunnige svenska män, 23 Vols. (Upsala & Örebro, 1835–1857).

Björkbom, Carl, "Ett projekt att bygga en ångmaskin i Sverige år 1725", *Dædalus* 1936.

— "A Proposal to Erect an Atmospheric Engine in Sweden in 1725", *Transactions of the Newcomen Society* 18 (1937–38).

Boëthius, Bertil, "Hammarkommissionerna på 1680- och 1720-talet. En studie över deras ställning i bergskollegiets brukspolitiska system", in: *En bergsbok till Carl Sahlin* (Stockholm, 1921).

— "Carl Daniel Burén", *Svenskt Biografiskt Lexikon*, Vol. 6 (Stockholm, 1926).

— "Trävaruexportens genombrott efter det stora nordiska kriget", *Historisk Tidskrift* 49 (1929).

— *Gruvornas, hyttornas och hamrarnas folk: Bergshanteringens arbetare från medeltiden till gustavianska tiden* (Den svenska arbetarklassens historia, Vol. 9: Stockholm, 1951).

Boëthius, Bertil & Kromnow, Åke, *Jernkontorets historia*, 3 Vols. (Stockholm, 1947–1968).

de Boisgelin, Louis, *Travels through Denmark and Sweden*, Vol. 2 (London, 1810).

Booker, Peter J., *A History of Engineering Drawings* (London, 1963).

Bossut, Charles, *Traité élémentaire d'hydrodynamique ouvrage dans lequel la théorie et l'expérience*, 2 Vols. (Paris, 1771).

Boyle, Robert, *Experimentorum novorum physico-mechanicorum* (London, 1680).

Brand, John, *History and Antiquities of Newcastle-upon-Tyne*, Vol. 2 (London, 1789).

Braun, Hans-Joachim, "The National Association of German-American Technologists and Technology Transfer between Germany and the United States, 1884–1930", *History of Technology* 8 (1983).

Braune, Hjalmar, "Om utvecklingen af den svenska masugnen", *Jernkontorets Annaler* n.s. 59 (1904).

Bring, Samuel E., "Bidrag till Christopher Polhems lefnadsteckning", *Christopher Polhem: Minnesskrift utgifven af Svenska Teknologföreningen* (Stockholm, 1911).

— ed., *Göta Kanals historia*, Vol. 2:1 (Uppsala, 1930).

British Museum General Catalogue, Vol. 238 (London, 1964).

von Bromell, Magnus, *Inledning til nödig kundskap at igenkiänna och upfinna allahanda bergarter* (Stockholm, 1730).

Brötje, Anne-Marie, "Samuel Worster. En frihetstida Stockholmsköpman", *Personhistorisk Tidskrift* 41 (1942–1945).

Burke's Genealogical and Heraldic History of the Landed Gentry (London, 1937).

Burke's Genealogical and Heraldic History of the Landed Gentry of Ireland (London, 1958, 4th ed.).

Burke's Peerage and Baronetage (London, 1978).

Bæckström, Arvid, "Kongl. Modellkammaren", *Daedalus* 1959.

Börjeson, Hjalmar, *Biografiska anteckningar om örlogsflottans officerare 1700–1799* (Stockholm, 1942).

Caledonian Mercury 1725, March 25, March 29, March 30, April 1, April 5, April 6, April 8.

Calvert, Albert F., *The Grand Lodge of England 1717–1917: Being an Account of 200 Years of English Freemasonry* (London, 1917).

Cardwell, Donald S. L., *From Watt to Clausius: The Rise of Thermodynamics in the Early Industrial Age* (Ithaca, 1971).

— *Technology, Science and History* (London, 1972).

Carlberg, Johan O., *Historiskt sammandrag om svenska bergverkens uppkomst och utveckling samt grufvelagstiftningen* (Stockholm, 1879).

Carlborg, Harald, "Om tramphjul och andra motorer i äldre tid vid svenska malmgruvor", *Med Hammare och Fackla* 25 (1967).

Carlgren, Wilhelm, *De norrländska skogsindustrierna intill 1800-talets mitt* (Norrländskt handbibliotek, Vol. 11: Uppsala, 1926).

— "Norrländsk trävarurörelse genom seklen", *Svenska kulturbilder*, n.s. Vol. 4, part 7–8 (Stockholm, 1937).

Carlqvist, Sten, "Kanonborrningsmaskin från 1700-talet", *Daedalus* 1934.

Carlsson, Alfred Bernhard, *Den svenska centralförvaltningen 1521–1809* (Stockholm, 1913).

— ed., *Uppsala universitets matrikel*, Vol. 2 (Uppsala, 1919–1923).

Carlsson, Bo et al., *Teknik och industristruktur—70-talets ekonomiska kris i historisk belysning* (IUI-publikation; IVA-meddelande, No. 218: Stockholm, 1979).

Carlsson, Sten, *Ståndssamhälle och ståndspersoner 1700–1865: Studier rörande det svenska ståndssamhällets upplösning* (Lund, 1973, rev. ed.).

— *Fröknar, mamseller, jungfrur och pigor: Ogifta kvinnor i det svenska ståndssamhället* (Studia Historica Upsaliensia, No. 90: Uppsala, 1977).

Cassel, Bo, *Havet, dykaren, fynden under 2000 år* (Stockholm, 1967).

— "Dykarkonstens utveckling i Sverige fram till

1850-talet", *Sjöhistorisk Årsbok* 1975–1976.

Celsius, Anders, ed., *Almanach* ... 1728, 1730, 1731, 1732, 1734, 1735, 1736. Copies in KB.

Christie, John R. R., "The Origins and Development of the Scottish Scientific Community, 1680–1760", *History of Science* 12 (1974).

Clark-Kennedy, A. E., *Stephen Hales, D.D., F.R.S.: An Eighteenth Century Biography* (Cambridge, 1929).

Clarke, Edward Daniel, *Travels in Various Countries of Europe Asia and Africa*, Vol. 2 (London, 1824).

Corin, Carl-Fredrik, "A. N. Edelcrantz och hans ångmaskinsprojekt år 1809", *Daedalus* 1940.

— "N. A. Edelcrantz och Eldkvarn", *Daedalus* 1961.

Cronstedt, Nils, "Polhems arbeten på byggnadskonstens områden", in: *Christopher Polhem: Minnesskrift utgifven av Svenska Teknologföreningen* (Stockholm, 1911).

Daedalus hyperboreus 1716.

Dahl, Helmer, *Teknikk Kultur Samfunn: Om egenarten i Europas vekst* (Oslo, 1983).

Dahlberg, Gunilla, "Datering av skrifter med hjälp av vattenmärken", *Samlaren* 95 (1974).

Dahlgren, E. W., ed., *Svenska Vetenskapsakademiens protokoll för åren 1739, 1740 och 1741*, Vol. 1 (Stockholm, 1918).

von Dalin, Olof, ed., *Then swänska Argus* 1732–1734.

Daumas, Maurice & Gille, Paul, "Methods of Producing Power", in: Maurice Daumas, ed., *A History of Technology and Invention: Progress Through the Ages*, Vol. 3 (New York, 1979). The French original: idem, *Histoire générale des techniques*, Vol. 3 (Paris, 1968).

Delius, Christoph Traugott, *Anleitung zu der Bergbaukunst* (Wien, 1773).

Descriptions des arts et métiers (Paris, 1761–1788).

Dickinson, Henry W., *A Short History of the Steam Engine* (Cambridge, 1939).

Dictionary of National Biography, Vol. 14 (London, 1888).

Dunér, Nils C., *Kungliga Vetenskaps Societetens i Upsala tvåhundraårsminne* (Uppsala, 1910).

Eck, Johann Georg, *Reisen in Schweden* (Leipzig, 1806).

Edelcrantz, Abraham Niclas, "Afhandling om nödige rättelser vid mätningen af ångornes spänstighet, och determination af deras kraft, uti ångmachiner, samt beskrifning på en förbättrad ångmätare", *KVAH* 1809.

Edén, Nils, *Den svenska centralregeringens utveckling till kollegial organisation i början af sjuttonde århundradet (1602–1634)* (Skrifter utgivna av Humanistiska Vetenskaps-Samfundet i Uppsala, No. 8.2: Uppsala, 1902).

Edquist, Charles & Edqvist, Olle, *Social Carriers of Techniques for Development* (Swedish Agency for Research Cooperation with Developing Countries, SAREC Report No. R.3: Stockholm, 1979). For an abridged version with the same title and by the same authors, see: *Journal of Peace Research* 16 (1979).

Eenberg, Johan, *Kort berättelse af de märkwärdigste saker som för de främmande äre at besee och förnimma uti Upsala stad* (Upsala, 1704).

Ehrensvärd, Ulla, "Gruvor på kartor", in: Nicolai Herlofson et al., eds., *Vilja och kunnande: Teknikhistoriska uppsatser tillägnade Torsten Althin på hans åttioårsdag den 11 juli 1977 av vänner* (Uppsala, 1977).

Ek, Sven B., *Väderkvarnar och vattenmöllor: En etnologisk studie i kvarnarnas historia* (Nordiska museets handlingar, No. 58: Stockholm, 1962).

Ekström, Gunnar, "När upptäcktes Östra Silverberget?", *Historisk Tidskrift* 68 (1948).

Elgenstierna, Gustaf, *Den introducerade svenska adelns ättartavlor*, 9 Vols. (Stockholm, 1925–1936).

Elmroth, Ingvar, *Nyrekryteringen till de högre ämbetena 1720–1809: En social-historisk studie* (Bibliotheca Historica Lundensis, No. 10: Lund, 1962).

Elvius, Pehr, *Mathematisk tractat om effecter af vatndrifter, efter brukliga vatn-värks art och lag* (Stockholm, 1742).

— "Theorien om vatten-drifter jämförd med försök", *KVAH* 1743.

— "Rön vid trampkvarnar", *KVAH* 1744.

Erdman, Axel, "Dannemora jernmalmsfält i Upsala län, till dess geognostiska beskaffenhet skildradt", *KVAH* 1850.

Erici, Einar, *Inventarium över bevarade äldre kyrkorglar i Sverige* (Stockholm, 1965).

Ericsson, Ernst & Rabe, Gustaf, *Kungl. Fortifikationens historia*, Vol. 4:1 (Stockholm, 1930).

Eriksson, Gunnar, *Elias Fries och den romantiska biologien* (Lychnos-Bibliotek, No. 20: Uppsala, 1962).

— "Motiveringar för naturvetenskap: En översikt av den svenska diskussionen från 1600-talet till första världskriget", *Lychnos* 1971–1972.

— "Om idealtypbegreppet i idéhistorien", *Lychnos* 1979–1980.

— "Den nordströmska skolan", *Lychnos* 1983.

Eriksson, Gösta A., *The Iron Mines of Dannemora and Dannemora Iron* (Meddelanden från Uppsala universitets geografiska institution, No. 161: Uppsala, 1961).

Erixon, Sigurd, "Spjället, en exponent för svensk bostadsteknik", *Svenska Kulturbilder*, n. s. Vol. 5 (Stockholm, 1937).

Fairbairn, William, *Treatise on Mills and Millwork*, Vol. 1 (London, 1861).

Fant, Johan Eric & Låstbom, August Theodor, *Upsala ärkestifts herdaminne*, Vol. 1 (Upsala, 1842).

Farey, John, *A Treatise on the Steam Engine: Historical, Practical, and Descriptive* (London, 1827).

Ferguson, Eugene S., "The Origins of the Steam Engine", *Scientific American* 210 (1964).

— "Technical Museums and International Exhibitions", *Technology and Culture* 6 (1965).

— *Bibliography of the History of Technology* (Cambridge, Mass., 1968).

— "The Measurement of the 'Man-Day'", *Scientific American* 225 (1971).

— "Toward a Discipline of the History of Technology", *Technology and Culture* 15 (1974).

— "The Mind's Eye: Nonverbal Thought in Technology", *Science* 197 (1977).

Flinn, Michael W., "The Travel Diaries of Swedish Engineers of the Eighteenth Century as Sources of Technological History", *Transactions of the Newcomen Society* 31 (1957–1958 and 1958–1959).

Foucault, Michel, *Discipline and Punish: The Birth of the Prison* (Harmondsworth, 1979).

Forsstrand, Carl, "Mårten Triewald och hans Stockholmsträdgård", *Svenska Linné-Sällskapets Årsskrift* 10 (1927).

Franzén, Olle, "Pehr Lagerhjelm", *Svenskt Biografiskt Lexikon*, Vol. 22 (Stockholm, 1979).

Fries, Theodor M., *Linné: Lefnadsteckning*, Vol. 1 (Stockholm, 1903).

Frunck, Gudmund, *Bref rörande nya skolans historia, 1810–1811* (Skrifter utgifna af svenska litteratursällskapet: Upsala, 1891).

Frängsmyr, Tore, "Den gudomliga ekonomin: Religion och hushållning i 1700-talets Sverige", *Lychnos* 1971–1972.

— *Wolffianismens genombrott i Uppsala: Frihetstidens universitetsfilosofi till 1700-talets mitt* (Acta Universitatis Upsaliensis, No. C.26: Uppsala, 1972).

— "Christian Wolff's Mathematical Method and Its Impact on the Eighteenth Century", *Journal of the History of Ideas* 36 (1975).

— "Vetenskapsmannen och samhället i historisk belysning", in: *Vetenskapsmannen och samhället: Symposier vid Kungl. Vetenskapssamhället i Uppsala 1976–1977* (Acta Academiæ Regiæ Scientiarum Upsaliensis, No. 19: Stockholm, 1977).

— "Vetenskapens roll i historien", *Lychnos* 1983.

— "History of Science in Sweden", *Isis* 74 (1983).

Funck, Alexander, *Beskrifning om tjäru- och kolugnars inrättande* (Stockholm, 1748, 2nd ed. 1772).

Furuskog, Jalmar, *De värmländska järnbruken: Kulturgeografiska studier över den värmländska järnhanteringen under dess olika utvecklingsskeden* (Filipstad, 1924).

Gadd, Pehr Adrian, "Rön och försök med murbruk och ciment-arter", *KVAH* 1770.

— "Rön, om skiffergångarna i Finland, och takskiffer i dem", *KVAH* 1780.

Garney, Johan Carl, *Handledning uti svenska masmästeriet* (Stockholm, 1791).

Georgii, Johan F., *Jordens alla mått och vigter* (Linköping, 1842).

Gerland, E., "Die erste in Deutschland in dauernden Betrieb genommen Dampfmaschine", *Zeitschrift des Vereins Deutscher Ingenieure* 49 (1905).

Gillmor, C. Stewart, *Coulomb and the Evolution of Physics and Engineering in Eighteenth-Century France* (Princeton, 1971).

Glacken, Clarence J., *Traces on the Rhodian Shore: Nature and Culture in Western Thought from Ancient Times to the End of the Eighteenth Century* (Berkeley, 1976).

Glete, Jan, "Teknikhistoria—viktig i ekonomisk och historisk forskning", *Daedalus* 49 (1980).

Gough, J. B., "René-Antoine Ferchault de Réaumur", in: Charles Coulston Gillispie, ed., *Dictionary of Scientific Biography*, Vol. 11 & 12 (New York, 1981).

Granström, Gustaf A., "Svensk bergsmannagärning i Spanien under 1700-talet", *Med Hammare och Fackla* 1 (1928).

Gren, A. G., "Beschreibung der wesentlichen Einrichtung der neuern Dampf- oder Feuermaschinen nebst einer Geschichte dieser Erfindung …", *Neues Journal der Physik* 1 (1795).

Hahn, Roger, *The Anatomy of a Scientific Institution: The Paris Academy of Sciences, 1666–1803* (Berkeley, 1971).

Hall, A. Rupert, "John Theophilus Desaguliers", in: Charles C. Gillispie, ed., *Dictionary of Scientific Biography*, Vol. 3 & 4 (New York, 1981).

Halldin, Gustaf, ed., *Svenskt skeppsbyggeri: En översikt av utvecklingen genom tiderna* (Malmö, 1963).

Hallerdt, Björn, "Strumpvävstolar av Christopher Polhem", *Daedalus* 1951.

Hallman, Per, "Tillståndet i Uppland under det stora nordiska kriget", *Karolinska Förbundets Årsbok* 1919.

Hamberg, Erik, "Idéer kring och förslag till förbättrad skogshushållning i Sverige på 1700-talet" (mimeographed paper presented at the Department of History of Ideas and Science, University of Gothenburg, 1975).

Hammarskiöld, Ludvig, "Kopparkanoner i Sverige och deras tillverkning", *Med Hammare och Fackla* 18 (1949–1950).

Hammarskjöld, Hjalmar, *Om grufregal och grufegendom i allmänhet enligt svensk rätt* (Upsala, 1891).

Hansotte, Georges, "L'introduction de la machine à vapeur au pays de Liège (1720)", *La Vie Wallone* 24 (1950).

Harris, John R., "The Employment of Steam Power in the Eighteenth Century", *History* 52 (1967).

— "Recent Research on the Newcomen Engine and Historical Studies", *Transactions of the Newcomen Society* 50 (1978–79).

Hebbe, Per, "Anders Gabriel Duhres 'Laboratorium mathematico-oeconomicum'" *Kungl. Landtbruksakademiens Handlingar och Tidskrift* 72 (1933).

Heckscher, Eli F., "Jonas Alströmer", *Svenskt Biografiskt Lexikon*, Vol. 20 (Stockholm, 1918).

— "Den gamla svenska brukslagstiftningens betydelse", in: *En bergsbok till Carl Sahlin* (Stockholm, 1921).

— *Svenskt arbete och liv* (Stockholm, 1942).

— *Sveriges ekonomiska historia från Gustav Vasa*, 2 Vols. (Stockholm, 1936–1949).

Hederström, Hans, "Näsby socken i Östergötland, beskrifwen år 1755", *KVAH* 1757.

Heijkenskjöld, Carl, "Krigsfångar som arbetare i Sala gruva och annorstädes", *Med Hammare och Fackla* 8 (1937).

Heilbron, John L., *Electricity in the 17th and 18th Centuries: A Study of Early Modern Physics* (Berkeley, 1979).

— *Physics at the Royal Society during Newton's Presidency* (Los Angeles, 1983).

Heinke, F. W. & Davis, W. G., *A History of Diving from the Earliest Times to the Present Date* (London, 1873, 4th ed.).

Henriques, Pontus, *Skildringar ur Kungl. Tekniska Högskolans historia*, Vol. 1 (Stockholm, 1917).

Hildebrand, Bengt, "Till släkten Triewalds historia", *Personhistorisk Tidskrift* 39 (1938).

— *Kungl. Svenska Vetenskapsakademien: Förhistoria, grundläggning och första organisation* (Stockholm, 1939).

— "Pehr Elvius, d.y.", *Svenskt Biografiskt Lexikon*, Vol. 13 (Stockholm, 1950).

Hildebrand, Bror Emil, *Sveriges och svenska konungahusets minnespenningar, praktmynt och belöningsmedaljer beskrifna*, Vol. 1 (Stockholm, 1874).

Hildebrand, Karl-Gustaf, "Brukshistoria och brukskultur", *Blad för Bergshandteringens Vänner* 33 (1958).

— *Fagerstabrukens historia: Sexton- och sjuttonhundrata-*

len (Fagerstabrukens historia, Vol. 1: Uppsala, 1957).

Hills, Richard L., *Power in the Industrial Revolution* (Manchester, 1970).

— "A One-Third Scale Working Model of the Newcomen Engine of 1712", *Transactions of the Newcomen Society* 44 (1971–72).

Hindle, Brooke, "The Transfer of Power and Metallurgical Technologies to the United States, 1800–1880: Processes of Transfer, with Special Reference to the Role of the Mechanics", in: *L'acquisition des techniques par les pays non-initiateurs* (Colloques Internationaux du Centre National de la Recherche Scientifique, No. 538: Paris, 1973).

Hiärne, Urban, *En lijten oeconomisk skrifft om wedsparande, huru man i desse knappa tijder med weden som efter handen begynner at tryta, bätter omgås skall, och till wärmande anwända med bättre nytta och sparsamhet, dem oförmögnom till tröst och lindring* (Stockholm, 1696).

Hjelm, Peter Jacob, "Minerographiske antekningar om porphyrbergen i Elfdals-socken och Öster-Dalarna, samt deras gränsor i omkringliggande socknar", *KVAH* 1805.

Hjärne, Rudolf, ed., *Dagen före drabbningen eller nya skolan och dess män i sin uppkomst och sina förberedelser 1802–1810* (Stockholm, 1882).

Hodgson, John C., "An Alphabetical Catalogue of the Goldsmiths of Newcastle", *Archaeologica Aeliana*, 3rd. s., 11 (1914).

Hoffmann, Dietrich, "Die frühesten Berichte über die erste Dampfmaschine auf dem europäischen Kontinent", *Technikgeschichte* 41 (1974).

Hollister-Short, Graham J., "Leads and Lags in late Seventeenth-Century English Technology", *History of Technology* 1 (1976).

— "The Introduction of the Newcomen Engine into Europe", *Transactions of the Newcomen Society* 48 (1976–77).

— "A New Technology and Its Diffusion: Steam Engine Construction in Europe 1720–c.1780", *Industrial Archaeology* 13 (1978).

— "Antecedents and Anticipations of the Newcomen Engine", *Transactions of the Newcomen Society* 52 (1980–81).

Holmberg, Arne, ed., "Jac. Berzelius. Brev till medlemmar av familjen Brandel om resan till England 1812 och till Frankrike 1818/19", *Kungl. Svenska Vetenskapsakademiens Årsbok* 1955.

Holmkvist, Erik, *Bergslagens gruvspråk* (Uppsala, 1941).

— *Bergslagens hyttspråk* (Uppsala, 1945).

Holmstrand, Per Olov, "Dannemora gruvor 500 år", *Jernkontorets Annaler* 165 (1981).

von Horn, Lorentz Leopold, *Biografiska anteckningar*, Vol. 2 (Örebro, 1937).

Hubendick, Edvard, "'Konstige påfund' och tekniska frågor, dryftade på Kungl. Vetenskapsakademiens sammanträden under 1700-talet", *Kungl. Svenska Vetenskapsakademiens Årsbok* 1948.

Hughes, Edward, "The first Steam Engines in the Durham Coalfield", *Archaeologica Aeliana*, 4th s., 27 (1949).

— *North Country Life in the Eighteenth Century, the North East 1700–1750* (London, 1952).

Hughes, Thomas P., "Emerging Themes in the History of Technology," *Technology and Culture* 20 (1979).

Hunter, Louis C., *Waterpower: A History of Industrial Power in the United States, 1780–1930* (Charlottesville, 1979).

Hülphers, Abraham Abrahamsson, *Dagbok öfwer en resa igenom de, under Stora Kopparbergs höfdingedöme lydande lähn och Dalarne år 1757* (Wästerås, 1762).

— *Historisk afhandling om musik och instrumenter särdeles om orgwerks inrättningen i allmänhet, jemte kort beskrifning öfwer orgwerken i Swerige* (Westerås, 1773).

Hägglund, John, *Kinda kanals historia* (Linköping, 1966).

Högberg, Staffan, ed., *Anton von Swabs berättelse om Avesta kronobruk 1723* (Jernkontorets Bergshistoriska Skriftserie, No. 19: Stockholm, 1983).

Industriföretagens forsknings- och utvecklingsverksamhet 1977–1981 (Svenska Statistiska Centralbyrån, Statistiska meddelanden, series U: Stockholm, 1982).

Jackson, Melvin H. & de Beer, Carol, *Eighteenth Century Gunfounding* (Newton Abbot, 1973).

Jamison, Andrew, *National Components of Scientific Knowledge: A Contribution to the Social Theory of Science* (Research Policy Institute, University of Lund: Lund, 1982).

Jansson, Sam Owen, *Måttordbok: Svenska måttstermer före metersystemet* (Stockholm, 1950).

Jars, Antoine Gabriel, *Voyages métallurgiques, ou recherches et observations sur les mines & forges de fer*, Vol. 1 (Lyon, 1774).

Jeremy, David J., *Transatlantic Industrial Revolution: The Diffusion of Textile Technologies between Britain and America, 1790–1830* (Cambridge, Mass., 1981).

Johannisson, Karin, "Naturvetenskap på reträtt: En diskussion om naturvetenskapens status under svenskt 1700-tal", *Lychnos* 1979–1980.

Johnsen, Heinrich Philip, "Kort orgwerks-beskrifning", in: Abraham Abrahamsson Hülphers, *Historisk afhandling om musik och instrumenter särdeles om orgwerks inrättningen i allmänhet, jemte en kort beskrifning öfwer orgwerken i Swerige* (Westerås, 1773).

Jägerskiöld, Stig, *Sverige och Europa 1716–1718: Studier i Karl XII:s och Görtz' utrikespolitik* (Ekenäs, 1937).

Jöransson, Christian Ludvig, *Tabeller, som föreställa förhållandet emellan Sveriges och andra länders mynt, vigt och mått* (Stockholm, 1777).

Kalm, Pehr, *Menlöse tankar om brädsågning* (Åbo, 1772).

Kanefsky, John & Robey, John, "Steam Engines in 18th-Century Britain: A Quantitative Assessment", *Technology and Culture* 21 (1980).

Kerker, Milton, "Science and the Steam Engine", in: Thomas Parke Hughes, ed., *The Development of Western Technology since 1500* (New York, 1964). Originally published in *Technology and Culture* 2 (1961), 381–390.

Kjellander, Rune, "Carl Knutberg", *Svenskt Biografiskt Lexikon*, Vol. 21 (Stockholm, 1977).

Kjellberg, Sven T., *Ull och ylle: Bidrag till den svenska yllemanufakturens historia* (Lund, 1943).

Klemm, Friedrich, *Geschichte der naturwissenschaftlichen und technischen Museum* (Deutsches Museum, Abhandlungen und Berichte 41 (1973), Heft 2: München, 1973).

Knutberg, Carl, "Nytt påfund, vid väder-quarnars inrättning, at i lugnt väder malningen må kunna förrättas medelst hästvind", *KVAH* 1751.

— *Tal om nyttan af ett Laboratorium Mechanicum* (Inträdestal KVA: Stockholm, 1754).

— "Beskrifning, med bifogad ritning, på en finbladig såg-qvarn", *KVAH* 1769.

Kongl. stadgar, förordningar, privilegier och resolutioner, angående justitien och hushållningen wid bergwerken och bruken, med hwad som ther til hörer både inom och utom Bergslagerne uti Sweriges rike, och ther under lydande provincier, uppå Kongl. May:ts allernådigste befalning, til almän nytto och efterrättelse igenom trycket utgifne år 1736 (Stockholm, 1736).

Kongl. stadgar, förordningar, bref och resolutioner, angående justitien och hushållningen wid bergwerken och bruken. Första fortsättningen. Från och med år 1736 til och med år 1756 (Stockholm, 1786).

Kongl. stadgar, förordningar, bref och resolutioner, angående justitien och hushållningen wid bergwerken och bruken. Andra fortsättningen. Ifrån och med år 1757 til och med år 1791 (Stockholm, 1797).

Kraft, Jens, *Forelæsninger over mekanik med hosføiede tillæg* (Sorøe, 1763).

— *Forelæsninger over statik og hydrodynamik med maskinvæsenets theorier som den anden deel af forelæsningerne over mekaniken* (Sorøe, 1764).

Kromnow, Åke, "Övermasmästareämbetet under 1700-talet (1751–1805): Dess organisation och verksamhet samt betydelse för den svenska tackjärnstillverkningen", *Med Hammare och Fackla* 9 (1938) & 10 (1939).

Kronberg, Algot, *Pehr Lagerhjelm: Vetenskapsman, näringsidkare, politiker* (Stockholm, 1960).

Kuhn, Thomas S., "Energy Conservation as an Example of Simultaneous Discovery", in: idem, *The Essential Tension: Selected Studies in Scientific Tradition and Change* (Chicago, 1977). Originally published in: Marshall Clagett, ed., *Critical Problems in the History of Science* (Madison, 1959).

— "The Relations between History and the History of Science", in: idem, *The Essential Tension: Selected Studies in Scientific Tradition and Change* (Chicago, 1977). Originally published in *Daedalus*, the journal of the American Academy of Arts and Sciences, 100 (1971).

Kumlien, Kjell, ed., *Norberg genom 600 år: Studier i en gruvbygds historia* (Uppsala, 1958).

— "Bergsordningarnas tillkomst—en kulturbild från frihetstiden", *Med Hammare och Fackla* 23 (1963).

Kyhlberg, Bengt, "Orgelbyggarefamiljen Cahman, Hülphers och orgeln i Trefaldighetskyrkan i Kristianstad", *Svensk Tidskrift för Musikforskning* 27 (1945).

König, Carl Henrik, *Inledning til mecaniken och bygnings-konsten, jämte en beskrifning öfwer åtskillige af framledne commerce-rådet och commenduren af Kongl. Nordstierneorden hr. Polhem opfundne machiner* (Stockholm, 1752).

Lagerhjelm, Pehr; af Forselles, Jacob Henrik & Kallstenius, Georg Samuel, *Hydrauliska försök, antällda vid Fahlu grufva, åren 1811–1815*, 2 Vols. (Stockholm, 1818–1822).

Lagerhjelm, Pehr, *Tal om hydraulikens närvarande tillstånd* (Presidii tal KVA: Stockholm, 1837).

— "Jacob Henrik af Forselles", *Lefnadsteckningar öfver Kongl. Svenska Vetenskaps Akademiens efter år 1854 aflidna ledamöter*, Vol. 1 (Stockholm, 1869–1873).

Lagerholm, Julius, ed., *Södermanland-Närkes nation: Biografiska och genealogiska anteckningar om i Uppsala studerande södermanlänningar och närkingar 1595–1900* (Sunnansjö, 1933).

Lamm, Martin, "Samuel Triewalds lif och diktning", *Samlaren* 28 (1907).

Landes, David S., *The Unbound Prometheus: Technological Change and Industrial Development in Western Europe from 1750 to the Present* (London, 1970).

de Latocnaye, De Bougrenet, *Promenade d'un Français en Suède et en Norvège* (Brunswick, 1801).

Laurel, Lars, *Åminnelsetal öfwer Mårten Triewald* (Stockholm, 1748).

Layton, Jr., Edwin T., *The History of Technology as an Academic Discipline* (Stockholm Papers in the History and Philosophy of Technology, TRITA-HOT-1003: Stockholm, 1981).

Leide, Arvid, *Fysiska institutionen vid Lunds universitet* (Acta Universitatis Lundensis, Sectio I, Theologica Juridica Humaniora 8: Lund, 1968).

Leijel, Adam, "Narratio accurata de cadavere humano in fodina Cuprimontana ante duos annos reperto", *Acta literaria Sveciae* 1 (1720–1724).

Lejonmark, Gustaf Adolf, "Tilläggning til föregående afhandling, eller en jämförelse imellan den coniska och den cylindriska linkorgen", *KVAH* 1796.

Leupold, Jacob, *Theatrum Machinarum Hydraulicarum. Tomus I. Oder: Schau-Platz der Wasser-Künste, Erster Teil* (Leipzig, 1724).

Levander, Lars, *Brottsling och bödel* (1933), facsimile edition (Stockholm, 1975).

Lewenhaupt, Adam, *Karl XII:s officerare: Biografiska anteckningar*, Vol. 2 (Stockholm, 1921).

Lidén, Johan Hinric, ed., *Brefwäxling imellan ärkebiskop Eric Benzelius den yngre och dess broder, censor librorum Gustaf Benzelstierna* (Linköping, 1791).

Liljedahl, Gösta, "Om vattenmärken och filigranologi", *Historisk Tidskrift* 76 (1956).

— "Om vattenmärken i papper och vattenmärkesforskning (filigranologi)", *Biblis* 1970.

Liljencrantz, Axel, "Polhem och grundandet av Sveriges första naturvetenskapliga samfund jämte andra anteckningar rörande Collegium curiosorum", *Lychnos* 1939 & 1940.

— ed., *Christopher Polhems brev* (Lychnos-Bibliotek, No. 6: Uppsala, 1941–46).

Liljendahl, Lennart, "Upplands bruksdammar", *Uppland* 1979.

— "Dannemora gruvor först med ångmaskin i Sverige", *Upsala Nya Tidning* 1982, August 7.

Lilley, Samuel, "Technological Progress and the Industrial Revolution 1700–1914", in: Carlo M. Cipolla, ed., *The Fontana Economic History of Europe: The Industrial Revolution* (The Fontana Economic History of Europe, Vol. 3: Huntington, 1973).

Lindberg, Bo & Nilsson, Ingemar, "Sunt förnuft och inlevelse. Den nordströmska traditionen", in: Tomas Forser, ed., *Humaniora på undantag: Humanistiska forskningstraditioner i Sverige* (Stockholm, 1978).

Lindbom, Gustaf Aron, "Beskrifning på en ny hästvind vid Persberget", *KVAH* 1796.

— "Om coniska hästvindar, at nyttja til upfordring vid grufvor", *KVAH* 1798.

Lindborg, Rolf, *Descartes i Uppsala: Striderna om "nya filosofien" 1663–1689* (Lychnos-Bibliotek, No. 22: Uppsala, 1965).

Lindgren, Torild, "Från högbarock till senbarock. Stilstudier i Cahmanskolan", *Svensk Tidskrift för Musikforskning* 38 (1956).

Lindqvist, Svante, "Nya bidrag till ångmaskinens historia", *Daedalus* 1976.

— *Teknikhistoria som läroämne vid universiteten i Storbritannien* (Stockholm Papers in History and Philosophy of Technology, TRITA-HOT-5001: Stockholm, 1976).

— "The Work of Martin Triewald in England", *Transactions of the Newcomen Society* 50 (1978–79).

— "The Impact of the Introduction of Steam Power Technology on the Society of Dannemora Mines—A Case Study in Transfer of Technology", in: Sigvard Strandh, ed., *Technology and its Impact on Society* (Tekniska Museet Symposia, No. 1: Stockholm, 1979).

— Teknikhistoria—motiv och mål", *Daedalus* 49 (1980).

— *The Teaching of History of Technology in USA: A Critical Survey in 1978* (Stockholm Papers in History and Philosophy of Technology, TRITA-HOT-5003: Stockholm, 1981).

— "Projektet 'Det medeltida tramphjulet'—en övningsuppgift i teknikhistoria på KTH", *Daedalus* 50 (1981).

— "Discussion: An Engineer Is an Engineer Is an Engineer?", in: Carl Gustaf Bernhard et al., eds., *Science, Technology and Society in the time of Alfred Nobel* (Nobel Symposia, No. 52: Oxford, 1982).

— "Vad är teknik?", in: Bo Sundin, ed., *Teknik för alla: Uppsatser i teknikhistoria* (Institutionen för idéhistoria, Umeå Universitet, Skrifter, No. 17: Umeå, 1983).

— "Natural Resources and Technology: The Debate about Energy Technology in Eighteenth-Century Sweden", *Scandinavian Journal of History* 8 (1983).

Lindroth, Sten, "Urban Hiärne och Laboratorium chymicum", *Lychnos* 1946–1947.

— *Christopher Polhem och Stora Kopparberget: Ett bidrag till bergsmekanikens historia* (Uppsala, 1951).

— *Gruvbrytning och kopparhantering vid Stora Kopparberget intill 1800-talets början*, 2 Vols. (Uppsala 1955).

— *Kungl. Svenska Vetenskapsakademiens historia 1739–1818*, 2 Vols. (Stockholm, 1967).

— *Svensk lärdomshistoria*, Vols. 2–3 (Stockholm, 1975–1978).

— *Uppsala universitet 1477–1977* (Uppsala, 1976).

— "The Two Faces of Linnaeus", in: Tore Frängs-myr, ed., *Linnaeus: The Man and His Work* (University of California Press: Berkeley, 1983).

von Linné, Carl, *Skånska resa, på höga öfwerhetens befallning förrättad år 1749 ...* (Stockholm, 1751).

Linnaeus (von Linné), Carl, *Lapplandsresa år 1732* (Stockholm, 1975).

Louis, Henry, "Early Steam-Engines in the North of England", *Transactions of the Institution of Mining Engineers* 82 (1932).

Lundequist, Nils, *Stockholms stads historia, från stadens anläggning till närwarande tid*, Vol. 3 (Stockholm, 1829).

Lundqvist, Bo V:son, ed., *Västgöta nation i Uppsala från år 1595*, Vol. 1 (Uppsala, 1928–1946).

Macey, Samuel C., *Clocks and the Cosmos: Time in Western Life and Thought* (Hamden, 1980).

Machines et inventions approuvées par l'Académie royale des sciences, depuis son établissement jusqu'à present; avec leur description, 6 Vols. (Paris, 1735).

Mahoney, Michael S., "Edmé Mariotte", in: Charles G. Gillispie, ed., *Dictionary of Scientific Biography*, Vol. 9 & 10 (New York, 1981).

Malmeström, Elis & Uggla, Arvid Hj., eds., *Vita Caroli Linnæi: Carl von Linnés självbiografier* (Stockholm, 1957).

Malmsten, Karl, "Ingenjörens titel och tradition", *Med Hammare och Fackla* 11 (1940–1941).

Malmström, Oscar, *Nils Bielke och kriget mot turkarna, 1684–1687* (Stockholm, 1895).

— *Nils Bielke såsom generalguvenör i Pommern, 1687–1697* (Stockholm, 1896).

— *Högmålsprocessen mot Nils Bielke* (Stockholm, 1899).

Malone, Patrick M., "Teaching the History of Technology in Museums", *Humanities Perspectives on Technology* (Curriculum Newsletter of the Lehigh HPT Program), 1978, No. 9.

Mannix, Daniel P., *Black Cargoes: A History of the Atlantic Slave Trade 1518–1865* (London, 1963).

Matschoss, Conrad, "Die ersten Dampfmaschinen ausserhalb Englands", *Zeitschrift des Vereins Deutscher Ingenieure* 49 (1905).

— *Die Entwicklung der Dampfmaschine*, Vol. 1 (Berlin, 1908).

Mattsson, Leif & Stridsberg, Einar, *Det industriinriktade skogsbruket sett ur ett historiskt perspektiv* (Kulturgeografiskt seminarium, 8/79: Stockholm, 1979).

Meinander, Nils, *En krönika om vattensågen* (Helsingfors, 1945).

Meijer, Gerhard, "Beskrifning och ritning på en mindre bårr-machine til massift gutne canoner", *KVAH* 1782.

Menlös, Daniel, *Kort beskrifning af den hydrostatiske wåg-balken* (Stockholm, 1728).

Merchant, Carolyn, "Mining the Earth's Vomb", in: Joan Rotschild, ed., *Machina Ex Dea: Feminist Perspectives on Technology* (New York, 1983).

Meyerson, Åke, "Rationaliseringssträvanden vid svenska gevärsfaktorier under 1700-talets mitt", *Daedalus* 1937.

— "Carl Knutberg", *Daedalus* 1937.

Millqvist, Folke, "Bomullens tidiga historia och spinningens mekanisering", *Från Borås och De Sju Häraderna* 34 (1981).

de Miranda, Francisco, *Archivo del General Miranda*, Vol. 3 (Caracas, 1929).

Molander, Bo, "Ryska härjningar under Nordiska kriget knäckte svensk järnindustri", *Daedalus* 48 (1978–1979).

Molin, Harry, *Karlholms bruks bok: En krönika kring ett upplandsbruk* (Karlholm, 1950).

Montelius, Sigvard; Utterström, Gustaf & Söderlund, Ernst, *Fagerstabrukens historia: Arbetare och arbetarförhållanden* (Fagerstabrukens historia, Vol. 5: Uppsala, 1959).

Moran, Bruce T., "German Prince-Practitioners: Aspects in the Development of Courtly Science, Technology, and Procedures in the Renaissance", *Technology and Culture* 22 (1981).

Morton Briggs, J., "Antoine Parent", in: Charles C. Gillispie, ed., *Dictionary of Scientific Biography*, Vol. 9 & 10 (New York, 1981).

Mumford, Lewis, *Technics and Civilization* (New York, 1963).

Musson, A. E. & Robinson, Eric, *Science and Technology in the Industrial Revolution* (Manchester, 1969).

Nedoluha, Alois, "Kulturgeschichte des technischen Zeichnens", *Blätter für Technikgeschichte* 19 (1957), 20 (1958), 21 (1959).

Needham, Joseph, *Science and Civilization in China*, Vol. 4, Part 2 (Cambridge, 1965).

— "The Pre-Natal History of the Steam-Engine", in: idem, *Clerks and Craftsmen in China and the West: Lectures and Addresses on the history of Science and Technology* (Cambridge, 1970).

Neumeyer, Friedrich, "Christopher Polhem och hydrodynamiken", *Arkiv för matematik, astronomi och fysik*, Band 28 A (1942), No. 15.

Newcastle Courant 1723–1724.

Nilsson, Stig, *Terminologi och nomenklatur: Studier över begrepp och deras uttryck inom matematik, naturvetenskap och teknik. I* (Lundastudier i Nordisk Språkvetenskap, series A No. 26: Lund, 1974).

— "Materieuppfattningens termer i svenskan fram till år 1800", *Lychnos* 1975–1976.

Nisser, Marie, "Byggnadsteknisk debatt och utbildning i Sverige under 1600- och 1700-talen" (mimeographed fil. lic. dissertation at the Department of History of Art, University of Uppsala, 1966).

— ed., *Industriminnen: En bok om industri- och teknikhistoriska bebyggelsemiljöer* (Stockholm, 1979).

Norberg, Johan Eric, "Rön öfver den effect, som af manskap kan användas medelst handkraft, å machiner, som sättas i rörelse genom hvef", *KVAH* 1799.

Norberg, Jonas, *Inventarium öfver de machiner och modeller, som finnas vid Kungl. Modell-Kammaren i Stockholm, belägen uti gamla Kongshuset på K. Riddareholmen* (Stockholm, 1779).

Norberg, Jonas Adolf, "Inventarium öfwer de nyare machiner och modeller som finnas på Kongl. Modell-kammaren i Stockholm", *Magazin för svenska hushållningen och konsterne*, Vol. 1, No. 6 (Stockholm, 1801).

Nordenberg, Anders Johan, "Rön om kakelugnar och deras omslagning", *KVAH* 1739.

Nordenmark, Nils Viktor Emanuel, *Anders Celsius: Professor i Uppsala, 1701–1744* (Lychnos-Bibliotek, No. 1: Uppsala, 1936).

Nordmark, Zacharias, "Om slut-följdens giltighet, att, af försök med hydrotechniska modeller i smått, dömma om maschiners verkan i stort", *KVAH* 1813.

Nordwall, Erik, *Afhandling rörande mechaniquen, med tillämpning i synnerhet till bruk och bergverk*, Vol. 1 (Stockholm, 1800).

Näslund, Oskar Johannes, *Sågar: Bidrag till kännedomen om sågarnas uppkomst och utveckling* (Stockholm, 1937).

Odqvist, Folke K. G., "Hållfasthetsläran som förutsättning för materialprovning, särskilt i Sverige", *Daedalus* 1977.

O'Kelly d'Aghrim, Jean Prosper Désiré, *Annales de la Maison d'HyMancy, issue des anciens rois d'Irlande et connu depuis le XIe siècle sous le nom de O'Kelly* (La Haye, 1830).

Olsson, Carl-Axel, *Teknikhistoria som vetenskaplig disciplin* (Meddelande från Ekonomisk-historiska institutionen, Lunds Universitet, No. 11: Lund, 1980).

Olsson, Hugo, *Kemiens historia i Sverige intill år 1800* (Lychnos-bibliotek, No. 17.4: Uppsala, 1971).

(Oxley, Joseph), *Oxley's Patent A. D. 1768 No. 795: Machinery for Drawing Coals out of Pits, &c.* (London, 1855).

Philipp, Wolfgang, "Physicotheology in the Age of Enlightenment: Appearance and History", *Studies on Voltaire and the Eighteenth Century* 57 (1967).

Pipping, Gunnar, *The Chamber of Physics: Instruments in the History of Sciences Collections of the Royal Swedish Academy of Sciences* (Stockholm, 1977).

Le Play, Frederic, "Mémoire sur la fabrication de l'acier en Yorkshire, et comparasion des principaux groupes d'aciéries européennes, *Annales des Mines*, quatrième série, Tome III, 1843 (Paris, 1843).

— "Om ståltillverkningen i Yorkshire, samt jemförelse mellan de förnämsta ståltillverkningssorter i Europa", *Tidskrift för Svenska Bergshandteringen* 3 (1845).

Poda, Nicolaus & von Born, Ignaz Edlen, *Kurzgefaßte Beschreibung der, bei dem Bergbau zu Schemnitz in Nieder-Hungarn, errichteten Maschinen* (Prag, 1771).

Poggendorff, J. C., *Biographisch-Literarisches Handwörterbuch*, Vol. 1 (Leipzig, 1863).

Polhem 1 (1983).

Christopher Polhem: The Father of Swedish Technology, trans. by W. A. Johnsson (Hartford, Conn., 1963).

Polhem, Christopher, *Kort berättelse om de förnämsta mechaniska inventioner, som tid efter annan af Commercierådet Christopher Polhem blifvit påfundna* (Stockholm, 1729).

— "Theoriens och practiquens sammanfogning i mechaniquen, och särdeles i ström-wärk", *KVAH* 1741.

— "Fortsättning om theoriens ock practiquens sammanlämpning i mechaniquen", *KVAH* 1742.

Praetorius, Michael, *Syntagma musicum*, Vol. 2 (Wolfenbüttel, 1619).

Price, Derek J. de Solla, "Sealing Wax and String: A Philosophy of the Experimenter's Craft and its Role in the Genesis of High Technology" (George Sarton lecture at the Annual Meeting of the American Association for the Advancement of Science in 1983). Published under the title "Of Sealing Wax and String" in *Natural History* 93 (1984), No. 1.

Pursell, Jr., Carroll W., "History of Technology", in: Paul T. Durbin, ed., *A Guide to the Culture of Science, Technology, and Medicine* (New York, 1980).

— "The History of Technology and the Study of Material Culture", *American Quarterly* 35 (1983).

Qvarfort, Ulf, *Sulfidmalmshanteringens början vid Garpenberg och Öster Silvberg* (Jernkontorets Bergshistoriska Utskott, No. H 20: Stockholm, 1981).

Qvistgaard, Harald, "Frimureriska namnteckningar från Gustaf III:s tid till våra dagar", *Meddelanden från Svenska Frimurare Orden* 34 (1962), No. 5.

— "Frimureriska namnteckningar och deras ursprung", *Meddelanden från Svenska Frimurare Orden* 46 (1974), No. 1.

Rabenius, Theodor, "Om eganderätt till grufvor", *Uppsala Universitets Årsskrift* 1863.

Raistrick, Arthur, "The Steam Engine on Tyneside, 1715–1778", *Transactions of the Newcomen Society* 17 (1936–37).

— *Dynasty of Ironfounders: The Darbys and Coalbrookdale* (Newton Abbot, 1970).

Raymond, R. W., ed., *Glossary of Terms used in the Coal Trade of Northumberland and Durham* (Newcastle upon Tyne and London, 1849).

de Réaumur, René Antoine Ferchault, *L'art de convertir le fer forgé en acier, et l'art d'adoucir le fer fondu* (Paris, 1722).

Réaumur's Memoirs on Steel and Iron (1722), translated by Annelise Grünhaldt Sisco, with an introduction and notes by Cyril Stanley Smith (Chicago, 1956).

The Record of the Royal Society of London (London, 1912).

Retzius, Anders Jahan, "Berättelse om de försök som blifvit gjorda med åtskilliga utländska träd och buskarter", *KVAH* 1798.

Reynolds, Terry S., "Scientific Influences on Technology: The Case of the Overshot Waterwheel", *Technology and Culture* 20 (1979).

— *Stronger Than a Hundred Men: A History of the Vertical Water Wheel* (Baltimore, 1983).

Rider, Robin E., *The Show of Science* (Keepsakes issued by the Friends of the Bancroft Library, No. 31: Berkeley, 1983).

Viscountess Ridley, ed., *Cecilia: The Life and Letters of Cecilia Ridley 1819–1845* (London, 1968).

Rinman, Sven, *Försök till järnets historia, med tillämpning för slögder och handtwerk* (Stockholm, 1782).

— *Bergwerks lexicon*, 2 Vols. (Stockholm, 1788–1789).

— *Afhandling rörande mechaniquen, med tillämpning i synnerhet till bruk och bergverk*, Vol. 2 (Stockholm, 1794).

Robertson, T., ed., *A Pitman's Notebook: Hope and Success at Houghton, Co. Durham. The Diary of Edward Smith. Houghton Colliery Viewer 1749–1751* (Newcastle upon Tyne, 1970).

Robinson, Eric H., "The Early Diffusion of Steam Power", *The Journal of Economic History* 34 (1974).

Rolt, Lionel T. C. & Allen, John S., *The Steam Engine of Thomas Newcomen* (Hartington, 1977).

Ronan, Colin A., *Edmond Halley: Genius in Eclipse* (London, 1970).

Rosenberg, Nathan, *Technology and American Economic Growth* (New York, 1972).

— "Selection and Adaptation in the Transfer of Technology: Steam and Iron in America—1800–1870", in: *L'acquisition des techniques par les pays non-initiateurs* (Colloques Internationaux du Centre National de la Recherche Scientifique, No. 538: Paris, 1973).

— *Perspectives on Technology* (Cambridge, 1976).

— *Inside the Black Box: Technology and Economics* (Cambridge, 1982).

Rowlands, Marie B., "Stonier Parrott and the Newcomen Engine", *Transactions of the Newcomen Society* 41 (1968–69).

Rudenschöld, Ulric, *Tal om skogarnes nytjande och vård* (Stockholm, 1748).

Runeberg, Ephraim Otto, "Beskrivning öfver Lajhela Socken i Österbotten", *KVAH* 1758.

Runeberg, Edvard F., *Åminnelsetal öfver Gerhard Meijer* (Stockholm, 1798).

Rydberg, Sven, *Svenska studieresor till England under frihetstiden* (Lychnos-Bibliotek, No. 12: Uppsala, 1951).

— *Dannemora genom 500 år* (Fagersta, 1981).

Rydén, Stig, ed., *Miranda i Sverige och Norge 1787: General Francisco de Mirandas dagbok från hans resa september-december 1787* (Stockholm, 1950).

Rüdling, Johann Georg, *Supplement till thet i flor stående Stockholm* (Stockholm, 1740).

Rönnow, Sixten, *Wedevågs bruks historia* (Stockholm, 1944).

Sahlin, Carl, "När började den svenska järnhandteringen använda torf som bränsle?", *Blad för Bergshandteringens Vänner* 14 (1913–15).

— "Länshållningen i Dannemora grufvor i äldre tider", *Teknisk Tidskrift: Kemi och Bergsvetenskap* 48 (1918).

— "Dannemora-litteratur", *Teknisk Tidskrift: Kemi och bergsvetenskap* 48 (1918).

— "Historien om den förstenade gruvarbetaren i Falun och denna berättelses användning som diktmotiv", *Jerkontorets annaler*, n.s. 75 (1920).

— "Ett nytt bidrag till historien om den förstenade gruvarbetaren i Falu gruva", *Jernkontorets Annaler*, n.s. 83 (1928).

— "Det svenska bergsbruket i Georg Agricolas skrifter", *Med Hammare och Fackla* 1 (1928).

— *Svenskt stål före de stora götstålsprocessernas införande* (Stockholm, 1931).

— "Det svenska järnets världsrykte, I: Från äldsta tider till omkring år 1850", *Daedalus* 1932.

— "Föremål av guld och silver, förfärdigade av metall från svenska bergverk", *Med Hammare och Fackla* 6 (1935).

Sahlstedt, Abraham, *Swensk ordbok* (Stockholm, 1773).

Salander, Erik, *Systematiske nöd-hielps tankar, eller okulstöteliga grund-saker til wälgång för höga och låga i et fattigt land, der penningen har rymt, näringen ligger öde och winsten är förswunnen; begrundade af En Som önskar allas wälmågo!* (Göteborg, 1730).

Samuelsson, Kurt, *De stora köpmanshusen i Stockholm 1730–1815: En studie i den svenska handelskapitalismens historia* (Skrifter utgivna av Ekonomisk-historiska institutet i Stockholm: Stockholm, 1951).

Sandblad, Henrik, ed., *Christopher Polhems efterlämnade skrifter*, Vol. 1 (Lychnos-bibliotek, No. 10.1: Uppsala, 1947).

Sandel, Samuel, "Rön angående malm- och bergsprengning", *KVAH* 1769.

Sandklef, Albert, "Äldre biskötsel i Sverige och Danmark: Bidrag till kännedomen om sydskandinavisk biskötsel före mitten av 1800-talet", *Göteborgs Kungl. Vetenskaps- och Vitterhets-Samhälles Handlingar*, 5:e följd., ser. A, bd. 6, No. 3 (Göteborg, 1937).

Sandqvist, Inga-Britta, "Elfdalsporfyr—idé och utformning: En svensk konstindustri 1788–1856" (mimeographed fil. lic. dissertation at the Department of History of Art, University of Stockholm, 1972).

— *Litteratur om svenska industriföretag* (IVA-Meddelanden, No. 227: Stockholm, 1979).

Sanner, Lars-Erik, "Musiklitteratur i Kungl. Tekniska Högskolans Bibliotek, Stockholm", *Nordisk Tidskrift för Bok- och Biblioteksväsen* 54 (1967).

Savery, Thomas, *The Miners Friend ...* (1702), facsimile edition (Edinburgh, 1979).

(Savery, Thomas), "An Account of Mr. Tho. Savery's Engine for Raising Water by the Help of Fire", *Philosophical Transactions* 21 (1699), 228.

Scheutz, Per Georg, "Pehr Lagerhjelm. Assessor i Kongl. Bergskollegium", in: *Lefnadsteckningar öfver Kongl. Svenska Vetenskaps Akademiens efter år 1854 aflidna ledamöter*, Vol. 1 (Stockholm, 1869–1873).

Schieche, Emil, *400 Jahre Deutsche St. Gertruds Gemeinde im Stockholm 1571–1971* (Stockholm, 1971).

Schiller, Bernt, "Technology – History – Social Change: A Methodological Comment and an Outline of a Nordic Account", *Scandinavian Journal of History* 8 (1983), 71–82.

Schmidt, Johann Wilhelm, *Reise durch einige schwedische Provinzen* (Hamburg, 1801).

Schück, Henrik, ed., *Bokwetts gillets protokoll* (Uppsala, 1918).

Sege, Carl A:son, "Bidrag till kännedomen om Sala silververks vattenkraftsanläggningar", *Blad för Bergshandteringens Vänner* 20 (1931–1932).

Seitz, Heribert, "Den stora förödelsen år 1719 och striderna vid Södra Stäket", *Karolinska Förbundets Årsbok* 1960.

Selling, Gösta, "Bruken som kulturmiljö", *Blad för Bergshandteringens Vänner* 33 (1958).

— "Den svenska kakelugnens tvåhundraårsjubileum", *Saga och Sed* 1967.

Shapin, Steven, "The Audience for Science in Eighteenth Century Edinburgh", *History of Science* 12 (1974).

— "Social Uses of Science", in: G. S. Rousseau & Roy Porter, eds., *The Ferment of Knowledge: Studies in the Historiography of Eighteenth-Century Science* (Cambridge, 1980).

Sjöberg, Sven, *Rysshärjningarna i Roslagen* (Stockholm, 1981).

Sjögren, Hjalmar, "Några ord om Swedenborgs manuskript: 'Nya anledningar til grufvors igenfinnande' etc.", *Geologiska Föreningens Förhandlingar* 29 (1907).

Sjöstrand, Wilhelm, *Den militära undervisningen i Sverige intill år 1792* (Uppsala, 1941).

— *Pedagogikens historia*, vol. 3:1 (Malmö, 1961).

Slott och herresäten i Sverige: Uppland, Vol. 2 (Malmö, 1967).

Smith, Alan, "Steam and the City—The Committee of the Proprietors of the Invention for Raising Water by Fire", *Transactions of the Newcomen Society* 49 (1977–78).

— "The Newcomen Engine at Passy, France, in 1725: A Transfer of Technology Which Did Not Take Place", *Transactions of the Newcomen Society* 50 (1978–79).

Smith, Merritt Roe, ed., *Military Enterprise and Technological Change: Perspectives on the American Experience* (to be published by MIT Press).

Smith, Norman F., *Man and Water: A History of Hydro-Technology* (London, 1976).

Solders, Severin, *Älvdalens sockens historia, Del II: Gamla porfyrverket* (Dalarnas fornminnes och hembygds förbunds skrifter, No. 8: Stockholm, 1939).

Solitander, Axel, *Några anteckningar rörande träförädlingens historia i Finland* (Helsingfors, 1930).

Sommarin, Emil, *Bidrag till kännedom om arbetareförhållanden vid svenska bergverk och bruk i äldre tid till omkring år 1720* (Lund, 1908).

— "Det svenska bergsregalets ursprung", *Statsvetenskaplig Tidskrift* 13 (1910).

Sondén, Per, *Nils Bielke och det svenska kavalleriet 1674–1679* (Stockholm, 1883).

Staf, Nils, ed., *Borgarståndets riksdagsprotokoll från frihetstidens början*, Vol. 4 (Stockholm, 1958).

Staudenmaier, John M., "What SHOT Hath Wrought and What SHOT Hath Not: Reflections on 25 years of the History of Technology" (paper presented at the Annual Meeting of the Society for the History of Technology in 1983, to be published in *Technology and Culture*).

de Stein d'Altenstein, Isidore, *Annuaire de la Noblesse de Belgique* (Bruxelles, 1859).

Stockholmske Post Tidningar 1727, 1728, 1729, 1734.

Stockholms Posten 1816, July 22.

Stridsberg, Einar & Mattsson, Leif, *Skogen genom tiderna: Dess roll för lantbruket från forntid till nutid* (Stockholm, 1980).

Stråle, Gustaf H., *Alingsås manufakturverk* (Stockholm, 1884).

Ström, Carl Fredrik, *Karta öfver landsvägarna uti Sverige och Norrige* (Stockholm, 1846).

Stuart, Robert, *Historical and Descriptive Anecdotes of the Steam Engine*, 2 Vols. (London, 1829).

Sundin, Bo, *Ingenjörsvetenskapens tidevarv: Ingenjörsvetenskapsakademin, Pappersmassekontoret, Metallografiska institutet och den teknologiska forskningen i början av 1900-talet* (Acta Universitatis Umensis, Umeå Studies in the Humanities, No. 42: Umeå, 1981).

Svedenstierna, Eric Thomas, *Ödmjukt yttrande på anmodan af Bruks-Societetens Herrar Fullmäktige inlemnadt till Jern-Kontoret* (Stockholm, 1805).

— *Försök till hytte-ordning för Nora, Lindes och Ramsbergs Bergslager* (Stockholm, 1819).

Svenska Akademiens Ordbok, 28 Vols. (Lund, 1898–1981).

Svenska Män och Kvinnor, 8 Vols. (Stockholm, 1942–1955).

Svenska slott och herresäten vid 1900-talets början: Uppland (Stockholm, 1909).

Svenskt Biografiskt Lexikon, 23 Vols. (Stockholm, 1918–1981).

Sveriges riksbank 1668–1924: Bankens tillkomst och verksamhet, Vol. 5 (Stockholm, 1931).

Swab, Anton von, *Om grufvebrytning i Sverige* (Upsala, 1780).

Swar på Kongl. Wettenskaps Academiens fråga: hwilka författningar äro de bästa, at underhålla tilräckelig tilgång på skog här i landet? (Stockholm, 1768).

Swedenborg, Emanuel, "Beskrifning öfver svenska masvgnar och theras blåsningar", *Noraskogs arkiv*, Vol. 4 (Stockholm, 1901–1903).

— *Opera quaedam aut inedita aut obsoleta de rebus naturalibus*, Vol. 1 (Stockholm, 1907).

Sylwan, Otto, *Svenska pressens historia till statshvälfningen 1772* (Lund, 1896).

Söderbaum, Henrik Gustaf, *Jac. Berzelius: Lefnadsteckning*, Vol. 1 (Uppsala, 1929).

Söderberg, Tom, *Bergsmän och brukspatroner i svenskt samhällsliv* (Det levande förflutna, Svenska Historiska Föreningens folkskrifter, No. 12: Stockholm, 1948).

Söderlund, Ernst, *Stockholms hantverkarklass 1720–1772: Sociala och ekonomiska förhållanden* (Monografier utgivna av Stockholms kommunalförvaltning: Stockholm, 1943).

Tandberg, John G., "Die Triewaldsche Sammlung am Physikal. Institut der Universität zu Lund und die Original-Luftpumpe Guerickes", *Lunds Universitets årsskrift*, n. s. avd. 2 bd. 16, No. 9 (Lund, 1920).

— "Historiska instrument i Lund", *Kosmos* 1922.

Tann, Jennifer, ed., *The Selected Papers of Boulton & Watt: Volume I, The Engine Partnership 1775–1825* (London, 1981).

Technikgeschichte 50 (1983), No. 3.

Teich, Mikuláš, "Diffusion of Steam, Water and Air-Power to and from Slovakia during the 18th Century and the Problem of the Industrial Revolution", in: *L'acquisition des techniques par les pays non-initiateurs* (Colloques Internationaux du Centre National de la Recherche Scientifique, No. 538: Paris, 1973).

— "The Early History of the Newcomen Engine at Nová Baňa (Königsberg): Isaac Potter's Negotiations with the *Hofkammer* and the Signing of the Agreement of 19 August 1721", *East-Central Europe* 9 (1982).

Thäberg, Carl-Th., "Föreningen Tekniska Museet under år 1932", *Daedalus* 1933.

Tilas, Daniel, "Mineral-historia öfwer Osmundsberget uti Rättwiks sochn och Öster-Dalarne", *KVAH* 1740.

— "Dag bok öfwer Hans Kongl. Höghet Cronprintsen Prints Gustavs resa igenom bergslagerne in sept: 1768", *Noraskogs arkiv*, Vol. 6: 1 (Stockholm, 1928).

Tillhagen, Carl-Herman, *Järnet och människorna: Verklighet och vidskepelse* (Stockholm, 1981).

Timoshenko, Stephen P., *History of Strength of Materials* (New York, 1953).

Todhunter, Isaac, *A History of the Theory of Elasticity and of the Strength of Materials from Galilei to the Present Time*, Vol. 1 (Cambridge, 1886).

Triewald, Mårten, *Notification, om trettijo publique föreläsningar, öfwer nya naturkunnigheten, illustrerad genom mechaniske, hydrostatiske, aërometriske och optiske experimenter och försök, som på Riddare-huset i Stockholm komma at taga sin början den 15:de Octobris innewarande åhr 1728* (Stockholm, 1728).

— *Nödig tractat om bij, deras natur, egenskaper, skiötzel och nytta, utur åtskillige nyares wittra anmärkningar samt egne anstälte rön och försök sammanfattad, jemte beskrifning på en ny inrättning av bijstockar* (Stockholm, 1728).

— "Queries Concerning the Cause of Cohesion of the Parts of Matter, Proposed in a Letter to Dr. Desaguliers, F.R.S. By Fr. Triewald, Director of Mechaniks in the Kingdom of Sweden", *Philosophical Transactions* 36 (1729–30).

— "An Extraordinary Instance of the Almost Instantaneous Freezing of Water; and giving an Account of Tulips, and such bulbous Plants, Flowering much Sooner when Their Bulbs are Placed upon Bottles Filled with Warm Water, than when Planted in the Ground", *Philosophical Transactions* 37 (1731–32).

— *Kort beskrifning, om eld- och luft-machin wid Dannemora grufwor* (Stockholm, 1734). Also published in an English translation by the Newcomen Society, London, in 1928 as Extra Publication No. 1, see: *Mårten Triewald's Short Description of the Atmospheric Engine. Published at Stockholm, 1734. Translated from the Swedish with Foreword, Introduction and Notes* (London, 1928).

— *Konsten at lefwa under watn eller en kort beskrifning om de påfunder, machiner och redskap hwarpå dykeri- och bärgnings-societetens privilegier äro grundade, hwarmed de anstält profwen under twenne riks-dagar för Sweriges rikes högl. ständers herrar deputerade* (Stockholm, 1734 b).

— "Nova inventio Follium Hydravlicorum", *Acta literaria et scientiarum Sveciae*, 4 (1735–1739).

— *Föreläsningar öfwer nya naturkunnigheten*, 2 Vols. (Stockholm, 1735–1736, 2nd ed. Vol. 1 1758).

— "An Improvement of the Diving Bell", *Philosophical Transactions* 39 (1735–36).

— "A Description of a New Invention of Bellows, Called Water-Bellows", *Philosophical Transactions* 40 (1737–1738).

— "Om alt hwad som länder till kundskapen om stenkol efter många års rön framgifwit", *KVAH* 1739.

— "Fortsättning om stenkolswettenskapen", *KVAH* 1740.

— "Ytterligare fortsättning om stenkåls wetenskapen", *KVAH* 1740.

— "Och sidsta fortsättningen af stenkåls wetenskapen. Beskrifning af et påfund, hwarigenom den dödeliga luften på en kort tid, utur et skakt blef dragen", *KVAH* 1741.

— "Walk- eller klädemakarlers grufwornes beskaffenhet uti Bedfordshire i England", *KVAH* 1742.

— "En ny ler-älta särdeles tiänlig för taktegel bruk", *KVAH* 1742.

— *Nyttan och bruket af wäderwäxlings-machin på Kongl. Maj:ts och riksens örlogs flotta* (Stockholm, n. d. 1742 ?).

— "Väderväxlings machin påfunnen och ingifven", *KVAH* 1744.

— "Rön och försök angående möjligheten at Svea Rike kunda äga egit rådt silke", *KVAH* 1745.

(Triewald, Mårten), *Triewald's Patent A. D. 1722 No. 449: Engine for Drawing Coals and Waters from Mines, Supplying Water to Towns, &c.* (London, 1857).

Tunander, Ingemar & Wallin, Sigurd, eds., *Anders Tidströms resa genom Dalarna 1754* (Falun, 1954).

Tunborg, Andreas Nicolaus, *Om bergs-domstolar* (Upsala, 1799).

von Tunzelmann, G. N., *Steam Power and British Industrialization to 1860* (Oxford, 1978).

af Uhr, Carl David, *Berättelse om kolnings-försök åren 1811, 1812 och 1813. På Bruks-Societetens bekostnad anstälde* (Stockholm, 1814).

— *Handbok för kolare* (Stockholm, 1814, 2nd ed. 1814, 3rd ed. 1823).

— *Betänkande till Herrar Fullmägtige i Jern-Contoret, innehållande en öfversigt af bergs-mechanikens tillstånd* (Stockholm, 1817).

Unnerbäck, R. Axel, "Leufsta bruks-orgeln restaurerad", *Kyrkomusikernas Tidning* 30 (1964).

Usher, Abbot Payson, *A History of Mechanical Inventions*, (Cambridge, Mass., 1954, rev. ed.).

Vallerö, Rolf, "Henric Kalmeter", *Svenskt Biografiskt Lexikon*, Vol. 20 (1975).

Vincenti, Walter G., "The Air-Propeller Tests of W. F. Durand and E. P. Lesley: A Case Study in Technological Methodology", *Technology and Culture* 20 (1979).

Voorn, Henk, *De papiermolens in de provincie Noord-Holland* (De geschiedenis der Nederlandse papierindustrie, No. 1: Haarlem, 1960).

Wahlund, Johan, *Dannemora grufvor: Historisk skildring* (Stockholm, 1879).

Waldén, Bertil, "Några ämbetsstavar från svenska bergslager", *Med Hammare och Fackla* 10 (1939).

Waldenström, Erland, "Näringslivet och den teknikhistoriska forskningen", *IVA-NYTT* 1983, No. 1.

Wallerius, Johan Gottschalk, "Undersökning om den jords beskaffenhet, som fås af vatten, vegetabilier och mineralier. Fjerde stycket, Om kalkjordens skillnad ...", *KVAH* 1760.

Wallner, Magnus Edvardi, *Kolare konsten uti Swerige, korterligen beskrifwen* (Stockholm, 1746).

Watkins, George M., "The Development of the Steam Winding Engine", *Transactions of the Newcomen Society* 50 (1978–79).

Weber, Wolfhard, *Innovationen im frühindustriellen deutschen Bergbau und Hüttenwesen: Friedrich Anton von Heynitz* (Studien zu Naturwissenschaft, Technik und Wirtschaft im Neunzehnten Jahrhundert, No. 6: Göttingen, 1976).

Webster, Charles, *The Great Instauration: Science, Medicine and Reform 1626–1660* (New York, 1975).

Wenström, Jonas & Granström, Gustaf A., "Electriciteten i grufhandteringens tjenst", *Jernkontorets Annaler*, n.s. 47 (1892).

Wester, Bertil, "Studier i svensk orgelkonst under 1600- och 1700-talen", *Svensk Tidskrift för Musikforskning* 13 (1931).

— "Orgelverket i Leufsta bruks kyrka", *Fornvännen* 28 (1933).

White, Jr., Lynn, "Technology Assessment from the Stance of a Medieval Historian", in: idem, *Medieval Religion and Technology* (Berkeley, 1978). Originally published in *American Historical Review* 79 (1974).

Wieselgren, Sigfrid, *Sveriges fängelser och fångvård från äldre tider till våra dagar* (Stockholm, 1895).

Wieslander, Gösta, "Skogsbristen i Sverige under 1600- och 1700-talen", *Svenska Skogsvårdsföreningens Tidskrift* 34 (1936).

Wilcke, Johan Carl, "Försök til en ny inrättning af luftpumpar, förmedelst kokande vattu-ångor", *KVAH* 1769.

Wittrock, Georg, "Förräderipunkten i Nils Bielkes process 1704–1705", *Karolinska Förbudets Årsbok* 1917.

— "Nils Bielkes underhandling i Brandenburg 1696", *Karolinska Förbundets Årsbok* 1918.

— "Nils Bielke", *Svenskt Biografiskt Lexikon*, Vol. 4 (Stockholm, 1924).

Wohlert, Klaus, "Svenskt yrkeskunnande och teknologi under 1800-talet. En fallstudie av förutsättningar för kunskapstransfer", *Historisk Tidskrift* 99 (1979).

Wollin, Nils G., *Från ritskola till konstfackskola: Konstindustriell undervisning under ett sekel* (Stockholm, 1951).

Woodcroft, Bennett, *Alphabetical Index of Patentees of Inventions* (1854), facsimile ed. (London, 1969).

Zenzén, Nils, "Från den tid, då vi skulle transmutera järn till koppar och få lika mycket silver i Sverige som gråberg", *Med Hammare och Fackla* 7 (1936).

Åberg, Alf, "Gustaf Horn", *Svenskt Biografiskt Lexikon*, Vol. 19 (Stockholm, 1971–1973).

Åkerrehn, Olof, *De lege Kepleriana, comparandi distantias planetarum medias a sole* (Uppsala, 1784).
— *Utkast til en practisk afhandling om vattenverk, grundad på physiske mechaniske lagar* (Örebro, 1788).
— *Svenska blåsverkens historia, från år 1786; med tillägg, om jernberedningen i allmänhet* (Stockholm, 1805).
— "Grufve-konst, till vatten-pumpning för dragare", *KVAH* 1806.

Åström, Sven-Erik, "Technology and Timber Exports from the Gulf of Finland, 1661–1740", *Scandinavian Economic History Review* 23 (1975).

Ährling, Ewald, ed., *Carl von Linnés svenska arbeten i urval*, Vol. 2 (Stockholm, 1880).

Index